Lecture Notes in Networks and Systems

Volume 755

AF831312

Series Editor

Janusz Kacprzyk, Systems Research Institute, Polish Academy of Sciences, Warsaw, Poland

Advisory Editors

Fernando Gomide, Department of Computer Engineering and Automation—DCA, School of Electrical and Computer Engineering—FEEC, University of Campinas—UNICAMP, São Paulo, Brazil

Okyay Kaynak, Department of Electrical and Electronic Engineering, Bogazici University, Istanbul, Türkiye

Derong Liu, Department of Electrical and Computer Engineering, University of Illinois at Chicago, Chicago, USA
Institute of Automation, Chinese Academy of Sciences, Beijing, China

Witold Pedrycz, Department of Electrical and Computer Engineering, University of Alberta, Alberta, Canada
Systems Research Institute, Polish Academy of Sciences, Warsaw, Poland

Marios M. Polycarpou, Department of Electrical and Computer Engineering, KIOS Research Center for Intelligent Systems and Networks, University of Cyprus, Nicosia, Cyprus

Imre J. Rudas, Óbuda University, Budapest, Hungary

Jun Wang, Department of Computer Science, City University of Hong Kong, Kowloon, Hong Kong

The series "Lecture Notes in Networks and Systems" publishes the latest developments in Networks and Systems—quickly, informally and with high quality. Original research reported in proceedings and post-proceedings represents the core of LNNS.

Volumes published in LNNS embrace all aspects and subfields of, as well as new challenges in, Networks and Systems.

The series contains proceedings and edited volumes in systems and networks, spanning the areas of Cyber-Physical Systems, Autonomous Systems, Sensor Networks, Control Systems, Energy Systems, Automotive Systems, Biological Systems, Vehicular Networking and Connected Vehicles, Aerospace Systems, Automation, Manufacturing, Smart Grids, Nonlinear Systems, Power Systems, Robotics, Social Systems, Economic Systems and other. Of particular value to both the contributors and the readership are the short publication timeframe and the world-wide distribution and exposure which enable both a wide and rapid dissemination of research output.

The series covers the theory, applications, and perspectives on the state of the art and future developments relevant to systems and networks, decision making, control, complex processes and related areas, as embedded in the fields of interdisciplinary and applied sciences, engineering, computer science, physics, economics, social, and life sciences, as well as the paradigms and methodologies behind them.

Indexed by SCOPUS, INSPEC, WTI Frankfurt eG, zbMATH, SCImago.

All books published in the series are submitted for consideration in Web of Science.

For proposals from Asia please contact Aninda Bose (aninda.bose@springer.com).

Anurag Mishra · Deepak Gupta · Girija Chetty
Editors

Advances in IoT and Security with Computational Intelligence

Proceedings of ICAISA 2023, Volume 1

Springer

Editors
Anurag Mishra
Department of Electronics
Deen Dayal Upadhyaya College
University of Delhi
New Delhi, India

Deepak Gupta
Department of Computer Science
and Engineering
MNNIT Allahabad
Prayagraj, India

Girija Chetty
Faculty of Science and Technology
University of Canberra
Bruce, ACT, Australia

ISSN 2367-3370 ISSN 2367-3389 (electronic)
Lecture Notes in Networks and Systems
ISBN 978-981-99-5084-3 ISBN 978-981-99-5085-0 (eBook)
https://doi.org/10.1007/978-981-99-5085-0

© The Editor(s) (if applicable) and The Author(s), under exclusive license to Springer Nature Singapore Pte Ltd. 2023

This work is subject to copyright. All rights are solely and exclusively licensed by the Publisher, whether the whole or part of the material is concerned, specifically the rights of translation, reprinting, reuse of illustrations, recitation, broadcasting, reproduction on microfilms or in any other physical way, and transmission or information storage and retrieval, electronic adaptation, computer software, or by similar or dissimilar methodology now known or hereafter developed.
The use of general descriptive names, registered names, trademarks, service marks, etc. in this publication does not imply, even in the absence of a specific statement, that such names are exempt from the relevant protective laws and regulations and therefore free for general use.
The publisher, the authors, and the editors are safe to assume that the advice and information in this book are believed to be true and accurate at the date of publication. Neither the publisher nor the authors or the editors give a warranty, expressed or implied, with respect to the material contained herein or for any errors or omissions that may have been made. The publisher remains neutral with regard to jurisdictional claims in published maps and institutional affiliations.

This Springer imprint is published by the registered company Springer Nature Singapore Pte Ltd.
The registered company address is: 152 Beach Road, #21-01/04 Gateway East, Singapore 189721, Singapore

Preface

Along with the advancement of technologies in cyber-physical systems, Internet of things, cloud computing and big data, challenges are traditionally solved by optimization of organizations, business processes and information systems at the whole-enterprise level. This is increasingly requiring an extension of this perspective to the societal level. We live in society and deal with it. Without this, we cannot sustain technological advances. In particular, the innovation that originated in Industry 4.0 has got transformed into Society 5.0, which is expected to play an important role in bringing about change not only in the enterprises but also in society. Therefore, how to design, develop, and manage enterprise and societal architectures and systems by using information technology gain more attention than ever before.

With this objective in mind, the International Conference on Advances in IoT, Security with AI (ICAISA-2023) was organized by Deen Dayal Upadhyaya College, University of Delhi, New Delhi, India, in collaboration with University of Canberra, Canberra, Australia, and NIT, Arunachal Pradesh, Itanagar, Arunachal Pradesh, India, during March 24–25, 2023. We are thankful to our contributors, participants and sponsors—STPI Chennai, REC Limited and Power Finance Corporation Limited who have supported this event wholeheartedly.

This conference has been organized having thirteen parallel technical sessions besides inaugural and valedictory sessions as the most suitable tracks capable of serving the electronics, IT and software industries to be specific. Few presentations by Indian and international industry specialists have been done in these sessions. This is done with a view to establishing a connection between academia and the industry, and both of them can get fruitful ideas from each other. We are really thankful to all our overseas sector and academic experts who have either joined us physically or in online mode.

We are particularly grateful to Dr. Rajendra Pratap Gupta, Mr. Animesh Mishra, Mr. M. S. Bala and Prof. Balram Pani who blessed us in the inaugural session. We are also thankful to Mr. N. K. Goyal for his presence in the valedictory session. We are extremely grateful to Springer Nature, especially Dr. Aninda Bose who agreed to publish two volumes of conference proceedings in the prestigious series of *Lecture Notes in Networks and Systems*.

New Delhi, India
Prayagraj, India
Bruce, Australia

Anurag Mishra
Deepak Gupta
Girija Chetty

Contents

Editors and Contributors

About the Editors

Prof. Anurag Mishra has bachelor's and master's in Physics from the University of Delhi. He completed his M.E. in Computer Technology and Applications and Ph.D. in Electronics also from the University of Delhi. He has extensive experience of teaching B.Sc. (Hons.), M.Sc., B.Tech. and M.Tech. programs in Electronics and Computer Science. He has about 28 years of experience as a teacher and as an active researcher. He has been a consultant for offshoot agencies of the Ministry of Education, Government of India. Presently, he is nominated as a visitor's nominee in a central university by the Government of India. He has 65 refereed papers in highly cited journals, international conferences and book chapters, three authored, one edited book and two patents to his credit. He has recently entered into developing medical applications using deep convolutional neural networks. He is an active reviewer of papers for Springer, Elsevier and IEEE Transactions. He is a member of IEEE and also holds membership of the Institute of Informatics and Systemics (USA).

Dr. Deepak Gupta is an assistant professor in the Department of Computer Science and Engineering at Motilal Nehru National Institute of Technology Allahabad, Prayagraj, India. Previously, he worked in the Department of Computer Science and Engineering at the National Institute of Technology Arunachal Pradesh. He received a Ph.D. in Computer Science and Engineering from the Jawaharlal Nehru University, New Delhi, India. His research interests include support vector machines, ELM, RVFL, KRR and other machine learning techniques. He has published over 70 referred journal and conference papers of international repute. His publications have more than 1384 citations with an h-index of 22 and an i10-index of 45 (Google Scholar, 21/06/2023). He is currently a member of an editorial review board member of *Applied Intelligence*. He is the recipient of the 2017 SERB-Early Career Research Award in Engineering Sciences which is the prestigious award of India at the early career level. He is a senior member of IEEE and currently an active member of many scientific societies like IEEE SMC, IEEE CIS, CSI and many more. He has served as a reviewer of many scientific journals and various national and international conferences. He was the general chair of the 3rd International Conference on Machine Intelligence and Signal Processing (MISP-2021) and associated with other conferences like IEEE SSCI, IEEE SMC, IJCNN, BDA 2021, etc. He has supervised three Ph.D. students and guided 15 M.Tech. projects. He is currently the principal investigator (PI) or a co-PI of two major research projects funded by the Science and Engineering Research Board (SERB), Government of India.

Dr. Girija Chetty has a bachelor's and master's degrees in Electrical Engineering and Computer Science from India and Ph.D. in Information Sciences and Engineering from Australia. She has more than 38 years of experience in industry, research and teaching from Universities and Research and Development Organisations from India and Australia and has held several leadership positions including the head of Software Engineering and Computer Science, the program director of ITS courses, and the course director for Master of Computing and Information Technology Courses. Currently, she is a full professor in Computing and Information Technology at School of Information Technology and Systems at the University of Canberra, Australia, and leads a research group with several Ph.D. students, post-docs, research assistants and regular international and national visiting researchers. She is a senior member of IEEE, USA; a senior member of Australian Computer Society; and ACM member, and her research interests are in multimodal systems, computer vision, pattern recognition, data mining and medical image computing. She has published extensively with more than 200 fully refereed publications in several invited book chapters, edited books, and high-quality conferences and journals, and she is in the editorial boards, technical review committees and a regular reviewer for several Springer, IEEE, Elsevier and IET journals in the area related to her research interests. She is highly interested in seeking wide and interdisciplinary collaborations, research scholars and visitors in her research group.

Contributors

S. K. Abdul Alim Central Institute of Technology, Kokrajhar, Assam, India

Matthew O. Adigun Department of Computer Science, University of Zululand, Richards Bay, South Africa

Owais Ahmad Thoucentric, Bangalore, India

Sunday Adeola Ajagbe Department of Computer & Industrial Engineering, First Technical University, Ibadan, Nigeria

Dagani Anudeepthi Koneru Lakshmaiah Education Foundation, Andhra Pradesh, India

Sandhya Avasthi Department of CSE, ABES Engineering College, Ghaziabad, UP, India

Joseph Bamidele Awotunde Department of Computer Science, University of Ilorin, Ilorin, Nigeria

Surendiran Balasubramanian National Institute of Technology Puducherry, Karaikal, India

Prasanta Baruah Central Institute of Technology Kokrajhar, Assam, India

Siddhant Bhalla Symbiosis Institute of Technology, Symbiosis International (Deemed) University, Pune, India

Prachi Bhanaria Department of Electronic Science, University of Delhi, South Campus, Delhi, India

Bharat Bhargava Purdue University, West Lafayette, IN, USA

Vallam Reddy Bhargavi Reddy Koneru Lakshmaiah Education Foundation, Andhra Pradesh, India

Jyoti Bharti Maulana Azad National Institute of Technology, Bhopal, India

Monika Bhattacharya Acharya Narendra Dev College, University of Delhi, Delhi, India

Mamta Bisht Department of CSE, ABES Engineering College, Ghaziabad, UP, India

Girija Chetty Faculty of Science and Technology, University of Canberra, Bruce, Australia

Kartikeya Choudhary Department of CSE, ABES Engineering College, Ghaziabad, UP, India

Nurul Amin Choudhury Department of Computer Science and Engineering, National Institute of Technology Silchar, Silchar, Assam, India

Sudakshina Das School of Commerce, Finance and Accountancy, CHRIST (Deemed to Be University), Bengaluru, India

Sneha Darade Symbiosis Institute of Technology, Symbiosis International (Deemed) University, Pune, India

Nirmallya Dey Central Institute of Technology, Kokrajhar, Assam, India

Anubhav Gagare Symbiosis Institute of Technology, Symbiosis International (Deemed) University, Pune, India

Abhinav Gaikwad Symbiosis Institute of Technology, Symbiosis International (Deemed) University, Pune, India

Pitamber Gaire Dr. A.P.J. Abdul Kalam University, Indore, Madhya Pradesh, India

Vrinda Garg School of Commerce, Finance and Accountancy, CHRIST (Deemed to Be University), Bengaluru, India

Vibha Gaur Acharya Narendra Dev College, University of Delhi, Delhi, India

Summi Goindi Chandigarh University, Gharuan, Mohali, Punjab, India

Somya Goyal Manipal University Jaipur, Jaipur, Rajasthan, India

Avdhesh Gupta Ajay Kumar Garg Engineering College, Ghaziabad, India

Ritu Gupta Department of Information Technology, Bhagwan Parshuram Institute of Technology, IP University, New Delhi, India

Shivam Gupta Department of Information Technology, Ajay Kumar Garg Engineering College, Uttar Pradesh, Ghaziabad, India

Sonam Gupta Department of Information Technology, Ajay Kumar Garg Engineering College, Uttar Pradesh, Ghaziabad, India

Vishan Kumar Gupta Sir Padampat Singhania University, Udaipur, Rajasthan, India

Harikrishanan Symbiosis Institute of Technology, Symbiosis International (Deemed) University, Pune, India

Naveen Kumar Jain Zakir Hussain College, University of Delhi, Delhi, India

Sarika Jain National Institute of Technology, Kurukshetra, India

Akshansh Jha Bhaba University, Bhopal, Madhya Pradesh, India

Prem Shankar Jha Department of Statistics, Patna College, Patna University, Bihar, India

Jerush John Joseph School of Commerce, Finance and Accountancy, CHRIST (Deemed to Be University), Bengaluru, India

Mrignainy Kansal Department of Information Technology, Ajay Kumar Garg Engineering College, Uttar Pradesh, Ghaziabad, India

Amita Kapoor NePeur, Delhi, India

Divneet Singh Kapoor Chandigarh University, Gharuan, Mohali, Punjab, India

Ravneet Kaur Acharya Narendra Dev College, University of Delhi, Delhi, India

Vallidevi Krishnamurthy Vellore Institute of Technology, Chennai, Tamil Nadu, India

Anoop Kumar Banasthali Vidyapith, Tonk, Rajasthan, India

Arun Kumar Department of Computer Science and Engineering, SRM Institute of Science and Technology, Delhi-NCR Campus, Ghaziabad, Uttar Pradesh, India

Dinesh Kumar KIET Group of Institutions, Ghaziabad, India

Vimal Kumar Bennett University, Greater Noida, India

Prakirty Kumari Department of Energy System Engineering, National Institute of Technology, Jamshedpur, Jharkhand, India

Taruna Kumari Discipline of Statistics, School of Sciences, Indira Gandhi National Open University, New Delhi, India

Snehal V. Laddha Department of Electronics Engineering, Shri Ramdeobaba College of Engineering and Management, Nagpur, Maharashtra, India

Shreya Maheshwari Department of Information Technology, Ajay Kumar Garg Engineering College, Uttar Pradesh, Ghaziabad, India

Hussain Kaide Johar Manasi Maulana Azad National Institute of Technology, Bhopal, India

Maneesha Maharaja Agrasen College, University of Delhi, Delhi, India

Mihir Mendse Department of Electronics Engineering, Shri Ramdeobaba College of Engineering and Management, Nagpur, India

Gouri Morankar Department of Electronics Engineering, Shri Ramdeobaba College of Engineering and Management, Nagpur, India

Om Mishra Symbiosis Institute of Technology, Symbiosis International (Deemed) University, Pune, India

Nibedita Priyadarsini Mohapatra NIST Institute of Science and Technology (Autonomous), Berhampur, Odisha, India

Sunil K. Muttoo Department of Computer Science, University of Delhi, Delhi, India

Manjushree Nayak NIST Institute of Science and Technology (Autonomous), Berhampur, Odisha, India

Ademola Temidayo Opadotun Oyo State College of Agriculture and Technology, Igbo-Ora, Nigeria

Surya Kant Pal Department of Mathematics, Sharda School of Basic Sciences and Research, Sharda University, Greater Noida, India

Gaurav Pandey Netaji Subhas University of Technology, Delhi, India

Maneesha Pandey Hindu College, University of Delhi, Delhi, India

Praveen Kant Pandey Maharaja Agrasen College, University of Delhi, Delhi, India

Rakesh Kumar Pandey Kirori Mal College, University of Delhi, Delhi, India

Rupal Paliwal Department of Electronics Engineering, Shri Ramdeobaba College of Engineering and Management, Nagpu, India

Manikant Paswan Department of Mechanical Engineering, National Institute of Technology, Jamshedpur, Jharkhand, India

Deepak Parashar Symbiosis Institute of Technology, Symbiosis International (Deemed) University, Pune, India

Anupam Pathak Department of Statistics, Ramjas College, University of Delhi, Delhi, India

D. Patteshwari Department of Cognitive Neurosciences, School of Life Sciences, JSS Academy of Higher Education and Research, Mysuru, Karnataka, India

Ramakrishna Pinninti Central Institute of Technology, Kokrajhar, Assam, India

Priyanka Banasthali Vidyapith, Tonk, Rajasthan, India

Oma Junior Raffik Department of Mathematics, Sharda School of Basic Sciences and Research, Sharda University, Greater Noida, India

Santhosh Rajaram National Institute of Technology Puducherry, Karaikal, India

Aditi Ramprasad Sri Sivasubramania Nadar College of Engineering, Kalavakkam, Chennai, India

Rita Roy Department of Computer Science and Engineering, GITAM Institute of Technology, GITAM (Deemed to be University), Visakhapatnam, Andhra Pradesh, India

Anuj Sahani Manipal University Jaipur, Jaipur, Rajasthan, India

V. Sakthivel Konkuk University, Seoul, South Korea

T. Santhi Sri Koneru Lakshmaiah Education Foundation, Andhra Pradesh, India

Vikas Saxena Department of Computer Science and Information Technology, Jaypee Institute of Information Technology, Noida, India

Harkesh Sehrawat Maharishi Dayanand University, Rohtak, Haryana, India

S. Selvabharathy National Institute of Technology Puducherry, Karaikal, India

Aditi Sharan School of Computer and System Sciences, Jawaharlal Nehru University, New Delhi, India

Abhisek Sharma National Institute of Technology, Kurukshetra, India

Meera Sharma SSN College, University of Delhi, Delhi, India

Neha Sharma Department of Computer Science and Information Technology, Jaypee Institute of Information Technology, Noida, India

Rashmi Sindhu Maharishi Dayanand University, Rohtak, Haryana, India

Abhilasha Singh Department of Computer Science and Engineering, SRM Institute of Science and Technology, Delhi-NCR Campus, Ghaziabad, Uttar Pradesh, India

Narotam Singh Department of Cognitive Neurosciences, School of Life Sciences, JSS Academy of Higher Education and Research, Mysuru, Karnataka, India

Pancham Singh Department of Information Technology, Ajay Kumar Garg Engineering College, Ghaziabad, Uttar Pradesh, India

Pankaj Pratap Singh Central Institute of Technology, Kokrajhar, Assam, India

Sakshi Singh Department of Computer Science and Engineering, National Institute of Technology Silchar, Silchar, Assam, India

Shatakshi Singh Department of Information Technology, Ajay Kumar Garg Engineering College, Uttar Pradesh, Ghaziabad, India

V. B. Singh School of Computer & Systems Sciences, JNU, Delhi, India

Amartya Sinha Acharya Narendra Dev College, University of Delhi, Delhi, India

Vikas Siwach Maharishi Dayanand University, Rohtak, Haryana, India

Sonam Department of Electronics & Communications Engineering, Indraprastha Institute of Information Technology Delhi, New Delhi, India

Badal Soni Department of Computer Science and Engineering, National Institute of Technology Silchar, Silchar, Assam, India

Govind Soni School of Computer and System Sciences, Jawaharlal Nehru University, New Delhi, India

Neha Soni University of Delhi South Campus, Delhi, India

Khushal Thakur Chandigarh University, Gharuan, Mohali, Punjab, India

Neha Tomar Department of Electronics Engineering, Shri Ramdeobaba College of Engineering and Management, Nagpur, India

Rishikesh Bhupendra Trivedi Manipal University Jaipur, Jaipur, Rajasthan, India

Monica Uttarwar Faculty of Science and Technology, University of Canberra, Bruce, Australia

Mohit Varshney Department of CSE, ABES Engineering College, Ghaziabad, UP, India

Gyanendra K. Verma Department of Information Technology, National Institute of Technology Raipur, Raipur, India

Shikha Verma School of Computer and System Sciences, Jawaharlal Nehru University, New Delhi, India

Vinita Verma Deen Dayal Upadhyaya College, University of Delhi, Delhi, India

Gayathri Vutla Koneru Lakshmaiah Education Foundation, Andhra Pradesh, India

Vivek Wanve Department of Electronics Engineering, Shri Ramdeobaba College of Engineering and Management, Nagpur, India

Mamta Yadav Department of CSE, ABES Engineering College, Ghaziabad, UP, India

An Automatic Decision Support System for Assessing SDG Contributions

Monica Uttarwar and Girija Chetty

Abstract In this paper, we propose an innovative computer-based decision support scheme based on Artificial Intelligence, consisting of novel text mining and machine learning techniques, to assess an organizational commitment to sustainable practices, with a granular analysis of the text content in their documents, and evaluate their conformance and alignment to one or more of the UN's sustainable development goals. The proposed decision support system can help assist an organization to self-assess their business practices and marketing messages, in terms of presence in the social media channels, company documents, and refine their corporate vision and responsibility statements appropriately.

Keywords Machine learning · Text processing · SDG

1 Introduction

The impact of climate change continues to threaten human life globally, and entire scientific community is unequivocal in professing that global warming and climate change are due to lack of commitment from individuals, organizations, companies and institutions towards environmentally sustainable activities [1]. Some of the eminent research studies [2, 3] have estimated that this trend could cost around 3.6% of a nations GDP by end of this decade, and a global economic impact of 4% loss, due to 4 degrees Celsius rise in the average temperature. Apart from climate change, several other factors could be attributed to risks associated with several developmental measures, such as the economic growth, overall welfare, inequalities, discrimination and corruption. To mitigate the impact of these threats in destabilizing the human society and global ecosystems, organizations and institutions need to be increasingly conscious of how they operate and demonstrate their commitment to sustainable business practices. Particularly, financial institutions need to be

M. Uttarwar · G. Chetty (✉)
Faculty of Science and Technology, University of Canberra, Bruce, Australia
e-mail: girija.chetty@canberra.edu.au

© The Author(s), under exclusive license to Springer Nature Singapore Pte Ltd. 2023

A. Mishra et al. (eds.), *Advances in IoT and Security with Computational Intelligence*,
Lecture Notes in Networks and Systems 755,
https://doi.org/10.1007/978-981-99-5085-0_1

highly responsive to address these challenges, since several studies have indicated that socially responsible companies can achieve higher return on their investments, especially to attract high net worth investors and millennial investors, who are very conscious about global good, and like to ensure that their investments create positive impact [4, 5]. Companies need to adapt to this new reality and revisit their commitments and actions towards CSR (corporate social responsibility) models, which ensures the commitment from an organization towards its impact on society, be accountable and take responsibility towards their actions, policies and procedures [4, 6]. Recently, there has been a bit of criticism on CSR model, being perceived as too soft a model, vulnerable to greenwashing, and some of the organizations showed preference to ESG (environmental, social and governmental) model, which consists of certain new measures to assess the institutions' business and stakeholder commitments [7]. Despite this promising change, there are still some gaps in the model, particularly certain non-binding and suggestive guidelines to reporting compliance to ESG criteria that can be exploited by companies, leading to irregularities and challenging interpretations. This has led to adoption of UN's SDG framework by many companies to report their CSR/ESG efforts [8].

The introduction of United Nation's Sustainable Development Goals (UN's SDG) framework in 2015 has turned out to be better and more relevant in guiding companies in their actions towards their contribution to sustainable practices, and report on the targets achieved. UN's SDG provides 17 goals for achieving sustainable development globally by 2030 and contributes towards a sustainable society. Figure 1 shows a snapshot of these 17 SDG goals to be achieved by 2030. It also helps investors to better identify the companies and business on assessing their commitment towards achieving some of the SDG goals, which in turn can fuel up the companies' efforts towards sustainable development in future [8]. These proactive efforts from companies then get reflected in their vision statements, organizational culture, company reports, social media channels websites, recruitment campaigns and job postings.

Fig. 1 Matrix of 17 SDGs (sustainable development goals) [9]

One of the issues that has been hampering these efforts is malicious attempts by certain entities, who have been plaguing the global efforts towards achieving environmentally sustainable business practices, which is the problem of "greenwashing" [9]. Greenwashing involves a company's attempt to mislead the stakeholders by falsely claiming to be more sustainable than they are in actual practice and create false impression. While most of the organizations and institutions globally have been responsible and tried to honestly revisit their business practices and actions to this new reality (of a need to adopt sustainable business models), there has been some attempts of deception or false impression created by some companies and business to demonstrate their commitment. Greenwashing is one such deceptive practice used by certain organizations to respond to rising awareness to sustainable business practices. This is hard to detect unless special and appropriate data sources and in-depth, objective analysis techniques are used to detect such deceptive and fraudulent practices, and refrain from making any false accusations on a company's genuine attempts as greenwashing attempts accidentally. Much of this abundant self-reported data from companies is available mainly in textual form. It is challenging for humans to dig deeper into this massive text data and get a holistic assessment of company's business practices and their intentions and actions to contribute to UN SDG indicators. Some of the recent advances in computer-based decision support approaches have been promising, particularly in text analysis, involving sophisticated natural language processing (NLP) and machine learning for analysis of data in textual form in an unbiased manner. Further, based on semantic relationships, these approaches provide an assessment in terms of an objective measure, such as the sustainability index or a rating score, which would be a useful measure, and can help guide the organisations in identification and tracking of their eforsts towards SDG contributions [10].

2 NLP Machine Learning Scheme

The NLP/text mining and machine learning scheme developed to process the raw text corresponding to different text phrases and label them to 15 different SDG labels is shown in Fig. 2. The details of each processing module are as described next.

2.1 NLP/Text Mining Dataset

The text data sources used for building the model for classifying the text into different SDG labels involve the use of a publicly available dataset called OSDG-CD Community Dataset [11], made available as an open-source initiative, with an aim to integrate various existing attempts to classify research according to Sustainable Development Goals, and making this process open, transparent and user-friendly. The dataset

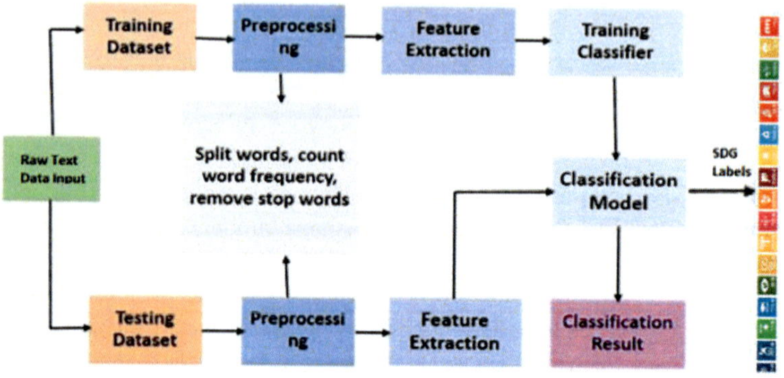

Fig. 2 Block schematic proposed text mining-machine learning scheme

consists of agreement scores in addition to ground truth labels, where average agreement score is computed for each text document was assigned by users, in terms of feedback from several users on the ground truth validity.

Further, the dataset was somewhat unbalanced with more negative ground truth labels (does not assign correctly the text into any of the 15 SDG labels), and fewer positive ground truth labels (classifies the text into one of the 15 SDG categories). Hence this required some pre-processing, where agreement score available in the dataset was used to balance the dataset for building multiclass SDG classifier model. The input text/sentences in the dataset with an agreement score of more than 60% were used for building the model, and any input data with less than 60% agreement score was discarded. This step resulted in total input size of 17,233 text records. Figure 3 shows the snapshot of a sample text record from the database.

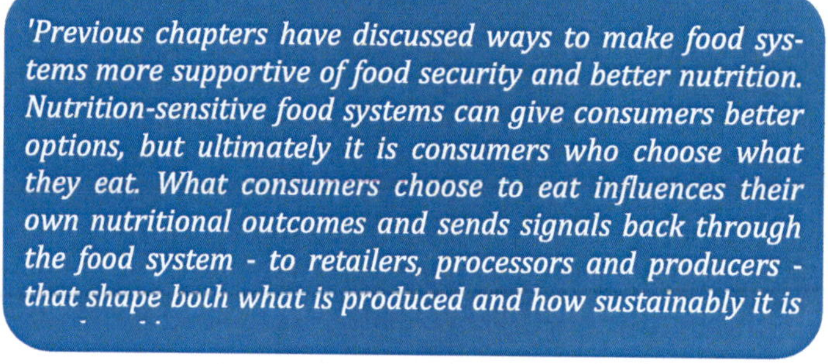
'Previous chapters have discussed ways to make food systems more supportive of food security and better nutrition. Nutrition-sensitive food systems can give consumers better options, but ultimately it is consumers who choose what they eat. What consumers choose to eat influences their own nutritional outcomes and sends signals back through the food system - to retailers, processors and producers - that shape both what is produced and how sustainably it is

Fig. 3 Snapshot of input text record from the OSDG-CD dataset

2.2 NLP Feature Extraction and Machine Learning Model Building

The input text goes through some of the basic NLP pre-processing steps, such as removal of stop words, punctuation, tokenization and stemming before extraction of text features. Several NLP text features were extracted from the pre-processed text such as the N-gram (unigram and bigram) text features and tf-idf statistical text features. The text feature vectors were then used for building and validating different machine learning models based on supervised learning techniques, by splitting the feature vector sets into training subset and test subset (discussed in experimental results section). For this research, five different machine learning algorithms were used for building the classifier model in a supervised learning mode, namely, the Naïve Bayes model, logistic regression model, SVM model, random forest model and gradient boost model. Each of these machines learning algorithmic models are briefly described next.

Naïve Bayes Model: Naive Bayes is a probability-based machine learning algorithm based on Bayes' theorem with the assumption of "naive" independence between the variables (features) and serves as a baseline reference for assessing the classifier performance. The Naive Bayes algorithms are most useful for classification problems and predictive modelling. A Naive Bayes text classifier is based on the Bayes' theorem, which allows computation of conditional probabilities of occurrence of two events based on the probabilities of occurrence of each individual event. For example, knowing that the probabilities of appearance of words user and interface in texts within the category Ease of Use of a feedback classifier for SaaS products are higher than the probabilities of appearance within the Pricing or Features categories, which is used by Naïve Bayes classifier to predict how likely it is for an unknown text that contains those words to belong to either category. Further details of Naïve Bayes algorithm are available elsewhere [12].

Logistic Regression model: Logistic regression is a modelling approach, where prior observations of a dataset are used to predict a binary outcome, such as yes or no. For a logistic regression model, the relationship between one or more of the existing independent variables is analysed to predict the dependent data variable. The simplest form of logistic regression model produces binary outcomes, allowing straightforward decisions between two alternatives. For example, a logistic regression model for determining whether a high school student will receive the acceptance to a college involves taking into consideration multiple input criteria (independent variables) and the logistic function build could consider factors such as the student's grade point average, SAT score and number of extracurricular activities. Based on historical data about earlier outcomes involving the same input criteria, it then scores new cases on their probability of falling into one of two outcome categories, about whether the high school student will get admission into a particular college (the dependent variable). Similar model can be developed for another setting, involving whether a political candidate will win or lose an election (a dependent variable), based on several independent variables used to construct the logistic function. In the machine learning field, logistic regression model is an important tool, often used as baseline reference model (like Naïve Bayes model), for comparing with other advanced machine learning models. In machine learning applications, the logistic regression function modelled with historical data allows predicting the outcome of unseen input data, and with increase in historical data, the predictive power improves. Further details of logistic regression modelling approach are given in [13].

Support Vector Machine model: A Support Vector Machine (SVM) model, based on a supervised machine learning algorithm, can be employed for both classification and regression purposes, although more often used for classification problems. SVM model development is based on the concept of finding a hyperplane that best divides a dataset into two classes, using support vectors (data points or feature sets) nearest to the hyperplane. The support vectors are critical elements of the data/feature set, and by altering or removing any of them, the position of dividing hyperplane gets altered. For example, a classification task involving just two features, a hyperplane could be thought of as a line that linearly separates and classifies the data or feature set, and further than the position of feature set from the hyperplane, more is the confidence that the data has been correctly classified. So, ideally the data points/ feature sets need to be far away, while still on the correct side of hyperplane. Also, with the new unseen test data getting added, the hyperplane determines the class that is assigned to it, depending on which side of the hyperplane the dataset ends up with. The right shape and location of hyperplane is determined during training phase (or model building phase), where different optimization algorithms can be used to maximize the margin between any of the data point and hyperplane surface. Further details of the SVM modelling approach are available in [14].

Random Forest model: Random Forest model involves an ensemble of multiple decision trees and is based on a powerful and versatile supervised machine learning algorithm that involves spreading and combining multiple decision trees in a random fashion to create a "forest." It finds use in both classification and regressing modelling. The decision tree algorithm is based on a simple decision-making technique for classifying the data points into one or other class, based on comparison with a threshold, and contributes to the structure of random forest tree with different thresholds. In Random Forest classification, multiple decision trees are merged for a more accurate prediction. It is based on the logic that that multiple uncorrelated models (the individual decision trees) perform much better as an ensemble or group than they do alone. In this approach, each tree gives a vote or classification decision, and the forest chooses the classification decision with majority vote. The important point here is the fact that there is low (or no) correlation between the individual models, that is, between the decision trees that make up the larger Random Forest model. While individual decision trees may produce errors, majority the group (assessed by votes) will be correct, thus moving the overall outcome in the right direction. The training of random forest model in supervised learning mode happens either with a bagging or boosting method. Further details of this powerful learning algorithm are provided here [15].

Gradient Boost model: Gradient Boost model is an ensemble machine learning model like random forest model, but relies on a different type of boosting strategy, to achieve the efficiencies. It assumes that the best possible model when combined with all previous models would have minimum overall prediction error. The key differentiator for this model involves setting the target outcome for the next best model, with an objective to minimize the error. The name gradient boosting was given because target outcomes for each case are set based on the gradient of the error with respect to the prediction. Each new modelling step proceeds in a direction to minimize prediction error, in the space of possible predictions for each training case. Here, the target outcome for each case in the data depends on how much changing that case's prediction impacts the overall prediction error. If a small change in the prediction for a case causes a large drop in error, then next target outcome of the case is boosted or has a high value. Predictions from the new model that are close to its targets will reduce the error. If a small change in the prediction for a case causes no change in error, then next target outcome of the case is zero. Further details of this efficient machine learning model are described in [16].

For comparing the performance of each machine learning model, different evaluation metrics were used such as the confusion matrix, micro-averaged and macro-averaged metrics, prediction accuracy, precision, recall and F1-measure, as described next.

2.3 *Evaluation Metrics*

Evaluation metrics allow objective assessment of a model's performance based on quality measures and help improve the state of the art though feedback and comparison based on target standards [17, 18]. The evaluation metrics selected depend on the classification task and domain, and metrics chosen for this study are described next.

Confusion Matrix

The confusion matrix metric allows evaluating the performance of a machine learning model for classification problems. For a binary classification problem, for instance, there are four possible classification decisions: True Positive (TP), False Positive (FP), False Negative (FN) and True Negative (TN). For TP, the values are correctly predicted as 1 and for FP mispredicted as 1. The same applies for the 0 predictions, where the correctly predicted ones are considered TN and the falsely predicted ones as FN [19]. For this study, we used multiclass classification, not binary classification, involving classification into 15 classes, corresponding to different SDG categories.

Accuracy

The accuracy evaluation metric involves measurement of the ratio of correct predictions over the total number of predictions, as shown in Eq. (1) [20]. It is the most popular evaluation metric, but indicative of correct performance only for balanced class distribution in the dataset, with equal distribution of labels in the data. With imbalanced classes, the metric does not truly reflect the performance of the classifier, particularly for minority class or under-represented class, and provides a misleading picture [21].

$$Accuracy = \frac{TP + TN}{TP + FP + TN + FN} \tag{1}$$

F1-Score

F1-score is an evaluation metric more suitable when there is a large imbalance in class distribution, and more representative of the performance of minority classes, which is the case with the proposed SDG classification from text input. This is due to problem with many of the sentences in the text input from the data subsets, which do not clearly fit into any of the SDG label thematically. The F1-score is created from the harmonic mean of another two metrics, precision and recall, which are described next.

Precision and Recall

Precision is calculated by dividing the true positives by anything that was predicted as a positive, as shown in Eq. (2).

$$Precision = \frac{TP}{TP + FP} \tag{2}$$

Recall on the other hand is calculated by dividing the true positives by anything that should have been predicted as positive, as depicted in Eq. (3).

$$Recall = \frac{TP}{TP + TN} \tag{3}$$

In other words, precision measure evaluates the number of instances that are relevant, out of the total instances the model retrieved, and tries to answer the question–*How many are relevant, out of the total instances the model retrieved*? Whereas Recall measure evaluates the number of instances which the model correctly identified as relevant, out of the total relevant instances, and answers the question–*How many retrieved items are relevant*? Recall is also called as True Positive Rate (TPR) in machine learning literature. With precision and recall introduced, the F1-score can now be defined as follows (Eq. (4)):

$$F1\,Score = 2 * \frac{Precision * Recall}{Precision + Recall} \tag{4}$$

Micro- and Macro-evaluation Metrics: Since the proposed SDG classification is a multiclass classification task, involving classification of input text into 15 different SDG classes, 2 more metrics are more representative of the model performance.

They are the micro- and macro-level scores and focus on different aspects of model performance. The macro-scores provide a weighted score on each class similarity, irrespective of its size, whereas the micro-score assigns proportionally more weight to majority classes than smaller ones. Using both macro- and micro-level scores, it is possible to assess the holistic performance, by evaluating through different perspectives. For instance, the Precision, Recall and F1-micro-scores are calculated as depicted in Eq. (5), where c is the number of classes.

$$\begin{aligned}
Precision_{micro} &= \frac{\sum_{n=1}^{c} TP_n}{\sum_{n=1}^{c} TP_n + \sum_{n=1}^{c} TN_n} \\
Recall_{micro} &= \frac{\sum_{n=1}^{c} TP_n}{\sum_{n=1}^{c} TP_n + \sum_{n=1}^{c} FP_n} \\
F1\,Score_{micro} &= 2 * \frac{\sum_{n=1}^{c} Precision_{micro_n} * \sum_{n=1}^{c} Recall_{micro_n}}{\sum_{n=1}^{c} Precision_{micro_n} * \sum_{n=1}^{c} Recall_{micro_n}}
\end{aligned} \tag{5}$$

The macro-level scores are calculated as binary metrics for each class, and the unweighted mean of the classes is formed. The lower performance of the model on minority classes is more weighted or emphasized, as depicted in Eq. (6).

$$F1\,Score_{macro} = \frac{\sum_{n=1}^{c} (2 * \frac{Precision*Recall}{Precision+Recall})n}{c} = \frac{\sum_{n=1}^{c} F1\,Score_n}{c} \tag{6}$$

The details of the experimental results obtained for different machine learning models and their performance assessment in terms of different evaluation metrics are presented next, and described in details in the next section.

3 Experimental Results

The NLP text feature vectors, consisting of unigram and bigram features, and tf-idf features, were split into training and test subsets, with around 12,924 features used for model building in training phase, and 4309 features for validation in the test phase. All the machine learning models discussed in Section II used a supervised learning strategy. Supervised learning mode involves model building with an annotated training dataset with ground truth labels. Since the model is trained using many different examples of various inputs and outputs, and thus learns how to classify any unseen, new input data, and predict future outcomes. For building machine learning models, different types of algorithms can be used to train and learn how to classify the data points from input sources into different groups or classes. For example, an email spam filter can be taught to distinguish between an email as spam and non-spam by learning to identify the patterns with an annotated training set. In real-world enterprise settings, the prediction capabilities of machine learning models based on different learning algorithms either in supervised or unsupervised (the dataset with no ground truth) can help provide useful computer-based decision support strategies needed. Further, supervised or unsupervised machine learning models can be built from data or feature vectors, for either regression task or classification task, where regression task involves predicting the continuous value, and the classification task involves predicting the category, label, or class, often a discrete number (for example, a spam/no spam for a binary spam classifier). For classification tasks, the class label is called dependent attribute, and the input feature vectors or data point the independent

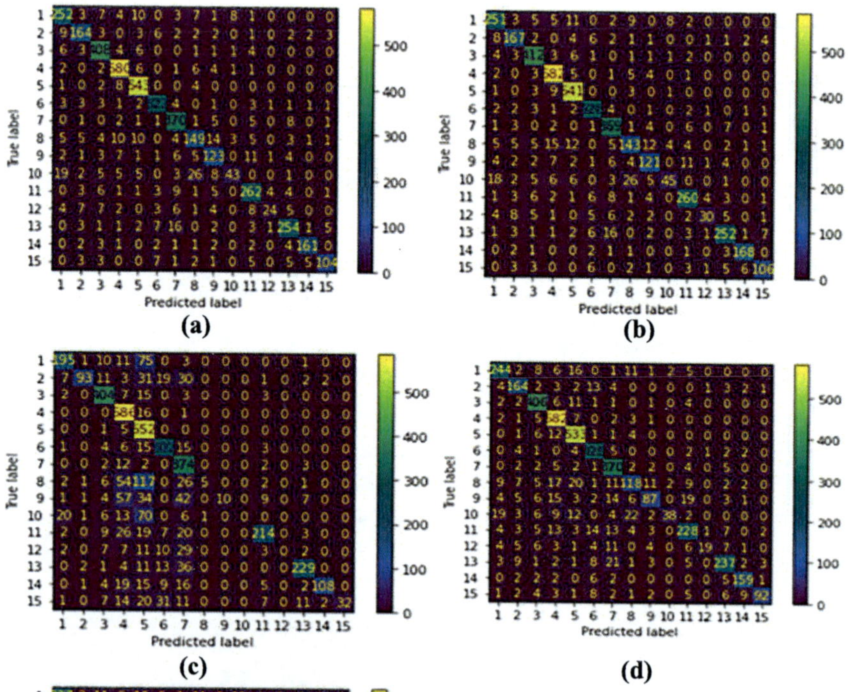

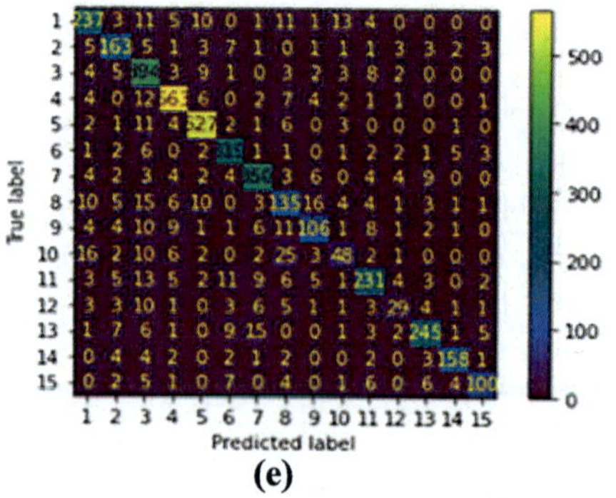

	Model	Score
1	SVM Classifier	0.875609
0	Logistic Regression	0.872360
4	Gradient Boosting	0.836621
3	Random Forest	0.835925
2	Naive Bayes	0.720353

(f)

Fig. 4 **a** Confusion matrix-Naïve Bayes model. **b.** Confusion matrix-logistic regression model. **c.** Confusion matrix-support vector machine model. **d.** Confusion matrix-random forest model. **e.** Confusion matrix-gradient boost model. **f.** Comparing different classifiers with accuracy metric

Fig. 5 Classification report for best performing classifier (SVM classifier) with all evaluation metrics (accuracy, precision, recall, F1-score

accuracy 0.875609190067301

	precision	recall	f1-score	support
1	0.83	0.85	0.84	296
2	0.81	0.84	0.82	199
3	0.90	0.95	0.93	434
4	0.92	0.97	0.94	603
5	0.91	0.97	0.94	558
6	0.91	0.94	0.92	346
7	0.87	0.93	0.90	395
8	0.72	0.67	0.69	214
9	0.77	0.73	0.75	165
10	0.76	0.38	0.51	117
11	0.88	0.87	0.87	300
12	0.77	0.42	0.55	71
13	0.89	0.85	0.87	296
14	0.94	0.94	0.94	179
15	0.87	0.78	0.82	136
accuracy			0.88	4309
macro avg	0.85	0.81	0.82	4309
weighted avg	0.87	0.88	0.87	4309

variables or attributes. The performance of each machine learning model in terms of Confusion Matrix metric is shown in Fig. 4a–e. Figure 4f shows the performance of each machine learning algorithm used in this study, in terms of micro-averaged F1-score. Figure 5 shows the detailed classification in terms of performance achieved for each of the 15 SDG classes, in terms of accuracy, precision, recall, F1-score, and micro- and macro-averaged metrics [27].

4 Conclusions and Further Plan

This paper reports the research in progress towards development of an innovative computer decision support system based on novel text mining and machine learning scheme for assessing the organizational commitment to sustainable business practices, using the UN SDG framework for assessment. An open-source publicly available OSDG text dataset (representing several corporate and marketing documents) was used for building the models, and a multiclass classifier based on different machine learning models and NLP text features extracted from the datasets. The performance of the 15-class SDG multiclass classifier assesses the classifier performance in terms of different evaluation metrics such as the confusion matrix, accuracy, precision and recall. Further research is in progress towards using larger datasets with different types of corporate documents and building advanced machine learning models based on deep learning and transformer networks.

References

1. United Nations Sustainable Development Goals https://sdgs.un.org/goals. Accessed 17 Dec 2022
2. Ackerman F, Stanton EA, Hope C, Alberth S., Fisher J, Biewald B, Martin E (2007) The cost of climate change: what we'll pay if global warming continues unchecked. Technical report
3. Bralower T, Bice D (2021) The economic costs of climate change. https://www.e-education.psu.edu/earth103/node/717. Accessed 17 Dec 2022
4. Friede G, Busch T, Bassen A (2015) ESG and financial performance: aggregated evidence from more than 2000 empirical studies. J Sustain Finan Invest 5(4):210–233
5. King G (2019) The importance of ESG investing for HNWIs. https://insight.factset.com/the-importance-of-esg-investing-for-hnwis. Accessed 17 Dec 2022
6. Fernando J (2021) Corporate social responsibility (CSR). https://www.investopedia.com/terms/c/corp-social-responsibility.asp. Accessed 17 Dec 2022
7. Heller C (2021) From CSR to ESG: how to kickstart your ESG program in 2021–NAVEX global. https://www.jdsupra.com/legalnews/from-csr-to-esg-how-to-kickstart-your-3703495/. Accessed 17 Dec 2022
8. Eccles RG, Krzus MP, Rogers J, Serafeim G (2012) The need for sector-specific materiality and sustainability reporting standards. J Appl Corp Finan 24(2):65–71
9. What is natural language processing? https://www.sas.com/en_us/insights/analytics/what-is-natural-language-processing-nlp.html. Accessed 17 Dec 2022
10. Resolutions adopted by the General Assembly on 6 July 2017. Technical report. https://unstats.un.org/sdgs/files/report/2017/thesustainabledevelopmentgoalsreport2017.pdf. Accessed 17 Dec 2022
11. Putting sustainability and social equity at the heart of global economic policies. https://mptf.undp.org/fund/pge00#:~:text=The%20work%20of%20Partnership%20for,annual%20contributions%20amounted%20to%20%245%2C250%2C667. Accessed 17 Dec 2022
12. OSDG Community Dataset (OSDG-CD) https://zenodo.org/record/5550238#.Y51cc3ZBw2y. Accessed 17 Dec 2022
13. Naïve Bayes Classifier. https://en.wikipedia.org/wiki/Naive_Bayes_classifier. Accessed 17 Dec 2022
14. Logistic Regression Classifier https://en.wikipedia.org/wiki/Logistic_regression. Accessed 17 Dec 2022
15. Support Vector Machine. https://en.wikipedia.org/wiki/Support_vector_machine. Accessed 17 Dec 2022
16. Random Forests. https://en.wikipedia.org/wiki/Random_forest. Accessed 17 Dec 2022
17. Gradient Boost. https://en.wikipedia.org/wiki/Gradient_boosting. Accessed 17 Dec 2022
18. DeepAI (2021) Evaluation metrics definition. https://deepai.org/machine-learning-glossary-and-terms/evaluation-metrics. Accessed 17 Dec 2022
19. Srivastava T (2019) 11 important model evaluation metrics for machine learning everyone should know. https://www.analyticsvidhya.com/blog/2019/08/11-important-model-evaluation-error-metrics/. Accessed 17 Dec 2022
20. Narkhede S (2018) Understanding confusion matrix. https://towardsdatascience.com/understanding-confusion-matrix-a9ad42dcfd62. Accessed 17 Dec 2022
21. Hossin M (2015) Article in International Journal of Data Mining & Knowledge Management Process March. Int J Data Min Knowl Manag Proc (IJDKP) 5(2)
22. Dreher A, Herzfeld T (2005) The economic costs of corruption: a survey and new evidence. SSRN Electron J
23. Ferrant G, Kolev A (2016) The economic cost of gender-based discrimination in social institutions. Technical Report
24. Mcadams RH (2010) Economic costs of inequality recommended citation. Technical Report 1
25. Kenton W (2022) Greenwashing definition. https://www.investopedia.com/terms/g/greenwashing.asp. Accessed 17 Dec 2022

26. Partnership for action on green economy link. Accessed 17 Dec 2022
27. Wood T (2021) F-score definition. https://deepai.org/machine-learning-glossary-and-terms/f-score. Accessed 17 Dec 2022

Removing Stegomalware from Digital Image Files

Vinita Verma, Sunil K. Muttoo, V. B. Singh, and Meera Sharma

Abstract Steganography, a data hiding technique, has trended into a lucrative means to hide malware within digital media and other files to avoid detection. Such malware hidden by means of steganography is known as stegomalware. Detecting stegomalware has been difficult and indeed removing such malware from the file is a big challenge. A tool has been created (Verma V, Muttoo SK, Singh VB Detecting stegomalware: malicious image steganography and its intrusion in windows. In: International conference on security, privacy and data analytics. Springer, Singapore (2022). 10.1007/978–981-16–9089-1_9) to detect such malware hidden within widely used JPEG file format. This paper introduces new and significant functionality to our tool (Verma V, Muttoo SK, Singh VB Detecting stegomalware: malicious image steganography and its intrusion in windows. In: International conference on security, privacy and data analytics. Springer, Singapore (2022). 10.1007/978–981-16–9089-1_9) to remove such malware from the file after detection, unlike the existing techniques limited to detection. The tool proposed in this paper has rendered the malicious image files benign with a success rate of 99.92%.

Keywords Malware · Steganography · Digital image · JPEG

V. Verma (✉)
Deen Dayal Upadhyaya College, University of Delhi, Delhi, India
e-mail: vinitaducs@gmail.com

S. K. Muttoo
Department of Computer Science, University of Delhi, Delhi, India
e-mail: drskmuttoo@gmail.com

V. B. Singh
School of Computer & Systems Sciences, JNU, Delhi, India
e-mail: vbsingh@mail.jnu.ac.in

M. Sharma
SSN College, University of Delhi, Delhi, India
e-mail: meerasharma@ss.du.ac.in

© The Author(s), under exclusive license to Springer Nature Singapore Pte Ltd. 2023
A. Mishra et al. (eds.), *Advances in IoT and Security with Computational Intelligence*, Lecture Notes in Networks and Systems 755,
https://doi.org/10.1007/978-981-99-5085-0_2

1 Introduction

Malware, short for malicious software, is intentionally created to affect computer systems and their functioning, and exploit data on systems via unauthorized access. On the other hand, steganography is a technique for hiding confidential data within digital media without raising suspicion. However, steganography has trended into the malicious use of hiding malware within the files. Such malware hidden using steganography is known as stegomalware.

The techniques for malware detection majorly comprise static and dynamic analysis. The static technique disassembles and analyzes the source code for suspicious operations while dynamic analysis executes the code and tracks malicious operations. The static analysis doesn't generally analyze steganography within images and hence anything suspicious in image files. On other hand, execution in the dynamic analysis is time- and resource-consuming. Therefore, malware developers find steganography a lucrative means to hide malware to evade detection. The stegomalware has been on the rise in recent years [1, 2]. Specifically, digital images are popularly used for they are massively used in cyberspace, contain redundancy, and therefore can hide data without degrading the visual quality.

Some methods have been discussed in the literature to detect malicious image steganography. There was detected unusual data appended at the end of GIF images [3] and the URL embedded in LSBs of images [4]. The DCT-based technique proposed [5] resulted in 80% protection against cyber-attacks on images and video streams. Kernel tracing-based detection [6] aims for a more programmatic approach to trace execution patterns for stegomalware. A study performed on favicons [7] using steganalysis and features that exploit flat areas detected malicious steganography in favicons.

Stegomalware in smartphones [8] was demonstrated through an app containing malign executable components within its assets. The results were not convincing, though revealed substantial data hidden within the app assets. The malicious JPEG/PNG images stored as resources in the application showed the ease of trivial hiding techniques to evade anti-malware [9]. Steganography was used as a threat model [10] implementing a malicious Android app. As per observation, the image chunks that fail to meet standard file format are ignored by the picture viewer and therefore used for inserting malicious codes. A technique for obfuscated malware component detection [11] was used for stegomalware detection within smartphone apps. It is based on analyzing behavioral differences between the original and the fault-injected app. The technique, however, utilizes a dynamic approach. A tool was created to detect stegomalware within widely used JPEG file format [12].

There has been found a lack of effective techniques for removing malware from files. This paper addresses this issue. The rest of the paper is structured as follows. Section 2 discusses related work. The procedure for the tool proposed is presented in Sect. 3. Section 4 describes the encoding scheme. Section 5 describes the experiment and the results are discussed in Sect. 6. Section 7 provides an analysis of the results, and Sect. 8 concludes the paper.

2 Related Work

A machine learning-based technique concerning stegomalware used file structure-based features for detecting malicious JPEG images [13]. It resulted in a false positive rate of 0.004 and a true positive rate of 0.951. Artificial immune system-based steganalysis [14] for JPEG images obtained a steganogram detection rate of 94.33% using horizontal coefficients in Haar wavelet while an average rate of 85.71% using vertical and diagonal coefficients. The image entropy-based technique [15] attained an accuracy of about 80% for the images using several JPEG hiding methods. A framework to identify malign JPEG images [16] over social networking sites used steganalysis, extracting embedded data and identifying file headers. The data were uploaded to VirusTotal [17] to verify the results. The technique, however, has not analyzed any real-world dataset and the data embedded may not necessarily contain a header, failing the approach. Fridrich et al. [18] discussed steganalysis based on features for JPEG images and its association with the design of stegosystems. A technique was proposed to differentiate between malicious and legitimate JPEG operations [19].

Several steganographic threats have been carried out in the wild particularly using JPEG file format. An incident was reported in July 2013 where a backdoor used PHP functions within the JPEG header to compromise a site [20]. The term 'stegosploit' was introduced at the 2015 Black Hat conference, which refers to an image-based exploit [21]. Such exploits are inserted into JPEG and PNG images with JavaScript and HTML code. It results in a polyglot that appears as an innocuous image though triggered in the victim's browser. November 2016 brought to the limelight a popular social networking site, Facebook, for the use of JPEG images to transmit Locky ransomware [22]. There was spread SyncCrypt ransomware in August 2017 through JPEG [23]. Further, Twitter made use of memes (JPEG images) in December 2018 for the purpose of communication with malware [24]. A malware named LokiBot came to light for its upgradation in August 2019 for hiding its source code within images [25]. In December 2019, the JPEG of Taylor Swift was reported for hiding a crypto mining botnet [26]. Given these escalating cyber attacks using JPEG, there is a need for such a technique that can not only detect but also wipe off the malware from the file. The tool presented in this paper performs detection. Alongside, it removes the hidden stegomalware from the image file.

3 Tool Proposed

We proposed a tool JPEGious [12] (created using Python) to classify the input image file into malicious or non-malicious. It detects stegomalware within JPEG format files. The tool performs detection based on locating malicious content within the image or deviation from the standard file format. This paper introduces a new functionality to our tool [12], which is to remove the stegomalware detected, to render the file benign. We name this tool JPEGious+ for it not only detects the stegomalware but also removes such malware from the file. The malicious content detected by JPEGious+ is produced as the output in text form, which is convenient for the user to comprehend unlike that of JPEGious [12] which outputs the content in both hexadecimal and text form. The working of JPEGious+ is described as follows:

a. JPEGious+ checks the beginning two bytes of the input image file. In case it finds the file starting with bytes other than SOI marker bytes 0xFFD8 (in hexadecimal notation), which indicates the start of the image [12], it classifies the image as malicious.
b. If the image is found non-malicious following the case mentioned above, JPEGious+ locates the header segment containing metadata using corresponding marker bytes 0xFFEn, n = 0...F. The tool then converts the bytes comprising metadata into ASCII format followed by further conversion into the respective text. JPEGious+ detects malicious content present in metadata using suspicious keywords such as 'eval', 'iframe', and 'script' [12]. The string 'eval' indicates the presence of a JavaScript function named eval(), which executes or evaluates the arguments passed to it. The argument can be any malicious JavaScript statements. Another string 'iframe' marks the presence of an HTML inline frame tag which is used to embed another document within the current HTML document. The presence of the 'script' string indicates the presence of another HTML script tag. It is used to specify a JavaScript or a PHP file to load. Detection of the presence of either of these strings or keywords results in JPEGious+ classifying the image file as malicious. Further, the tool removes such malicious metadata keeping the marker bytes for the header. Along with this, the two bytes following the marker bytes that indicate the length of data comprising the segment (including two bytes for itself) are altered to 0×0002 for now the metadata consists of just two length bytes.
c. If no malicious data is found within the header, JPEGious+ searches the comment segment indicated by 0xFFFE marker bytes for any suspicious content using the same keywords [12] in a similar manner as used for the header in case b. On detecting the malicious comment, JPEGious+ removes such comment keeping its two marker bytes and changing the following two length bytes to 0×0002 as done for the header.
d. If the tool doesn't find any malicious comment, it looks for the EOI marker bytes 0xFFD9 which indicates the end of the file [12]. In case the tool finds such bytes missing, it classifies the image as malicious for the image deviates from the standard file format.
e. In case of the EOI marker bytes are located, JPEGious+ searches for anomalous data appended at the end of the image file by locating data containing suspicious keywords present after 0xFFD9 bytes [12]. In case malicious content is detected, such trailing data are stripped from the file.

In case, the tool doesn't find the file malicious following the cases mentioned above, it classifies the file as non-malicious. Additionally, JPEGious+ contains another functionality to decrypt the encrypted malicious data detected. While detection, JPEGious+ locates encrypted data using words indicating encoding such as 'Base64' (discussed in Sect. 4). The tool decrypts the data and displays it as part of the output followed by removing such data. Figure 1 illustrates the working of the tool proposed.

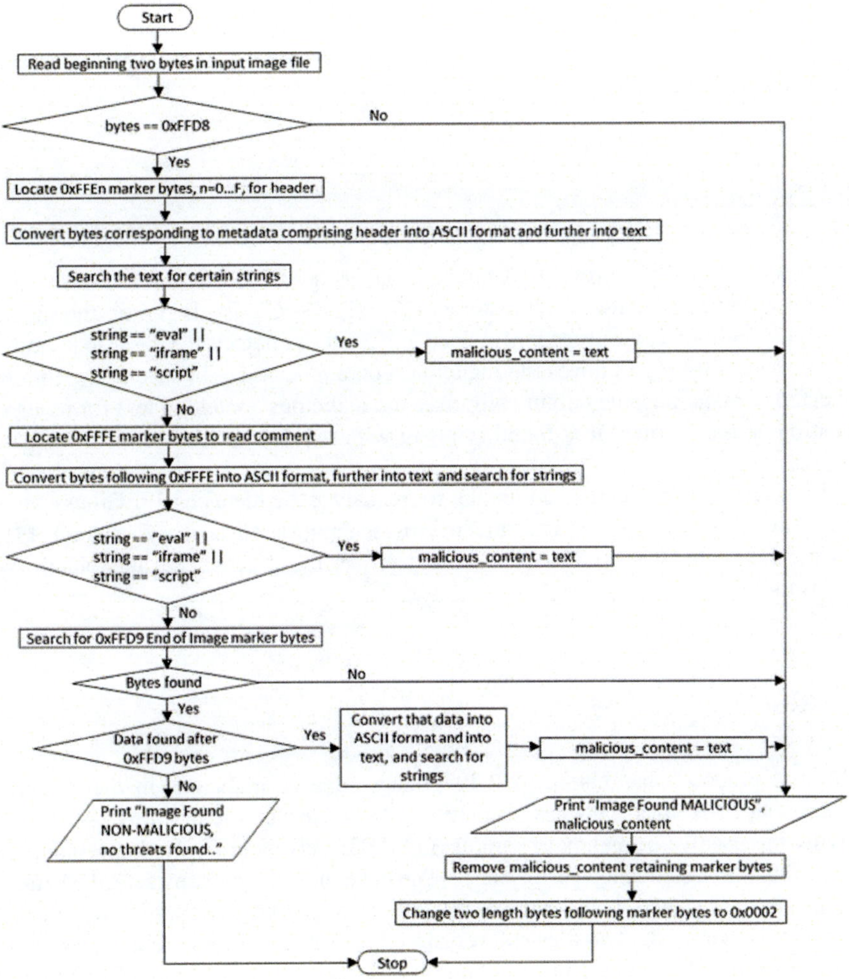

Fig. 1 Working of JPEGious+

4 Encoding Scheme

A binary-to-text encoding is the encoding of data in plain text. Base64 encoding is used to encode binary data into ASCII characters. It represents binary data (sequence of 8-bit bytes) as sequences of 24 bits that can be represented by four 6-bit Base64 digits. Base64 is designed to transmit data stored in binary format across channels that reliably only support text content. This encoding technique is commonly used in many applications such as to send e-mails, embed image files, or other binary assets inside textual assets, etc.

5 Experiment

The experimental data consist of a dataset created using 2,620 malicious JPEG images [12] collected from malware repositories [27, 28]. JPEGious+ is run on the input image file. Following the procedure in Sect. 3, JPEGious+ detects the file as malicious or non-malicious and removes the malicious content detected. In case of the file is detected as malicious, the output comprises the malicious content in text form, also mentioning the segment it is found in along with the decrypted data in cases that follow.

To ensure that JPEGious+ is reliable, we reanalyze the file using JPEGious+ and indeed also scan the file using a prominent virus scanning service/site, VirusTotal [17] that contains over 60 anti-malware scanners to check the effectiveness and reliability of our tool.

6 Result

The tool proposed has classified 2,397 image files as malicious in the dataset comprising 2,620 files. This has resulted in a False Negative Rate (rate of misclassifying the positive or malicious samples) of 0.08. This parameter is important for evaluating the classification performance. The lower the false negative rate, the better the performance. JPEGious+ has successfully removed stegomalware from every such file. However, the two files still remain malicious. The files have been verified using VirusTotal [17]. Therefore, JPEGious+ has rendered the malicious files benign with a success rate of 99.92%. The output of the tool, JPEGious+ , is presented in Figs. 2, 3 and 4 via screenshots corresponding to three different files referred to by their SHA-256 values. Figure 2 presents the malicious content appended to the end of the file. Figure 3 displays the malicious content detected in the metadata in the header. We find that such data is encrypted using the base64 encoding technique. The proposed tool possesses the ability to decrypt such data. Therefore, JPEGious+ presents the decrypted data along with the corresponding encrypted malicious data

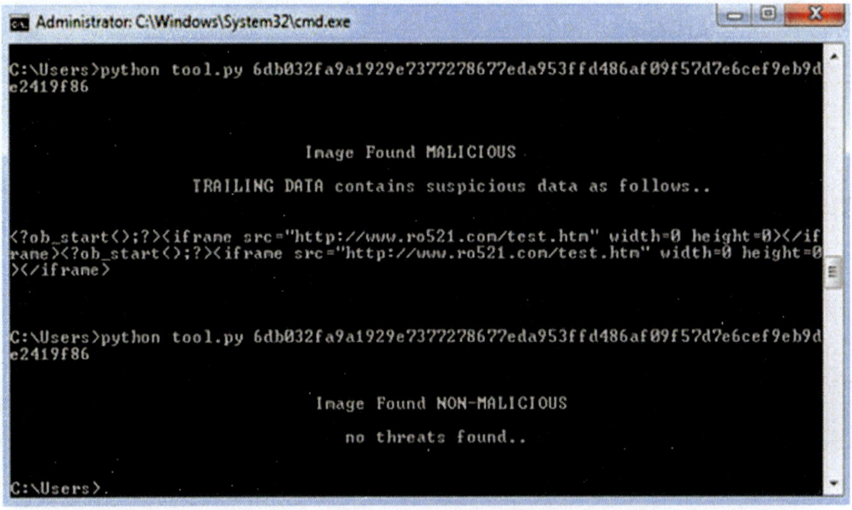

Fig. 2 Output for a file containing malware appended to end of the file

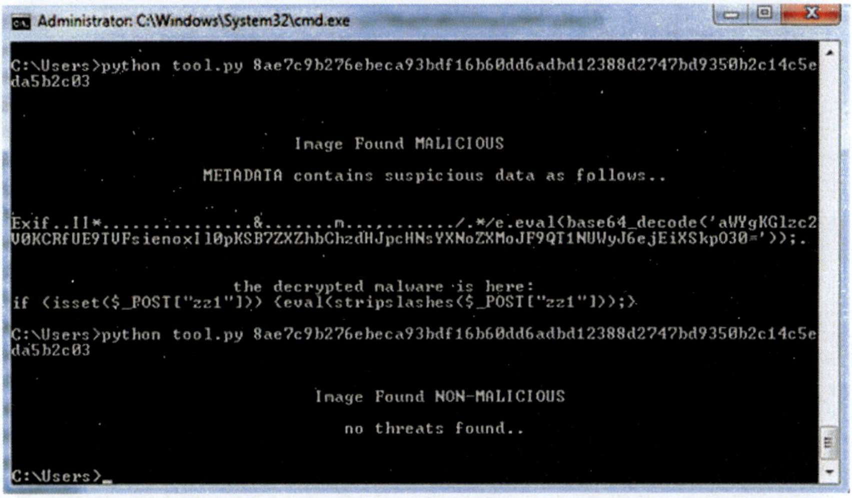

Fig. 3 Output for a file containing malware in header

detected as shown in Fig. 3. Figure 4 displays the detected malicious comment. All three figures show that JPEGious+ is run again after detection. It is to verify if the malware detected is removed successfully. JPEGious+ has resulted in a clean file on every second execution for the same file.

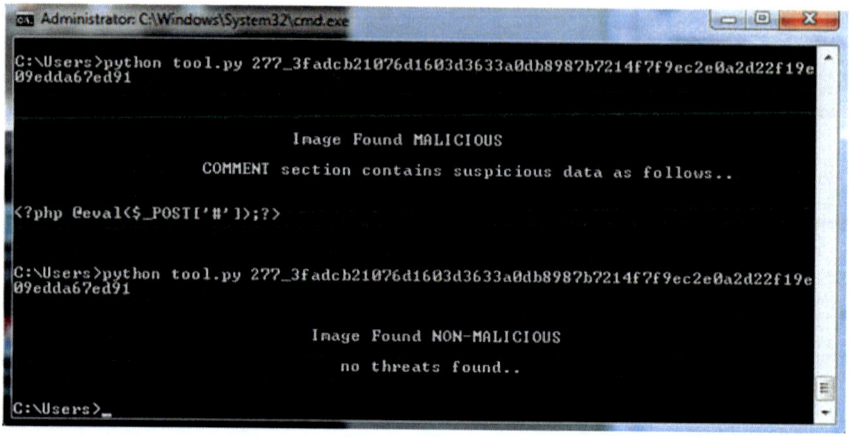

Fig. 4 Output for a file containing malware in comment

7 Analysis

Analyzing the results, we find that the proposed tool has resulted in a success rate of about 99% when it comes to removing the stegomalware from the file. This can be attributed to the functionality of the proposed tool for detection. It uses certain suspicious keywords for locating malicious content such as 'eval', 'script', and 'iframe'. Besides this functionality, the output of JPEGious+ is different from that of our previously introduced tool [12]. As mentioned in Sect. 3, the output of the tool [12] as depicted in Figs. 5 and 6 presents the content detected in both hexadecimal and text form while JPEGious+ focuses on presenting the content only in text form comprehensible to the user as provided in Figs. 2, 3 and 4. The figures provide a better view of the difference in the outputs of JPEGious+ and the tool proposed [12].

Fig. 5 Output of the tool [12]

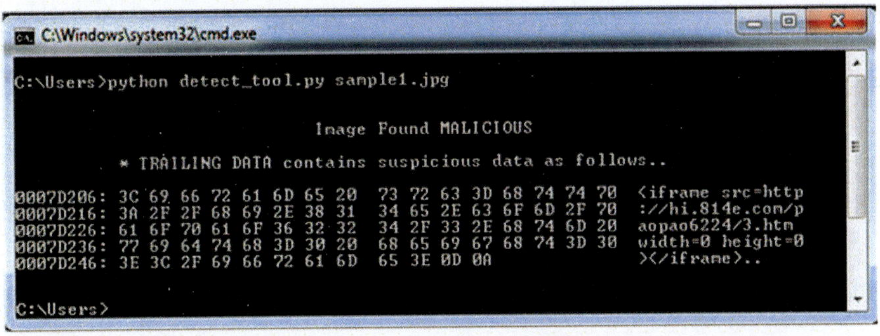

Fig. 6 Another output of the tool [12]

Figure 7 depicts the percentage distribution of the presence of these suspicious keywords within the malign images analyzed. We find the script tag in about 53% of images as depicted in the figure. 38% of images are embedded with the iframe tag with hidden visibility. On the other hand, about 6% of images contain both elements. The image files containing JavaScript eval function executing malicious string arguments comprise 3% of the distribution.

Therefore, we find that image files contain HTML and JavaScript code to carry out malicious purposes. The proposed tool uses certain keywords which can locate HTML and JavaScript elements embedded within the file and indeed, removes such contents upon detection which were causing the file to be malign, resulting in a good success rate.

Another analysis is illustrated in Fig. 8 presenting the threat distribution in malicious images. The figure shows that about 94% of images contain malicious content appended at the end of the file. The images containing malicious metadata in the

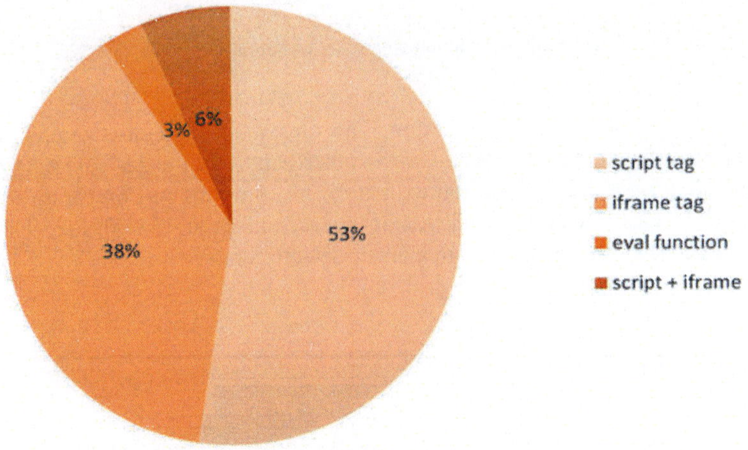

Fig. 7 Percentage distribution of suspicious keywords within malicious images

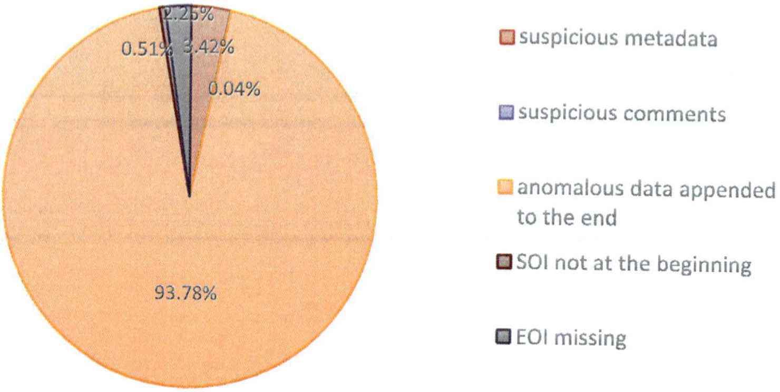

Fig. 8 Threat distribution in malicious images

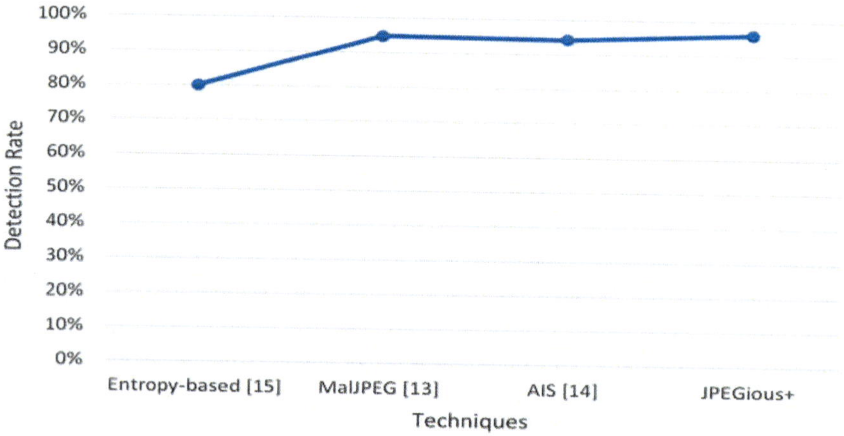

Fig. 9 Comparison of our result of detection with other techniques

header comprise 3% of the distribution. The files with malicious comments are comparatively fewer, about 0.04%. On the other hand, the images deviating from standard file format account for approximately 3%. Providing a comparative analysis of our result of detection with that of other techniques [13–15], Fig. 9 shows that the proposed tool has resulted in a better detection rate.

8 Conclusion

This paper has presented a tool named JPEGious+ containing functionality to remove the malware hidden by means of steganography (stegomalware) within a popularly used digital medium, JPEG format files. The image files are found to contain HTML and JavaScript code to carry out malicious purposes. The tool proposed is based on locating the malicious content within the file using certain suspicious HTML and JavaScript elements or keywords. The tool not only detects but also removes such malicious content from the file, retaining the file format. JPEGious+ is therefore found capable of rendering the malicious files benign with a success rate of 99.92%.

Acknowledgements The authors have contributed to the research without any conflict of interest. The authors are grateful to VirusShare.com [27] and Hybrid-Analysis.com [28] for providing access to their malware repositories. The research, to mention, has not received any grant from any funding agency in the commercial, public, or not-for-profit sectors.

References

1. Cabaj K, Caviglione L, Mazurczyk W, Wendzel S, Woodward A, Zander S (2018) The new threats of information hiding: the road ahead. IT Prof 20:31–39. https://doi.org/10.1109/MITP.2018.032501746
2. Mazurczyk W, Wendzel S (2017) Information hiding: challenges for forensic experts. Commun ACM 61:86–94. https://doi.org/10.1145/3158416
3. Puchalski D, Caviglione L, Kozik R, Marzecki A, Krawczyk S, Choraś M (2020) Stegomalware detection through structural analysis of media files. In: ARES '20: Proceedings of the 15th international conference on availability, reliability and security. ACM, pp 1–6. https://doi.org/10.1145/3407023.3409187
4. Aljamea MM, Iliopoulos CS, Samiruzzaman M Detection of URL in image steganography. In: ICC '16: Proceedings of the international conference on internet of things and cloud computing. ACM, pp 1–6. https://doi.org/10.1145/2896387.2896408
5. Amsalem Y, Puzanov A, Bedinerman A, Kutcher M, Hadar O (2015) DCT-based cyber defense techniques. In: Proc. SPIE 9599, applications of digital image processing XXXVIII. SPIE. https://doi.org/10.1117/12.2187498
6. Carrega A, Caviglione L, Repetto M, Zuppelli M (2020) Programmable data gathering for detecting stegomalware. In: 2020 6th IEEE conference on network softwarization (NetSoft). IEEE, pp 422–429. https://doi.org/10.1109/NetSoft48620.2020.9165537
7. Pevny T, Kopp M, Křoustek J, Ker AD (2016) Malicons: detecting payload in favicons. In: Electronic imaging symposium, media watermarking, security, and forensics. pp 1–9. https://doi.org/10.2352/ISSN.2470-1173.2016.8.MWSF-079
8. Suarez-Tangil G, Tapiador JE, Peris-Lopez P (2014) Stegomalware: playing hide and seek with malicious components in smartphone apps. In: International conference on information security and cryptology, LNCS. Springer, Cham, pp 496–515. https://doi.org/10.1007/978-3-319-16745-9_27
9. Badhani S, Muttoo SK (2018) Evading android anti-malware by hiding malicious application inside images. Int J Syst Assur Eng Manag 9:482–493. https://doi.org/10.1007/s13198-017-0692-7
10. Cao C, Zhang Y, Liu Q, Wang K (2015) Function escalation attack. In: International conference on security and privacy in communication networks, LNICST. Springer, Cham, pp 481–497. https://doi.org/10.1007/978-3-319-23829-6_33
11. Suarez-Tangil G, Tapiador JE, Lombardi F, Pietro RD (2016) Alterdroid: differential fault analysis of obfuscated smartphone malware. IEEE Trans Mob Comput 15:789–802. https://doi.org/10.1109/TMC.2015.2444847
12. Verma V, Muttoo SK, Singh VB (2022) Detecting stegomalware: malicious image steganography and its intrusion in windows. In: International conference on security, privacy and data analytics. Springer, Singapore. https://doi.org/10.1007/978-981-16-9089-1_9

13. Cohen A, Nissim N, Elovici Y (2020) MalJPEG: machine learning based solution for the detection of malicious JPEG images. IEEE Access 8:19997–20011
14. Pérez JDJS, Rosales MS, Cruz-Cortés N (2016) Universal steganography detector based on an artificial immune system for JPEG images. In: 2016 IEEE Trustcom/BigDataSE/ISPA. IEEE, pp 1896–1903. https://doi.org/10.1109/TrustCom.2016.0290
15. Natarajan V, Sheen S, Anitha R (2012) Detection of StegoBot: a covert social network botnet. In: SecurIT '12: Proceedings of the first international conference on security of Internet of Things. ACM, pp 36–41. https://doi.org/10.1145/2490428.2490433
16. Kunwar RS, Sharma P (2017) Framework to detect malicious codes embedded with JPEG images over social networking sites. In: 2017 International conference on innovations in information, embedded and communication systems (ICIIECS). IEEE, pp 1–4. https://doi.org/10.1109/ICIIECS.2017.8276144
17. Virus scanning website. https://www.virustotal.com
18. Fridrich J (2004) Feature-based steganalysis for JPEG images and its implications for future design of steganographic schemes. In: International workshop on information hiding, LNCS. Springer, Berlin, Heidelberg, pp 67–81. https://doi.org/10.1007/978-3-540-30114-1_6
19. Lin C-Y, Chang S-F (2001) A robust image authentication method distinguishing JPEG compression from malicious manipulation. IEEE Trans Circuits Syst Video Technol 11:153–168. https://doi.org/10.1109/76.905982
20. Cid DB (2013) Malware hidden inside JPG EXIF headers. Sucuri Blog, Website Security News. https://blog.sucuri.net/2013/07/malware-hidden-inside-jpg-exif-headers.html
21. Shah S (2015) Stegosploit–exploit delivery with steganography and polyglots. In: Briefings, Black Hat Conference. https://www.blackhat.com/eu-15/briefings.html
22. Khandelwal S (2016) Beware! Malicious JPG images on facebook messenger spreading locky ransomware. In: The Hacker News, Cybersecurity News and Analysis. https://thehackernews.com/2016/11/facebook-locky-ransomware.html
23. Abrams L (2017) SyncCrypt ransomware hides inside JPG files, appends .KK Extension. Bleeping computer, technology news website. https://www.bleepingcomputer.com/news/security/synccrypt-ransomware-hides-inside-jpg-files-appends-kk-extension
24. Zahravi A (2018) Malicious memes that communicate with Malware. Trend Micro, IT Security Company. https://www.trendmicro.com/en_us/research/18/l/cybercriminals-use-malicious-memes-that-communicate-with-malware.html
25. Osborne C (2019) LokiBot malware now hides its source code in image files. ZDNet, Technology News Website. https://www.zdnet.com/article/lokibot-information-stealer-now-hides-malware-in-image-files
26. Szappanos G, Brandt A (2019) MyKings botnet spreads headaches, cryptominers, and For share malware. Sophos, Cybersecurity Company. https://news.sophos.com/en-us/2019/12/18/mykings-botnet-spreads-headaches-cryptominers-and-forshare-malware
27. Malware repository. https://www.virusshare.com
28. Malware analysis service. https://www.hybrid-analysis.com

Entropic Analysis of Reservation Policy of Government of India

Rakesh Kumar Pandey and Maneesha Pandey

Abstract Analysis based on the entropy estimates of the society is carried out on a mathematical model in this paper, to make an assessment of reservation policy presently in operation in India. The reservation policy was provisioned in the constitution of India under the policy of affirmative action. The policy makers had identified some communities of India that were being discriminated socially. They identified them by noticing their dismal representation in positions of power and respect, within the society. It was argued that these communities are required to be treated differently to give them equal opportunity in the society. The policy makers had provisioned in the same policy that the effectiveness of this policy will be analysed at regular intervals to decide its continuation or improvisation. An attempt is made here to scientifically analyse the effectiveness of this policy and suggest possible improvements in the same. As per the understanding of entropy, a policy such as the reservation policy that is intended to encourage equality must be associated with an increase in the entropy of the society. After a careful mathematical analysis, it is shown here that the policy, in its present form, is designed to increase the entropy of the society in the short-term scenario but is an iso-entropic policy in the long term. A way has been suggested to complement the policy to make the entropy increase effectively and permanently.

Keywords Social entropy · Policy assessment · Policy performance · Policy analysis

R. K. Pandey (✉)
Kirori Mal College, University of Delhi, Delhi, India
e-mail: r.kr.pandey@gmail.com

M. Pandey
Hindu College, University of Delhi, Delhi, India

© The Author(s), under exclusive license to Springer Nature Singapore Pte Ltd. 2023

A. Mishra et al. (eds.), *Advances in IoT and Security with Computational Intelligence*, Lecture Notes in Networks and Systems 755,
https://doi.org/10.1007/978-981-99-5085-0_3

1 Introduction

Recently, attempts have been made to carry out entropy-estimate-based analysis of certain social scenarios [1–3]. Such studies are being carried out extensively under the field of Cybersyn and Cybernetics [4–6]. The importance of such studies while trying to explore the possibility of our dependence on machines for decision-making can be readily acknowledged [7]. An attempt was made in [8] to carry out such an entropic analysis of a simplified traffic intersection scenario. It is shown there that implementing traffic signals or constructing flyovers result in a decrease of entropy thereby transforming the situation into a systematic behaviour. It is expected that such estimates and analysis will help in understanding the social complexities in a better way through scientific and mathematical modelling [9–11]. Such analysis might turn into tools for better assessment, analysis and predictions of social behaviour and status [12]. While analysing these models on the basis of entropy estimates, it is believed that processes that bring a positive change in the entropy of a social system would lead to equality whereas negative change will make the system function systematically but at the cost of equality [13–15].

The constitution of India [16] has provisioned a Reservation Policy [17] to ensure equality in the Indian society. This paper attempts to make a scientific analysis of the policy, based on the entropy estimates of the society. Using the entropic calculations, the effectiveness of this policy is ascertained scientifically. An attempt is made to verify whether the social entropy that is expected to increase due to this policy will meet the promise or not.

The next section discusses the policy of reservation that is presently in operation in India. The need to have such a policy and the aim that it is looking to achieve are analysed briefly. Using the same, a mathematical formulation is developed in the subsequent section to identify and estimate the entropy of the society since the entire analysis is based on the changes that would be caused by the implementation of the policy. An example is then constructed with complete mathematical details in the next section to illustrate conclusively that the policy can increase the entropy of the society in a short term only till the time the criteria to identify discriminated sections are not reutilized to reidentify the communities.

2 The Need for a Policy of Reservation

After becoming independent in 1947, the government of India passed its own constitution that came into effect from January 26, 1949. The Constitution of India [16], in its articles 15(4) and 16(4), provisioned a sound basis to implement a policy under affirmative action that was called Reservation Policy [17]. The policy makers realized that there are several communities in India that are socially discriminated against sharing positions associated with power and policy making. These ignored but large sections of the society needed some affirmative action to get an equal opportunity as compared to those who were already better placed in the society. Based on this reality, a few categories were identified by grouping some communities together as Scheduled Castes (SC) and Scheduled Tribe (ST) at an initial stage, and later some additional communities were grouped under Other Backward Class (OBC). These communities were identified by their caste identity and were made eligible for the provisions of affirmative action.

These categories were identified by estimating their share in the existing structure of the society on the scales of power and respect. Instead of economy, the basis of identifying these categories were considered as per their say in the policy making, in drawing respect and power from the society.

The need for having this policy was felt because the share of these communities identified as SC, ST and OBC was not in the proportion to their share in the population. The policy, since then, has intended to achieve proportionate representation of all these communities in the social structure of power and respect.

The policy was adopted after concluding correctly that since the share in the power structure was far lesser by a large number of communities despite having larger share in the population, they were automatically being discriminated by others who were in minority but were enjoying the advantage. It was realized in principle by the policy makers that to eliminate such practices of discrimination, all communities must get proportionate representation in the power structure. And thus, the policy was conceptualized to correct this skewed imbalance [18]. The structure of power and respect in the society gets automatically built up around the government jobs. It was also realized in principle that government jobs dominantly contribute to the power structure and therefore the policy of reservation has been implemented in the government jobs belonging to all categories and stages. In particular, apart from those who could get into this social structure of respect and power without the help of the reservation policy, a minimum of 7.5% from the ST community, a minimum of 15% from SC and a minimum of 27% from the OBC community have been assured of the share in this structure. Since these are not the upper caps, the policy looks at facilitating these communities in sharing more than these percentages, if the policy is successfully implemented for a reasonably long duration.

3 The Mathematical Model

The present paper utilizes the idea of Social Entropy to carry out the analysis. As per the definition and the understanding of entropy, any change to achieve equality will necessarily be associated with an increase in entropy. An attempt is made in this paper to do the entropic analysis of this policy. It is well known that under the given conditions, the social entropy will be maximum when each community will get a representation in the power structure in proportion to their share in the population of India. To build a model, let the entropy of the society be defined in the following way.

Let us consider that there are a total of N positions available in the power structure of the society (loosely equal to the number of government jobs). As per the features of any power structure, let there be four sub-classes in this hierarchical structure having numbers as N_1, N_2, N_3 and N_4
where

$$\sum_{i=1}^{i=4}(N_i) = N \tag{1}$$

If the social power and respect index of these positions are denoted by $Pow(N_i)$, then let them be chosen in the following order:

$$Pow(N_1) > Pow(N_2) > Pow(N_3) > Pow(N_4) \tag{2}$$

According to the characteristics of any hierarchical structure, this must be associated with

$$N_1 < N_2 < N_3 < N_4 \tag{3}$$

Now, let us try to formulate the identification Criteria of Communities to be selected for differential treatment to make them equal with others under the affirmative action strategy.

In general, people belonging to several castes will be occupying the N positions in some way. India has a large list consisting of several hundreds of castes identified in the government records. In the next step, let us try to identify some categories of people by grouping these castes that are occupying these positions of power which is disproportionate to their share in the population. Let four such communities be identified by grouping these castes intelligently as C_1, C_2, C_3 and C_4 that occupy these N_1, N_2, N_3 and N_4 positions in a completely skewed manner as illustrated in the following observations.

1. That C_1, C_2, C_3 and C_4 have a%, b%, c% and d% share in the population, respectively.
2. Let a_{ij} denote the percentage share of the class C_j in the positions N_i.
3. That the percentage share in the most powerful positions denoted by N_1 is dominated by C_1 as compared to the other three C_2, C_3 and C_4. (Mathematically, a_{11} is very large as compared to a_{12}, a_{13} and a_{14}.)
4. That the percentage share in the second most powerful positions denoted by N_2 is dominated by C_2 as compared to the other three C_1, C_3 and C_4. (Mathematically, a_{22} is very large as compared to a_{21}, a_{23} and a_{24}.)
5. That the percentage share in the not so powerful positions denoted by N_3 is dominated by C_3 as compared to the other three C_1, C_2 and C_4. (Mathematically, a_{33} is very large as compared to a_{31}, a_{32} and a_{34}.)
6. That the percentage share in the least powerful positions denoted by N_4 is dominated by C_4 as compared to the other three C_1, C_2 and C_3. (Mathematically, a_{44} is very large as compared to a_{41}, a_{42} and a_{43}.)

To get the desirable proportionate representation in all the positions of power, the policy must result in changes in these parameters such that the following distribution is achieved:

$$a_{i1} : a_{i2} : a_{i3} : a_{i4} = a : b : c : d \tag{4}$$

for all $i = 1, 2, 3, 4$.

If the probability of C_j occupying one of the positions in N_i is given by P_{ij}, then

$$P_{ij} = N_{ij}/N \tag{5}$$

Sixteen such parameters denoted by P_{11}, P_{12}, P_{13}, P_{14}, P_{21}, P_{22}, P_{23}, P_{24}, P_{31}, P_{32}, P_{33}, P_{34}, P_{41}, P_{42}, P_{43} and P_{44} will represent a particular distribution of people belonging to different categories occupying different positions of power. The social entropy [19–21] of this distribution will then be given by

$$E = -\sum_{i=1}^{i=4}\sum_{j=1}^{j=4}(P_{ij}Log(P_{ij})) \tag{6}$$

It can be easily shown that when the condition given in Eq. (4) is satisfied, the entropy will attain its maximum value. In any skewed distribution therefore, the value of entropy will be less than that and the success of the policy can be assessed by estimating how close the entropy has reached to its maximum. The policy of reservation enforces proportionate representation in all the positions blindly irrespective of the hierarchy and thus promises to achieve proportionate representation in the entire power structure.

It is expected that in the new distribution, no community will be facing discrimination. But is this expectation realizable? Ironically, in an example discussed in the next section it is illustrated that when the proportionate distribution is attained, a new set of communities will start facing similar discrimination that was sought to be nullified through the policy. And so, with the identification of a new set of categories, the entropy will be still far from reaching the desirable maximum value. The entire exercise, therefore, becomes iso-entropic since it can never lead to a real increase in the entropy.

4 Illustrative Example

To get the complete insight about this strategy, we discuss a possible scenario here, by illustrating a perfectly ordinary example.

Let us consider the following scenario. Let there be 400 positions of power and respect in a particular society, and let these be distributed in four categories as

$$N_1 = 20, N_2 = 60, N_3 = 120, N_4 = 200 \tag{7}$$

Here, N_1 belongs to the most powerful positions, N_2 belongs to the next powerful ones, N_3 are not so powerful but N_4 are the least powerful positions. Let there be sixteen communities (castes) of people, denoted by

$$A, B, C, D, E, F, G, H, S, T, U, V, W, X, Y, Z$$

that are occupying these positions as shown in Fig. 1. Using the criteria to identify communities for affirmative action, it is noticed that

1. A, S, B and T form the community C_1 that occupies the majority share 85% in N_1.
2. C, Z, D and Y form the community C_2 that occupies the majority share 93% in N_2.
3. E, U, F and V form the community C_3 that occupies the majority share 87% in N_3.
4. G, X, H and W form the community C_4 that are left to occupy the majority share 96% in N_4.

This distribution is extremely skewed if compared with their population share as C_1, C_2, C_3, C_4 have 5%, 15%, 30% and 50% share in the population, respectively.

The entropy of this distribution is estimated in Table 1 using Eqs. (5) and (6) as 0.7278. When the policy of reservation is enforced, all the categories C_1, C_2, C_3, C_4 will get a proportionate representation separately in all stages of power. Therefore, we expect a distribution that is shown in Fig. 2. The entropy associated with this distribution is estimated as 0.9870 for such a distribution as shown in Table 1. This value is apparently very close to its maximum and therefore illustrates the success of the policy in achieving a situation wherein there is no discrimination.

To our surprise however, there is a twist in this understanding. If the criteria discussed in the earlier section is applied in this distribution as displayed in Fig. 2, one can identify four new categories by grouping the communities differently. To illustrate this, the following new categories are now identified with the features mentioned therein:

1. A, C, E and G form the community C_1' that occupies the majority share 75% in N_1.
2. S, Z, U and X form the community C_2' that occupies the majority share 93% in N_2.
3. B, D, F and H form the community C_3' that occupies the majority share 95% in N_3.
4. T, U, V and W form the community C_4' left to occupy the majority share 96% in N_4.

We see in Fig. 3 wherein the same distribution that exists in Fig. 2 becomes skewed with the new categories C_1', C_2', C_3', C_4' (C_4' now being the most discriminated set of communities). It is shown in Table 1 that entropy of this distribution becomes 0.5884, showing no improvement on the scale of equality from its earlier value of 0.6238. Obviously, the increase in entropy noticed earlier was only as we were ignoring

N_1=20	A	A	S	S	S	B	B	B	B	B	T	T	T	T	T	T	T	C	C	U
	T	T	C	C	Z	Z	Z	Z	Z	Z	Z	Z	D	D	D	D	D	D	D	D
N_2=60	D	D	D	D	D	D	D	D	D	D	D	Y	Y	Y	Y	Y	Y	Y	Y	Y
	Y	Y	Y	Y	Y	Y	Y	Y	Y	Y	Y	Y	Y	Y	Y	Y	Y	Y	E	E
	Y	Y	E	E	E	E	U	U	U	U	U	U	U	U	U	U	U	U	U	U
	U	U	U	U	F	F	F	F	F	F	F	F	F	F	F	F	F	F	F	F
N_3=120	F	F	F	F	F	F	F	F	F	F	F	F	F	F	F	F	F	F	V	V
	V	V	V	V	V	V	V	V	V	V	V	V	V	V	V	V	V	V	V	V
	V	V	V	V	V	V	V	V	V	V	V	V	V	V	V	V	V	V	V	V
	V	V	V	V	V	V	G	G	G	G	X	X	X	X	X	X	X	X	X	X
	V	V	V	V	V	V	V	V	V	G	G	G	G	G	G	G	X	X	X	X
	X	X	X	X	X	X	X	X	X	X	X	X	X	X	X	X	X	X	H	H
	H	H	H	H	H	H	H	H	H	H	H	H	H	H	H	H	H	H	H	H
	H	H	H	H	H	H	H	H	H	H	H	H	H	H	H	H	H	H	H	H
N_4=200	H	H	H	H	H	H	H	H	H	H	H	H	H	H	H	H	H	W	W	W
	W	W	W	W	W	W	W	W	W	W	W	W	W	W	W	W	W	W	W	W
	W	W	W	W	W	W	W	W	W	W	W	W	W	W	W	W	W	W	W	W
	W	W	W	W	W	W	W	W	W	W	W	W	W	W	W	W	W	W	W	W
	W	W	W	W	W	W	W	W	W	W	W	W	W	W	W	W	W	W	W	W
	W	W	W	W	W	W	W	W	W	W	W	W	W	W	W	W	W	W	W	W

Fig. 1 Initial given distribution of 16 different castes across 400 vacancies. **White Boxes** belong to C_1 community consisting of A, S, B, T castes. **Light Grey Boxes** belong to C_2 community consisting of C, Z, D, Y castes. **Dark Grey Boxes** belong to C_3 community consisting of E, U, F, V castes. **Black Boxes** belong to C_4 community consisting of G, X, H, W castes

and delaying to acknowledge the change in the power structure that the policy had brought in the society. The moment the changes were acknowledged in Fig. 3, the policy is realized as iso-entropic, promising no improvement on the scale of equality.

5 Conclusion and Solution

The policy apparently promises to achieve the proportionate representation of all the categories in all the positions to ensure that no category of people is left out in this sharing of power and respect in the society. However, a careful analysis as done in the previous section proves that the success lies in having ignored the new communities that will start satisfying the criteria for being recognized as enjoying power and/or are beginning to being denied their due share in the power structure. The moment the criteria is used for regrouping the communities, entropy will fail to show any appreciable increase. And thus, equality will never be achieved.

N_1=20	A	C	C	C	E	E	E	E	E	U	G	G	G	G	G	G	X	X	X	X
	A	Z	Z	Z	U	U	U	U	U	U	X	X	X	X	X	X	X	X	X	X
N_2=60	S	Z	Z	Z	U	U	U	U	U	U	X	X	X	X	X	X	X	X	X	X
	S	Z	Z	D	U	U	U	U	U	U	X	X	X	X	X	X	X	X	H	H
	S	D	D	D	F	F	F	F	F	F	H	H	H	H	H	H	H	H	H	H
	B	D	D	D	F	F	F	F	F	F	H	H	H	H	H	H	H	H	H	H
N_3=120	B	D	D	D	F	F	F	F	F	F	H	H	H	H	H	H	H	H	H	H
	B	D	D	D	F	F	F	F	F	F	H	H	H	H	H	H	H	H	H	H
	B	D	D	D	F	F	F	F	F	F	H	H	H	H	H	H	H	H	H	H
	B	D	D	D	F	F	F	F	V	V	H	H	H	H	H	H	H	W	W	W
	T	Y	Y	Y	V	V	V	V	V	V	W	W	W	W	W	W	W	W	W	W
	T	Y	Y	Y	V	V	V	V	V	V	W	W	W	W	W	W	W	W	W	W
	T	Y	Y	Y	V	V	V	V	V	V	W	W	W	W	W	W	W	W	W	W
	T	Y	Y	Y	V	V	V	V	V	V	W	W	W	W	W	W	W	W	W	W
N_4=200	T	Y	Y	Y	V	V	V	V	V	V	W	W	W	W	W	W	W	W	W	W
	T	Y	Y	Y	V	V	V	V	V	V	W	W	W	W	W	W	W	W	W	W
	T	Y	Y	Y	V	V	V	V	V	V	W	W	W	W	W	W	W	W	W	W
	T	Y	Y	Y	V	V	V	V	V	V	W	W	W	W	W	W	W	W	W	W
	T	Y	Y	Y	V	V	V	V	V	V	W	W	W	W	W	W	W	W	W	W
	C	Y	Y	E	V	G	G	G	G	G	W	W	W	W	W	W	W	W	W	W

Fig. 2 Distribution of 16 different castes across 400 vacancies after implementation of reservation policy. **White Boxes** belong to C_1 community consisting of A, S, B, T castes. **Light Grey Boxes** belong to C_2 community consisting of C, Z, D, Y castes. **Dark Grey Boxes** belong to C_3 community consisting of E, U, F, V castes. **Black Boxes** belong to C_4 community consisting of G, X, H, W castes

Let us try to find out the factor that is again bringing the entropy back to its initial levels. In fact, since the power associated with N_1, N_2, N_3 and N_4 is hierarchically ordered, the share of communities in these categories is bound to be skewedly different. Moreover, since only those who would have a larger share of presence in N_1 (smallest of all) can be considered powerful, it leads to a situation wherein a large number of communities will start feeling ignored.

The solution therefore exists in bridging the difference between the power substructures. For example, if the difference between the power and respect associated with N_2 and N_3 gets eliminated, we can expect a similar share in these positions by all the communities. This is reflected in Table 1 by proportionately distributing the numbers between N_2 and N_3. Entropy of this distribution wherein the parity between N_2 and N_3 is restored (see Appendix) is estimated as 0.7278 in Table 1 (under Case 4) which is a definite improvement over the distribution of Fig. 1. Moreover, when proportionate distribution is achieved by reducing/eliminating the difference

N_1=20	A	C	C	C	E	E	E	E	E	U	G	G	G	G	G	G	X	X	X	X
	A	Z	Z	Z	U	U	U	U	U	U	X	X	X	X	X	X	X	X	X	X
N_2=60	S	Z	Z	Z	U	U	U	U	U	U	X	X	X	X	X	X	X	X	X	X
	S	Z	Z	D	U	U	U	U	U	U	X	X	X	X	X	X	X	X	H	H
	S	D	D	D	F	F	F	F	F	F	H	H	H	H	H	H	H	H	H	H
	B	D	D	D	F	F	F	F	F	F	H	H	H	H	H	H	H	H	H	H
N_3=120	B	D	D	D	F	F	F	F	F	F	H	H	H	H	H	H	H	H	H	H
	B	D	D	D	F	F	F	F	F	F	H	H	H	H	H	H	H	H	H	H
	B	D	D	D	F	F	F	F	F	F	H	H	H	H	H	H	H	H	H	H
	B	D	D	D	F	F	F	F	V	V	H	H	H	H	H	H	H	W	W	W
	T	Y	Y	Y	V	V	V	V	V	V	W	W	W	W	W	W	W	W	W	W
	T	Y	Y	Y	V	V	V	V	V	V	W	W	W	W	W	W	W	W	W	W
	T	Y	Y	Y	V	V	V	V	V	V	W	W	W	W	W	W	W	W	W	W
	T	Y	Y	Y	V	V	V	V	V	V	W	W	W	W	W	W	W	W	W	W
N_4=200	T	Y	Y	Y	V	V	V	V	V	V	W	W	W	W	W	W	W	W	W	W
	T	Y	Y	Y	V	V	V	V	V	V	W	W	W	W	W	W	W	W	W	W
	T	Y	Y	Y	V	V	V	V	V	V	W	W	W	W	W	W	W	W	W	W
	T	Y	Y	Y	V	V	V	V	V	V	W	W	W	W	W	W	W	W	W	W
	T	Y	Y	Y	V	V	V	V	V	V	W	W	W	W	W	W	W	W	W	W
	C	Y	Y	E	V	G	G	G	G	G	W	W	W	W	W	W	W	W	W	W

Fig. 3 Distribution of 16 different castes across 400 vacancies after regrouping castes to form the four communities according to the criteria decided initially. **White Boxes** belong to C_1' community consisting of A, C, E, G castes. **Light Grey Boxes** belong to C_2' community consisting of S, Z, U, X castes. **Dark Grey Boxes** belong to C_3' community consisting of B, D, F, H castes. **Black Boxes** belong to C_4' community consisting of T, Y, V, W castes

between the power sub-structures, the revisiting of community-grouping will make no difference in the final estimate of the entropy.

Due to the long duration of sufferings caused by slavery and cultural invasions probably, the Indian society lost the sense to acknowledge dignity of labour and work. An urge to change one's profession merely to earn respect in the society originates from our collective loss of sense in respecting dignity of work. Unless this sense is restored, the reservation policy, in itself, will not be enough to achieve the aim of ensuring equality in the society. One of the ways to restore this sense could be to reduce the gap in the income associated with the stages denoted by N_1, N_2, N_3 and N_4 so that these positions do not get treated differently at least due to the financial status associated with these positions. As reflected in Table 1, even a restoration of parity between N_2 and N_3 has been able to improve the entropy from 0.6238 to 0.7278 that is closer to its maximum value of 0.9870. Just to help in choosing a strategy,

Table 1 Entries in the columns of this table are as follows. **Case 1** deals with the distribution shown in Fig. 1. **Case 2** deals with the distribution shown in Fig. 2. **Case 3** deals with the distribution shown in Fig. 3. **Case 4** deals with the distribution given in Fig. 1, modified on restoration of parity between N_2 and N_3. **Case 5** deals with the distribution given in Fig. 1, modified on restoration of parity between N_3 and N_4

		N_{ij}	P_{ij}	N_{ij}	P_{ij}	N_{ij}	P_{ij}	N_{ij}	P_{ij}	N_{ij}	P_{ij}
i	j	Case 1		Case 2		Case 3		Case 4		Case 5	
Index	Index	Number	Probability	Number	Probability	Number	Probability	Number	Probability	Number	Probability
1	1	17	0.0425	1	0.0025	15	0.0375	17	0.0425	17	0.0425
1	2	2	0.005	3	0.0075	5	0.0125	2	0.005	2	0.005
1	3	1	0.0025	6	0.015	0	0	1	0.0025	1	0.0025
1	4	0	0	10	0.025	0	0	0	0	0	0
2	1	2	0.005	3	0.0075	1	0.0025	1	0.0025	2	0.005
2	2	56	0.14	9	0.0225	56	0.14	19	0.0475	56	0.14
2	3	2	0.005	18	0.045	3	0.0075	35	0.0875	2	0.005
2	4	0	0	30	0.075	0	0	5	0.0125	0	0
3	1	0	0	6	0.015	0	0	1	0.0025	0	0
3	2	2	0.005	18	0.045	1	0.0025	39	0.0975	2	0.005
3	3	104	0.26	36	0.09	114	0.285	71	0.1775	42	0.105
3	4	14	0.035	60	0.15	5	0.0125	9	0.0225	77	0.1925
4	1	0	0	9	0.0225	7	0.0175	0	0	0	0
4	2	0	0	30	0.075	0	0	0	0	0	0
4	3	9	0.0225	56	0.14	0	0	9	0.0225	71	0.1775
4	4	191	0.4775	105	0.2625	193	0.4825	191	0.4775	128	0.32
Entropy			0.6238		0.9870		0.5884		0.7278		0.7625

one must note that restoring parity between N_3 and N_4 (see Appendix), instead of between N_2 and N_3, shows the promise of improvement in the entropy to 0.7625.

Acknowledgement, Data Availability and Ethical Compliance The authors thank Kirori Mal College and Hindu College to have provided them with the required environment to carry out this work.
No funds were received to carry out this work.
There was no conflict of interest involved, and ethical compliance was observed in this work.
All data generated or analysed during this study are included in this paper.

6 Appendix

To restore the parity between N_2 and N_3, the calculation is done in the following way: If a community C_i has a combined strength of N_c, then it is redistributed among N_2 and N_3 in the ratio 60:120 (or 1:2) since $N_2 = 60$ and $N_3 = 120$. According to Fig. 1, community C_1, C_2, C_3 and C_4 has a combined strength as 2, 58, 116 and 14, respectively. These are then proportionately distributed in N_2 and N_3 in the ratio 1:2 as given in Table 1. The two levels given by N_2 and N_3 are not made one to keep the system comparable with the initial state when the system has four levels of hierarchy.

Similarly, to restore the parity between N_3 and N_4, the following calculation is done: According to Fig. 1, community C_1, C_2, C_3 and C_4 has a combined strength as 0, 2, 113 and 205, respectively. These are then proportionately distributed in N_3 and N_4 in the ratio 120:200 (or 3:5) as given in Table 1.

References

1. Martinez S, Nicolas F, Pennini F, Plastino A (2000) Tsallis' entropy maximization procedure revisited. Phys A 286(3–4):489–502. https://doi.org/10.1016/S0378-4371(00)00359-9
2. Ortega PA, Daniel AB (2013) Thermodynamics as a theory of decision-making with information-processing costs. Proc R Soc A 469(3–4). https://doi.org/10.1098/rspa.2012.0683
3. Parrondo JMR, Horowitz JM, Sagawa T (2015) Thermodynamics of information. Nat Phys 11:131–139. https://doi.org/10.1038/nphys3230
4. Lavanderos L (2022) From cybersin to cybernet. Considerations for a cybernetics design thinking in the socialism of the XXI century. AI Soc 37:1279–1292. https://doi.org/10.1007/s00146-021-01354-2
5. Espejo R (2022) Cybersyn, big data, variety engineering and governance. AI Soc 37:1163–1177. https://doi.org/10.1007/s00146-021-01348-0
6. Araneda C (2022) Jaime Garretón's cybernetic theory of the city and its system: a missing link in contemporary urban theory. AI Soc 37:1179–1189. https://doi.org/10.1007/s00146-021-01349-z
7. Adnan R (2021) Entropic decision making. https://arxiv.org/abs/2001.00122v1
8. Pandey RK (2020) Entropic analysis to assess impact of policies on disorders and conflicts within a system: case study of traffic intersection as 12-qubit social quantum system. arXiv:2012.15012

9. Jacquemin AP, Berry CH (1979) Entropy measure of diversification and corporate growth. J Ind Econ 27(4):359–369
10. Perc M, Jordan JJ, Rand DG, Wang Z, Boccaletti S, Szolnoki A (2017) Statistical physics of human cooperation. Phys Rep 687:1–51
11. Lavertu S, Moynihan DP (2012) Agency political ideology and reform implementation: performance management in the bush administration. J Publ Adm Res Theory 23(3):521–549. https://doi.org/10.1093/jopart/mus026
12. Castellano C, Fortunato S, Loreto V (2009) Statistical physics of social dynamics. Rev Mod Phys 81:591
13. de Marchi S (2005) Computational and mathematical modelling in the social sciences. Cambridge University Press, New York
14. Jost L (2006) Entropy and diversity. Oikos 113(2):363–375. https://doi.org/10.1111/j.2006.0030-1299.14714.x
15. Overman ES (1996) The new science of management: chaos and quantum theory and method. J Publ Adm Res Theory 6(1):75–89. https://doi.org/10.1093/oxfordjournals.jpart.a024304
16. Government of India: Article 15(4) and Article of 16(4) given in the The Constitution of India can be accessed using the link. These articles provide the justification for taking affirmative action as carried out through the Reservation Policy. https://legislative.gov.in/sites/default/files/COI.pdf
17. Government of India: The basic idea of the Reservation Policy of Government of India can be understood through the document. https://dpe.gov.in/sites/default/files/Reservation_Brochure-2.pdf
18. Maheshwari SR (1997) Reservation policy in India: theory and practice. Indian J Publ Adm 43(3):662–679. https://doi.org/10.1177/0019556119970335
19. Shannon EC, Weaver W (1964) The mathematical theory of communication. University of Illinois Press, Urbana
20. Bailey KD (1990) Social entropy theory. State University of New York Press, Albany
21. Sargentis GF, Iliopoulou T, Dimitriadis P, Mamassis N, Koutsoyiannis D (2021) Stratification: an entropic view of society's structure. World 2(2):153–174. https://doi.org/10.3390/world2020011

Cybersecurity in the Supply Chain and Logistics Industry: A Concept-Centric Review

Sunday Adeola Ajagbe, Joseph Bamidele Awotunde, Ademola Temidayo Opadotun, and Matthew O. Adigun

Abstract A literature review assists authors in evaluating and analyzing relevant literature, identifying conceptual content in the subject, and contributing to theory building in a specific domain. As a result, the goal of this article is to conduct a concept-centric review of cybersecurity in logistics, and supply chain management (SCM) in order to identify the gap and recommend areas for improvement in this case study, based on security demands and technological innovation. The study follows the following steps: selection of the relevant studies in the subject area; carrying out a descriptive analysis of the selected papers and classifying the papers based on the type of key phrases used. The result shows limited attention has been focused on logistic management compared to the cybersecurity and supply chain (SC). The need to consider and improve the logistic infrastructure in human-dominated area like Saudi Arabia which hosts people from all works of life during Haji and Humira. We, therefore, suggest a redirection of research focus that will leverage AI and IoT, to provide efficient security for life and properties, during and after the Haji and other seasonal densely populated areas around the world.

Keywords Cybersecurity · Concept-centric review · Haji operation · Logistics · Supply chain management

S. A. Ajagbe (✉)
Department of Computer & Industrial Engineering, First Technical University, Ibadan 200255, Nigeria
e-mail: Sunday.ajagbe@tech-u.edu.ng

J. B. Awotunde
Department of Computer Science, University of Ilorin, Ilorin, Nigeria
e-mail: awotunde.jb@unilorin.edu.ng

A. T. Opadotun
Oyo State College of Agriculture and Technology, Igbo-Ora, Nigeria
e-mail: atopadotun@oyscatech.edu.com

S. A. Ajagbe · M. O. Adigun
Department of Computer Science, University of Zululand, Richards Bay 3886, South Africa
e-mail: adigunm@unizulu.ac.za

© The Author(s), under exclusive license to Springer Nature Singapore Pte Ltd. 2023
A. Mishra et al. (eds.), *Advances in IoT and Security with Computational Intelligence*, Lecture Notes in Networks and Systems 755,
https://doi.org/10.1007/978-981-99-5085-0_4

1 Introduction

People's ability to communicate with one another has been revolutionized by technological breakthroughs, which have made telework, telemedicine, and mobile commerce possible. Other technologies linked to the Industrial Internet of Things (IIoT) have also had an impact on the corporate world [1, 2]. Technological advancement has also contributed to fields of human endeavor, this includes but is not limited to commerce, education, supply chain (SC), security, and so on. The SC is becoming even more complex and global in scope. Even transportation service providers are using it to boost efficiency and coordination across the SC. IT is becoming more and more important in this industry. However, such advancements made the SC and logistics industry subject to cyberattacks [3, 4]. The existing SCM solutions are primarily concerned with improving SC company performance. Theories have emphasized how managers of industries can use a variety of tools, tactics, and management approaches to deal with demand uncertainty and market volatility, for example, while maximizing profits. Contract management, quality management, risk management, network reengineering, and make or purchase decisions, for example, are generally recognized by scholars as approaches to improve SC performance while reducing costs and negative environmental impacts [5, 6].

These and many more are in the front burner of research in recent times. Researchers are attempting to address the issues in different ways. The researchers' attempt to contribute to this domain includes preparing the accounts of happening in this area to recommend a possible way forward. Although others researchers simulated and implemented different solutions to a variety of problems, producing a survey of relevant literature can assist readers in evaluating and analyzing such literature, as well as locating conceptual context in the field and making contributions to the development of theory in a specific topic. In this paper, we aim to carry out a concept-centric review trying to analyze the current state of the art with respect to cybersecurity in logistics and SCM in order to identify the gap and suggest the area for improvement. The paper seeks to answer three research questions:

i. What is the current state of cybersecurity research in logistics and SCM?
ii. What is the contribution available in major databases?
iii. What types of research contributions are accessible on cybersecurity in logistics and SCM?

This study employs a descriptive approach for the study, focusing on the three prominent areas (Topic, Abstract, and the Keywords) of the research papers. We examine research on ways to improve cybersecurity research in logistics and SCM.

The remaining paper is organized as follows. Section 2 presents the approach for a concept-centric review of cybersecurity in logistics and SCM while Sect. 3 is the outcome and discussion of the study's descriptive analysis of the cybersecurity research in logistics and SCM. The open issues and future research points were discussed in Sect. 4. The conclusion, identified gaps, and recommendations are outlined in Sect. 5.

2 Adopted Research Approach

The research methodology used for conducting this review paper is described as follows. Due to the inter-disciplinary nature of the area of this paper (i.e., cybersecurity, SC, and logistics), the emergent synthesis method is used to synthesize a diverse corpus of literature, encompassing quantitative, qualitative, case studies, theoretical work, and conceptual frameworks [7]. This study used a concept-centric literature review process rather than an author-centric strategy [2, 8] to identify relevant publications. An emphasis was given to the classification of security methods which is the emphasis of this paper. The review process adopted in this study consists of the following steps:

A. Selection procedure: which paper was reviewed.
B. A description of the papers that were chosen.
C. Classification of the papers based on the type of key phrases used.

Based on the available time and space allotted for paper at this conference, 16 recent papers were downloaded and reviewed for this study, and all documents in English linked to "cybersecurity/cyberattack/cyberthreat", "logistics", and/or "SC" were found using the Google Scholar database. After experimenting with various keyword combinations, the search was narrowed down to all documents that focused on either "cybersecurity/cyberattack/cyberthreat", "logistics" or "SC" in the titles, abstract, or keywords of their studies. The main body of the papers was reviewed for contribution and the classification of the study.

3 The Outcome and Discussion

The outcome of this study is presented in Table 1 titled "Concept-centric report on cybersecurity in logistics and SCM". The studies focused on the three main concepts (cybersecurity research in logistics and SCM) were compared. Anyone that focused on any of the three concepts is denoted by "*" and the study that did not focus on any of the concepts is denoted by "-". Firstly, we centered the review of three concepts on three key areas of the publications, viz-a-viz the "Topic", "Abstract", and "Keyword". The concept-centric report shows that 13 papers considered the cybersecurity/cyberattack/cyberthreat in their studies, 2 papers considered logistics and 10 papers considered SC in their "Topic", 13 papers considered the cybersecurity/cyberattack/cyberthreat, 5 papers considered logistics and 13 papers considered SC in their "Abstract", 11 papers considered the cybersecurity/cyberattack/cyberthreat, 2 papers considered logistics, and 10 papers considered SC in their "Keyword". In addition, the databases where these studies were published were also considered in our review, we focused on the Elsevier, IEEE, and the studies that were not published in either of these databases were grouped with others.

Table 1 shows the result of the concept-centric review in this study, the three distinct concepts considered in this study were "cybersecurity/cyberattack/

Table 1 Concept-centric report of cybersecurity, logistic, and supply chain management

Study	Title, abstract, and keyword			Publisher/Database			Classification/ contributions
	Cybersecurity	Logistics	Supply chain	Elsevier	IEEE	Others	
[2]	* * *	***	* **	*	-	-	Review of literature
[3]	* * *	- - -	* * *	*	-	-	Systematic review
[9]	* * *	- - -	- - -	*	-	-	Review of literature
[10]	- * -	- - -	* * _	*	-	-	A complex framework that provides solutions to the supply chain problem was proposed
[11]	* * *	- - -	- - -	*	-	-	Integration of a cyberrisk assessment system for a container port was proposed
[12]	* _ *	- - -	* * *	*	-	-	Review of literature
[13]	* * *	- - -	* * *	*	-	-	A systematic review of literature
[14]	- - -	- - -	* * *	*	-	-	Development of blockchain strategy for sustainable supply chain management
[15]	* * *	* * *	- * -	*	-	-	A novel attacker–defender paradigm against a quantal response (QR) opponent was suggested
[16]	* * *	- - -	* * *	-	-	*	Qualitative analysis
[17]	* * *	- - -	* * *	-	-	*	Qualitative analysis
[18]	* * *	- * -	- * -	-	-	*	A conceptual framework to determine the level and area of cybersecurity was proposed

(continued)

Table 1 (continued)

Study	Title, abstract, and keyword			Publisher/Database			Classification/ contributions
	Cybersecurity	Logistics	Supply chain	Elsevier	IEEE	Others	
[19]	* * -	- - -	* * *	-	*	-	A technique for the detection of cyber supply chain issues was developed
[20]	- * -	- * -	- * *	-	*	-	An authentication strategy for weak elements in the supply chain was proposed
[21]	* - -	- * -	- - -	-	-	*	AI-IoT-based solution was proposed for the cybersecurity of a smart city
[22]	* * *	- - -	* * *	-	*	-	ML technique was proposed to predict cyberthreat intelligence

cyberthreat", "logistics", and "SC". Out of 16 papers reviewed, the Topic of 13 of the studies focused on cybersecurity/cyberattack/cyberthreat, 2 focused on logistics, and 10 focused on SCM. In the abstract of the reviewed studies, 13 of the studies focused on cybersecurity/cyberattack/cyberthreat, 5 focused on logistics, and 13 focused on SCM. Lastly on the conceptual review were the Keywords, 11 of the studies focused on cybersecurity/cyberattack/cyberthreat, 2 focused on logistics, and 10 focused on SCM. In the classification of the contributions of the reviewed scholars, the cyberrisk were earlier assessed and solutions to curtail security issues in the area through simulation. The Machine learning (ML) technique was proposed to predict cyberthreat intelligence in one of the reviewed papers, showing the contributions of AI to curtailing cyberthreat. Review of the database that published the reviewed papers, Elsevier published 9, IEEE published 3, and the third category published 4 papers.

The contributions of the reviewed studies to the academic world with respect to the concept of discussion were also reviewed and the focus of eight out of these studies were different implementations of various solutions, the majority of the solutions focused on cybersecurity, and SCM. Less attention has been focused on the technological solutions that addressed the logistics problem. Our finding reveals that there are limited studies that use the dataset to design and stimulate solutions for cybersecurity problems while there is none focusing on cybersecurity with respect to logistics and SCM. The available studies in logistics and SCM are review papers. In regard to our major findings, we suggest possible research directions on the way to improve cybersecurity in logistics, and SCM through the design and simulation of state-of-the-art that redefines cybersecurity in logistics and SCM in a heterogeneous densely populated area like Saudi Arabia. The results of the review and analysis are presented graphically in Figs. 1, 2 and 3. While Fig. 1 shows the concept-centric result of this study, Figs. 2 and 3 present the database for the reviewed papers and classifications of such contributions, respectively. In Fig. 1, the three key concepts were denoted with different colors for easy identification. The report of the three areas of focus in the study was shown in this graphical representation.

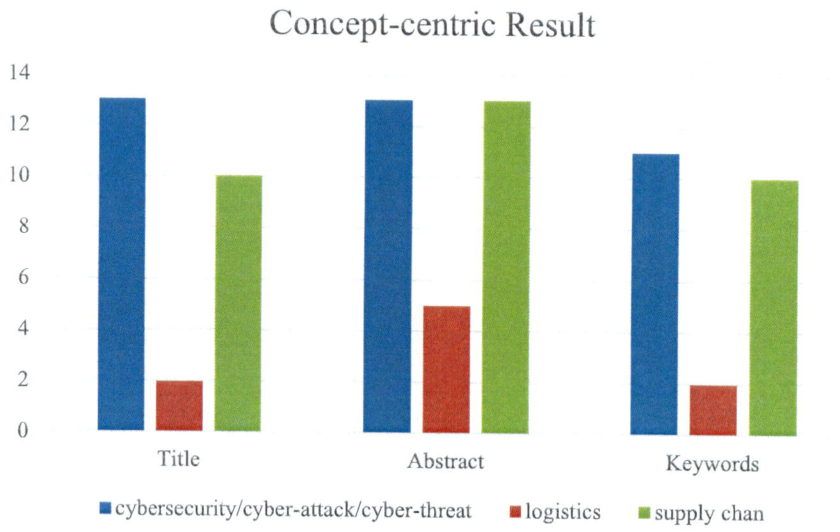

Fig. 1 Concept-centric result

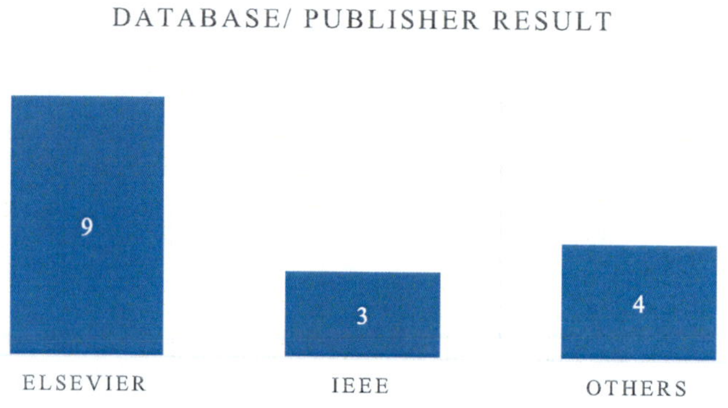

Fig. 2 Publisher report

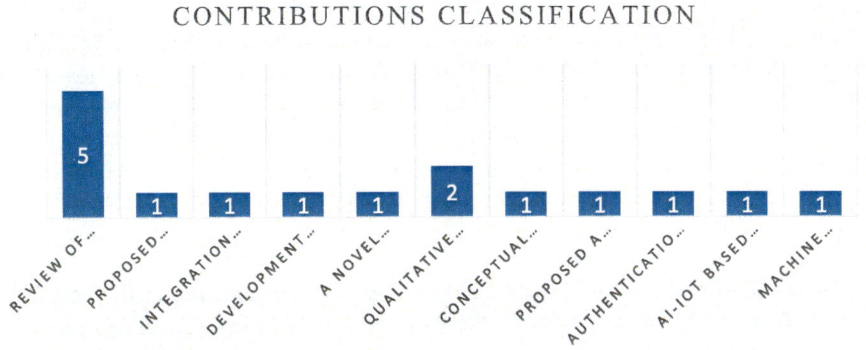

Fig. 3 Contributions classification

4 Open Issues and Future Research

Cybersecurity in logistics and SCM has witnessed a rise in the number of publications due to the acceleration in prominent cyberincidents in these areas.

In recent times, though the huge amount of existing literature in these areas has really enhanced the study of cybersecurity in various industries engaging in logistics and SCM, there is still need for further research to help in the enhancement of approaches and other measures to protect their operations against the growing number of cybersecurity in this sector. The potential research directions from the findings of this paper are stated as follows.

Absence of real cybersecurity data: Almost all the reviewed articles involve and present the development of conceptual architectures. This is due to little or no cybersecurity dataset that can be used in modeling various cyberattacks within the sectors. The existing privacy agreements or for possible reputational damage within clients and logistics and SCM sectors, there is a reluctance to release cybersecurity datasets that can be used for research. Hence, it becomes practically difficult to get a dataset for research purposes in logistics and SCM [2]. Hence, the lack of cybersecurity datasets for logistics and SCM limits research in these fields. The best way to overcome this issue is to collaborate between the researchers and the logistics industry for capturing cybersecurity incidents within the organizations [23, 24]. However, getting such collaboration maybe very difficult since various companies are not ready for such partnership, but substitutes like the use of secondary data or simulation models can be employed in this reaction [25–27]. The creation of hypothetical data from consulting cybersecurity experts in logistics and SCM companies such be of help too.

Deficiency of methodological multiplicity: The use of qualitative approaches which has been employed in the present studies reviewed, and the study of the authors in [28] show that present research is in that direction. The survey supports the use of the methods since this shows there is huge interest in the areas. In the academic field, the majority of qualitative studies reveal a lack of agreement on vital elements and the field's current state. The increase in the number of quantitative studies published reveals the maturity of an academic in the field. The study revealed that there has been an increase in the qualitative methodologies when compared to quantitative methodologies in procurement and delivery cybersecurity research, thus indicating that this field of study is still in its infancy. Hence, it's going to be easier to obtain cybersecurity data such as cyberbreach points, network attack vectors, cybernetworking protocols, and cyberincident investigation in quantitative studies in the area of logistics and SCM.

Insufficient research on cybersecurity in logistics management: Most existing studies focus on SCM, and in SCM, there has been very little research on cybersecurity. Despite the importance of logistics in SC, the significance of cybersecurity in logistics is sometimes underestimated. Third-party logistics (3PL) providers are frequently used by businesses to manage their logistics [29, 30]. A sense of security is produced when companies empower their 3PL partners to feel accountable for risk and vulnerability management is related to logistics. Because there is little research on cybersecurity in the logistics area, the success of this technique is unknown [29].

The infancy of Blockchain in the logistics and supply chains sector: The use of the blockchain-based model for security and privacy has been found in various fields like smart transportation, smart health care, smart agriculture, and smart city applications among others. Most reviewed studies in the areas of logistics and SCM support the use of blockchain that is rich in distributed control, anonymity, tamper-proofness, and immutability natures [31]. This technique can be used for authentication, verification, or data protection in sectors to protect cybersecurity [32]. In the luxury goods SC, blockchain technology-enabled a digital thumb-print mechanism to identify and validate diamond products. Blockchains can also update demand and supply information since they preserve trustworthy permanent recordings of data [33]. The use of blockchain has been shown to be effective in ensuring product quality, selling used goods on an Internet platform, and, more, thus improving the profit margins of the industries [34].

The increase of encryption models: Blockchain is a ledger to secure digital records that uses cryptographic technology to keep track of transactions [35]. Modern encryption methods rely on "one-way" theoretical equations that are simple to compute but difficult to reverse. A typical computer may quickly complete the Rivest–Shamir–Adleman approach by multiplying two very large prime numbers, but finding a pair of prime numbers for a single product could take years [36]. However, if quantum computing technologies advance, the backwards computation in a few hours, thousands of one-way mathematical equations may be solved. At that moment, distributed ledger innovation and its commonly used one-way encryption mechanisms will be vulnerable [37]. Due to the intricacy of quantum techniques, cryptosystem and quantum Internet could be used to mitigate this risk. Both academia and industry could do research on quantum computing-compatible cryptography techniques in the future, with an eye on their application in logistics and SCM.

Encouraging digital forensic and information security research: The advent of radio frequency identification (RFID) technologies has been at the forefront of logistics and SCM in recent years. This has increased warehouse and inventory control efficiency, accuracy, theft prevention, and visibility performance inside SCM [38]. To monitor and trace the physical movement of food in the SC accurately and efficiently, the implementation of RFID technologies becomes important and necessary to ensure food quality [39]. Undeniably, the development and implementation of RFID technologies and other modern technologies like autonomous vehicles, advanced robotics, and smart transportation among others for food tracking systems necessitate heaps of smart gadgets that can sense, collect, distribute, and analyze data to guarantee that SC runs smoothly [40]. In today's SC, real-time intercommunication of these smart devices is essential. The digitization of SC has been advanced by Industry 4.0. However, SCs are vulnerable to cyberattacks because they communicate through a variety of information and operating systems [41]. Manual intervention without Internet communications is beneficial in coping with cyberattacks on infrastructure in manufacturing SC [42]. However, as smart devices become more prevalent, manual intervention may become less successful, necessitating a greater emphasis on the significance of establishing information-sharing schemes across SC stakeholders [43].

5 Conclusion

Cybersecurity research in logistics and SC is not new globally. A study of relevant literature can help readers evaluate and analyze it, as well as locate conceptual background in the area and contribute to the development of theory in a certain issue. In this paper, we have conducted a concept-centric assessment in order to assess the existing state of the art in terms of cybersecurity in logistics and SCM, identify gaps, and offer areas for development. They are existing in various parts of human endeavor but the existing cybersecurity and SCM solutions are primarily concerned with improving SC company performance and enhancing the security in cyberspace. Despite its importance, logistic management has not gotten the desired attention compared to the cybersecurity and SC, and this is a big concern. Hence, the need to consider and improve the logistic infrastructure in the human-dominated area and heterogeneous environments like Saudi Arabia which hosts people from all works of life during Haji and Humira is important. We, therefore, suggest a redirection of research focus that will leverage AI and IoT, to provide efficient security for life and properties, during and after the Haji and other seasonal densely populated areas around the world. Our work expanded studies on cybersecurity in logistics and SCM by acknowledging the databases where current studies on the subject can be found. We also categorized the available studies to highlight the open issues for further studies. Despite the importance of SC and logistics sectors in the major assemblies of people like Haji operations, only limited publications from the scholars are found in the major scholarly databases, offering technical solutions within the concept of SC and logistics to human challenges and this is a big concern.

References

1. Culot G, Nassimbeni G, Orzes G, Sartor M (2020) Behind the definition of industry 4.0: analysis and open questions. Int J Prod Econ 107617
2. Cheung K-F, Bell MG, Bhattacharjya J (2021) Cybersecurity in logistics and supply chain management: an overview and future research directions. Transp Res Part E 146:102217
3. Latif MNA, Aziz NAA, Nussin NSN, Aziz ZA (2021) Cyber security in supply chain management: a systematic review. Sci J Log 49–57
4. Ajagbe SA, Adesina AO (2020) Design and development of an access control based electronic medical record (EMR). Centrepoint J (Science Edition) CPJ 26(1):98–119
5. Urciuoli L, Hintsa J (2017) Adapting supply chain management strategies to security—an analysis of existing gaps and recommendations for improvement. Int J Log Res Appl 20(3):276–295
6. Bamimore I, Ajagbe SA (2020) Design and implementation of smart home nodes for security using radio frequency modules. Int J Dig Sign Smart Syst 4(4):286–303
7. Schick-Makaroff K, MacDonald M, Plummer M, Burgess J, Neander W (2016) What synthesis methodology should I use? A review and analysis of approaches to research synthesis. AIMS Public Health 3(1):172
8. Webster J, Watson R (2002) Analyzing the past to prepare for the future: writing a literature review. MIS Q 26(2):13–23
9. Zarzuelo IP (2021) Cybersecurity in ports and maritime industry: reasons for raising awareness on this issue. Transp Policy 100:1–4
10. Aslama J, Saleema A, Khanb NT, Kim YB (2021) Factors influencing blockchain adoption in supply chain management practices: a study based on the oil industry. J Innov Knowl 6:124–134
11. Gunesa B, Kayisoglub G, Bolat P (2021) Cyber security risk assessment for seaports: a case study of a container port. Comput Secur 103:102196
12. Toppinga T, Dwyerb A, Michaleca O, Craggsa B, Rashida A (2021) Beware suppliers bearing gifts!: analysing coverage of supply chain cyber security in critical national infrastructure sectorial and cross-sectorial frameworks. Comput Secur 108:102324
13. Filhoa NG, Rego N, Claro J (2021) Supply chain flows and stocks as entry points for cyber-risks. In: International conference on project management/Hcist–international conference on health and social care information systems and technologies 2020
14. Kshetri N (2021) Blockchain and sustainable supply chain management in developing countries. Int J Inf Manage 60:102376
15. Cheung K-F, Bell MGH (2021) Attacker–defender model against quantal response adversaries for cyber security in logistics management: an introductory study. Eur J Oper Res 291:471–481
16. Vollmer BK (2021) NATO'S mission-critical space capabilities under threat: Cybersecurity gaps in the military space asset supply chain. SciencesPo, Paris
17. Wallis T, Johnson C, Khamis M (2021) Interorganizational cooperation in supply chain cybersecurity: a cross-industry study of the effectiveness of the UK implementation of the NIS directive. Inform Sec Int J 48(1):36–68
18. Konaszczuk W (2021) Cybersecurity threats in the sectors of oil, natural gas and electric power in the context of technological evolution. Stud Iurid Lublinensia 30(4):333–351
19. Yeboah-Ofori A, Boachie C (2019) Malware attack predictive analytics in a cyber supply chain context using machine learning. In: 2019 International conference on cyber security and Internet of Things (ICSIoT)
20. Martínez MM, Marin-Tordera E, Masip-Bruin X (2021) Scalability analysis of a blockchain-based security strategy for complex IoT systems. In: 2021 IEEE 22nd international conference on high performance switching and routing (HPSR)
21. Jun Y, Craig A, Shafik W, Sharif L (2021) Artificial intelligence application in cybersecurity and cyberdefense. In: Wireless communications and mobile computing. pp 1–10
22. Yeboah-Ofori A, Islam S, Lee SW, Shamszaman ZU, Muhammad K, Altaf M, Al-Rakham M (2021) Cyber threat predictive analytics for improving cyber supply chain security. IEEE Access 9:94318–94337

23. Fernández-Caramés TM, Blanco-Novoa O, Froiz-Míguez I, Fraga-Lamas P (2019) Towards an autonomous industry 4.0 warehouse: A UAV and blockchain-based system for inventory and traceability applications in big data-driven supply chain management. Sensors 19(10), 2394
24. Polatidis N, Pavlidis M, Mouratidis H (2018) Cyber-attack path discovery in a dynamic supply chain maritime risk management system. Comp Stand Interf 56:74–82
25. Zheng K, Albert LA (2019) A robust approach for mitigating risks in cyber supply chains. Risk Anal 39(9):2076–2092
26. Zheng K, Albert LA, Luedtke JR, Towle E (2019) A budgeted maximum multiple coverage model for cybersecurity planning and management. IISE Trans 51(12), 1303–1317
27. Yeboah-Ofori A, Islam S, Brimicombe A (2019) Detecting cyber supply chain attacks on cyber physical systems using Bayesian belief network. In: 2019 International conference on cyber security and Internet of Things (ICSIoT). IEEE, pp 37–42
28. Ghadge A, Wurtmann H, Seuring S (2020) Managing climate change risks in global supply chains: a review and research agenda. Int J Prod Res 58(1):44–64
29. Gkanatsas E, Krikke H (2020) Towards a pro-silience framework: a literature review on quantitative modelling of resilient 3PL supply chain network designs. Sustainability 12(10):4323
30. Ejem EA, Uka CM, Dike DN, Ikeogu CC, Igboanusi CC, Chukwu OE (2021) Evaluation and selection of Nigerian third-party logistics service providers using multi-criteria decision models. LOGI–Sci. J. Trans. Logist. 12(1):135–146
31. Adeniyi EA, Ogundokun RO, Misra S, Awotunde JB, Abiodun KM (2022) Enhanced security and privacy ıssue in multi-tenant environment of green computing using blockchain technology. In: Blockchain applications in the smart era. Springer, Cham, pp 65–83
32. Choi TM, Wen X, Sun X, Chung SH (2019) The mean-variance approach for global supply chain risk analysis with air logistics in the blockchain technology era. Transp Res Part E Logist Transp Rev 127:178–191
33. Shen B, Xu X, Yuan Q (2020) Selling secondhand products through an online platform with blockchain. Transp Res Part E Logist Transp Rev 142:102066
34. Zhai S, Yang Y, Li J, Qiu C, Zhao J (2019) Research on the application of cryptography on the blockchain. In: J Phys Conf Ser 1168(3):032077. IOP Publishing
35. Abdulraheem M, Awotunde JB, Jimoh RG, Oladipo ID (2020) An efficient lightweight cryptographic algorithm for IoT security. In: International conference on ınformation and communication technology and applications. Springer, Cham, pp 444–456
36. AbdulRaheem M, Balogun GB, Abiodun MK, Taofeek-Ibrahim FA, Tomori AR, Oladipo ID, Awotunde JB (2021) An enhanced lightweight speck system for cloud-based smart healthcare. In: International conference on applied ınformatics. Springer, Cham, pp 363–376
37. Biswal AK, Jenamani M, Kumar SK (2018) Warehouse efficiency improvement using RFID in a humanitarian supply chain: ımplications for Indian food security system. Transp Res Part E Logist Transp Rev 109:205–224
38. Shankar R, Gupta R, Pathak DK (2018) Modeling critical success factors of traceability for food logistics system. Transp Res Part E Logist Transp Rev 119:205–222
39. Tang CS, Veelenturf LP (2019) The strategic role of logistics in the industry 4.0 era. Transp Res Part E Logist Transp Rev 129:1–11
40. Abikoye OC, Bajeh AO, Awotunde JB, Ameen AO, Mojeed HA, Abdulraheem M, Oladipo ID, Shakirat AS (2021) Application of internet of thing and cyber physical system in Industry 4.0 smart manufacturing. In: Emergence of cyber physical system and ıot in smart automation and robotics. Springer, Cham, pp 203–217
41. Dynes S, Johnson ME, Andrijcic E, Horowitz B (2007) Economic costs of firm-level information infrastructure failures: estimates from field studies in manufacturing supply chains. Int J Logist Manag
42. Adeniyi JK, Adeniyi EA, Oguns YJ, Egbedokun GO, Ajagbe KD, Obuzor PC, Ajagbe SA (2022) Comparative analysis of machine learning techniques for the prediction of employee performance. Paradigmplus 3(3):1–15

43. Ajagbe SA, Misra S, Afe OF, Okesola KI (2022) Internet of Things (IoT) for secure and sustainable healthcare intelligence: analysis and challenges. In: Florez H, Gomez H (eds) Applied Informatics. ICAI 2022, Communications in computer and information science, vol 1643. Springer, Cham, pp 45–59

Machine Learning Methodology for the Recognition of Unsolicited Mail Communications

Surya Kant Pal, Oma Junior Raffik, Rita Roy, and Prem Shankar Jha

Abstract It is very demanding to classify and label data in machine learning algorithms manually. This works by analysing the probability of different words occurring in legit and spam mails and classifying them accordingly. This research used the notion of applying few labelled data to extrapolate and work on unlabelled data to attain high-accuracy classifiers. Many organizations, students and workplaces have recently used emails to communicate about current affairs, update people on events, etc. Unfortunately, spammers always find their way to exploit recipients by either flooding them with spam emails or sending spam messages periodically. Researchers started applying machine learning techniques to messages to detect if they were spam or not. This was met by the spammers fabricating standard mails. Most people regard spam mails as annoying emails repeatedly used for advertising products and brand promotion. Many such emails are blocked before they get into the user's inbox, but the spam menace is still prevalent. Some other spam mails pose a significant risk to users, like identity theft, containing malware and viruses used to hack people's gadgets, and spam that facilitates fraud. There is a need to create robust spam filters that can efficiently and correctly classify any incoming mail as either spam or not.

Keywords Machine learning · Spam mails · Spammers · Brand promoting

S. K. Pal (✉) · O. J. Raffik
Department of Mathematics, Sharda School of Basic Sciences and Research, Sharda University, Greater Noida 201306, India
e-mail: suryakantpal6676@gmail.com

R. Roy
Department of Computer Science and Engineering, GITAM Institute of Technology, GITAM (Deemed to be University), Visakhapatnam, Andhra Pradesh, India

P. S. Jha
Department of Statistics, Patna College, Patna University, Bihar 800005, India

© The Author(s), under exclusive license to Springer Nature Singapore Pte Ltd. 2023

A. Mishra et al. (eds.), *Advances in IoT and Security with Computational Intelligence*,
Lecture Notes in Networks and Systems 755,
https://doi.org/10.1007/978-981-99-5085-0_6

1 Introduction

In this digital era, emails are significantly used for personal, professional and commercial communication. Spam emails accounted for approximately 10% of the incoming emails to almost every corporate network. Due to the increase in demand and the user base (email users), there is a surge in spam (unwanted) mails. In some situations, people tend to receive more than 50% of spam mails than essential mails. If no countermeasures exist, it costs victims (including organizations) millions of dollars daily and undermines how mails are used [1]. Spam mails operate by foiling the user's internet bandwidth and increasing the time taken to process mails [2].

The birth of internet-based spam introduced the use of supervised machine learning algorithms to combat this malpractice. In this algorithm, a set of labelled data are fed into the required model (classifier), which is then trained to mark the messages accordingly (spam or non-spam). After testing the model and checking its accuracy, it is then deployed on unlabelled emails; if it also manages to attain the same accuracy rate as previously, it is then deployed for official use [3, 4].

Any mislabelling of the training data set due to obfuscating spam affects the entire learning process and its accuracy [5, 6]. This is because some people might find a mail beneficial, while identical mail can be considered spam to another person. People trying to overcome spammer's tactics have seen fewer people receiving spam-based emails. As noticed in the graph provided by 'Statista.com' below from 'Spam Statistics: spam email traffic' across the years, software engineers came up with reliable solutions that helped and kept on helping to reduce the malice act by a considerable percentage [7, 8].

2 Literature Review

With the use of supervised learning methods, the algorithms of this type were used in several text categorization tasks. The fascinating thing was they were also capable of classifying email into folders. Recently, tools to combat spam have become very necessary for many internet service providers in the quest to deal with the eternal phenomena of spam emails [9]. In 1998, Stanford researcher Mehran Shami and his Microsoft team published in the first connection between email classification and Naive Bayes on a project based on Spam Filtering. With the application of word-based features only, Shami achieved 97% precision, leading to Naive Bayes being a better method than the usual method of Keyword-Based Spam Filters [10]. A popular open-source software D-Spam is having a 99.5% precision with a Bayesian-based classification algorithm [11, 12].

Messaging Spam is a type of spam involving abusing mobile phone services used for messaging (texting). These spams usually possess advertisements and rarely target the user personally [13, 14]. These types of spam are more dangerous since they can also find their way to the user's Instant Message services (IM) like Telegram or WhatsApp and sometimes sends direct messages through website features present on social sites; like the Direct Message (DM) feature that is available on sites like Reddit and Twitter [15].

Link spam is when an entity embeds links in their webpages (websites) to increase the backlink count (number of links pointing to domains). Consideration of the contents of the links you are being directed to is not regarded (that is why it is called link spam) [16, 17]. Most links lead to unrelated sites since the main goal is to increase their presence in search engine indexes [18].

Unsolicited Bulk Emails (UBE) flood the inbox with many junk mails to many users simultaneously. The messages involve business adverts as well as malicious emails (Malspam). The malicious emails contain links for the receiver to click; once the link is clicked, the spammer injects the user's device with trojans, bots and sometimes ransomware [19]. The malspam are made very convincing to make the user believe (click) the story they have been provided [20, 21].

Any fraudulent activity received via a phone call is called such [22]. These SMS are mostly intrusive and less valuable to the recipient [23]. According to the information being disseminated, they can either be legal or illegal. This is because even network providers tend to send their customers these messages since they have the telephone numbers in their databases [24].

In general, it is called a fee scam spam. It is where spammers send messages that convince the receiver of a reward of money only if they agree to send some cash they regard as a processing fee [25]. If the victims believe the information they have received, they rush to make the payment; after the payment has been made, the victims do not receive the promised monetary reward, and the process becomes void [26].

3 Methodology

The research was based on a Multinomial Naive Bayes method. In this context, multinomial was applied since it is feasible to analyse text data (Categorical data). It calculates the probability of the occurrence of an event based on its prior knowledge of conditions related to the event.

$$P(A|\mathbf{B}) = P(A) * P(B|A)/P(B)$$

Calculating the probability of A given B.

P(B) = prior probability of B.

P(A) = prior probability of class A.

P(B|A) = occurrence of predictor B given class A probability.

Figure 1 shows the data set above (available in Python corpus) containing 5572 messages labelled as spam and non-spam (ham). The model was fed with this amount of data and allowed to study every instance accordingly; since there were no missing data and a need to remove stop words, data cleaning methods were not used. After the

	class	message	Unnamed: 2	Unnamed: 3	Unnamed: 4
0	ham	Go until jurong point, crazy.. Available only ...	NaN	NaN	NaN
1	ham	Ok lar... Joking wif u oni...	NaN	NaN	NaN
2	spam	Free entry in 2 a wkly comp to win FA Cup fina...	NaN	NaN	NaN
3	ham	U dun say so early hor... U c already then say...	NaN	NaN	NaN
4	ham	Nah I don't think he goes to usf, he lives aro...	NaN	NaN	NaN
...	...	...	...	...	...
5567	spam	This is the 2nd time we have tried 2 contact u...	NaN	NaN	NaN
5568	ham	Will ì_ b going to esplanade fr home?	NaN	NaN	NaN
5569	ham	Pity, * was in mood for that. So...any other s...	NaN	NaN	NaN
5570	ham	The guy did some bitching but I acted like i'd...	NaN	NaN	NaN
5571	ham	Rofl. Its true to its name	NaN	NaN	NaN

5572 rows × 5 columns

Fig. 1 Data set

algorithm analysed the entire data, it extracted two features that contained instances (class and message) crucial for the study of spam as well as non-spam to train the model to classify spam messages. The features extracted from the entire data set were the ones with the critical data type used in the Machine Learning model. Since the 'Class' feature classifies the type of mail (output), either spam or not, the message it belongs to. And the other element, 'Message,' tells the model how and what spam mail contains and specifies the types of legit messages. Depending on similar words existing in the news (if it exceeds the threshold), the model will classify the group it belongs. This formula helps calculate the probability of the tags present in the text. The algorithm finds keywords that are rampant in spam but rarely show up and uses them to classify future spam emails.

4 Experiment and Results

After applying the Naive Bayes-based method on the data set, which was composed of the data set that was previously split into training and test set that was used to detect spam emails, since selecting required features is an essential task as far as classification goes. The model turned out with an accuracy rate of 97%.

The model here was designed to take details about the message and compare it to the ones available in the data set for similar features on both spam and non-spam. When the model detects the similarity of phrases or words in the spam category, it sends a notification labelling it spam. The same applies when the given input does not pass the non-spam threshold (it gets classified as non-spam). It understands the text context of the received messages and analyses it for both probabilities and the one that surpasses the other. This helps reduce the load of manually filtering mails since spam emails are sent directly to the spam folder instead of a user sending them there. Hence increasing the user experience by not disturbing them with unwanted alerts [27].

```
from sklearn.naive_bayes import BernoulliNB

model = BernoulliNB().fit(X_train, y_train)
# Fit the model on train dataset
model.fit(X_train, y_train)
predicted_signal = model.predict(X_test)
```

```
from sklearn import metrics
print("Binomial Naive Bayes model accuracy(in %):", metrics.accuracy_score(y_test, predicted_signal)*100)
```

```
Binomial Naive Bayes model accuracy(in %): 97.82490483958674
```

Fig. 2 Timing and precision

In Fig. 2, finding the ideal balance between timing and precision is the key to the art of prediction. Since early predictions lack the luxury of 'history evidence,' getting them right requires a lot of intuition. Here, obtained 97.82%.

5 Evaluation

After finalizing the model, it is necessary to evaluate its performance to compare it with other models. This research was based on the Naive Bayes model, which had a performance of 97.8%. Using the same data set, a Support Vector Model was tested for its accuracy, which had an accuracy percentage of 97.6. There is a difference of 0.2% between these two models, with the Naive Bayes-based model being the first or best.

Figure 3 shows the most straightforward and fastest classification technique, Naive Bayes, which is appropriate for handling vast amounts of data. A statistical classification method based on the Bayes Theorem is called naive Bayes. One of the easiest ML techniques is supervised learning methods is this one. The fast, accurate and dependable algorithm is the Naive Bayes classifier. On large datasets, Naive Bayes classifiers perform quickly and accurately. It is about 97.824.

```
from sklearn.svm import SVC
classifier = SVC(kernel = 'rbf', random_state = 0)
classifier.fit(X_train, y_train)

print(classifier.score(X_test,y_test)*100)
```

```
97.66177270255574
```

```
from sklearn import metrics
print("Binomial Naive Bayes model accuracy(in %):", metrics.accuracy_score(y_test, predicted_signal)*100)
```

```
Binomial Naive Bayes model accuracy(in %): 97.82490483958674
```

Fig. 3 Simplest and fastest classification

Table 1 Matrices

		Predicted	
Actual		Negative	Positive
	Negative	TN	FP
	Positive	FN	TP

Table 1 shows that the matrices represent the predicted and actual values count, where TN means True Negatives. TN is a set of variables that have been correctly classified as negatives. TP represents True Positives; these values have been accurately classified as positives.

FP is False Positives; these are negative features or values that have been classified as positives. FN stands for False Negative; these values indicate the number of actual positives that have been wrongly classified as negatives. Models are checked for accuracy (with the use of a confusion matrix) by using the formula below:

$$\text{Accuracy} = \text{TN} + \text{TPTN} + \text{FP} + \text{FN} + \text{TP}$$

Accuracy can be false if applied to imbalanced datasets, hence the need to check the datasets and provide correct actual and predicted values to have a real accuracy rate.

After applying the confusion matrix on the datasets, 3238 mails were correctly classified as non-spam (all the instances of non-spam were correctly classified), while spam emails correctly classified were 486 out of 495 spam mails (9 turned out to be false negatives).

Figure 4 showed when a confusion matrix was applied on the test set that contained 1839 mails that were both spam and non-spam. It turned out with 1587 true negatives, mails that were accurately classified as belonging to the non-spam category while one mail was classified as belonging to the false positive (failed to be classified correctly). On spam mails, the matrix produced 42 false negatives (not correctly classified) but managed to make 210 spam emails that were accurately classified.

Figure 5 represents the precision, recall, F1-score and support of ham and spam mail; here, 0.97, 1.00, 0.99 and 1587 for ham mail and spam mail 1.00, 0.83, 0.91 and 252. Accuracy is 0.98, 0.92, 0.95 and 1839.

It turned out with 1587 true negatives, mails that were accurately classified as belonging to the non-spam category while one mail was classified as belonging to the false positive (failed to be classified correctly). On spam mails, the matrix produced 42 false negatives (not correctly classified) but managed to make 210 spam emails that were accurately classified.

Table 2 contains both readings of the train set and the test set of the entire data set of spam and non-spam emails (5572 rows * 2 columns). The training set had an accuracy rate of 99.75%, while the test set produced a 97.66% accuracy rate, providing an accuracy difference rate of 2.09%.

```
# Evaluating the model on the training data set
from sklearn.metrics import classification_report, confusion_matrix, accuracy_score
pred = classifier.predict(X_train)
print(classification_report(y_train, pred))
print()
print("Confusion Matrix: \n", confusion_matrix(y_train, pred))
print("Accuracy: \n", accuracy_score(y_train, pred)*100)
```

```
              precision    recall  f1-score   support

         ham       1.00      1.00      1.00      3238
        spam       1.00      0.98      0.99       495

    accuracy                           1.00      3733
   macro avg       1.00      0.99      0.99      3733
weighted avg       1.00      1.00      1.00      3733


Confusion Matrix:
 [[3238    0]
 [   9  486]]
Accuracy:
 99.7589070452719
```

Fig. 4 Confusion matrix

```
# Evaluating the model on the training data set
from sklearn.metrics import classification_report, confusion_matrix, accuracy_score
pred = classifier.predict(X_test)
print(classification_report(y_test, pred))
print()
print("Confusion Matrix: \n", confusion_matrix(y_test, pred))
print("Accuracy: \n", accuracy_score(y_test, pred)*100)
```

```
              precision    recall  f1-score   support

         ham       0.97      1.00      0.99      1587
        spam       1.00      0.83      0.91       252

    accuracy                           0.98      1839
   macro avg       0.98      0.92      0.95      1839
weighted avg       0.98      0.98      0.98      1839


Confusion Matrix:
 [[1586    1]
 [  42  210]]
Accuracy:
 97.66177270255574
```

Fig. 5 Precision, recall and support for the spam mail

Table 2 Precision, recall and F1-score

Sets	Precision	Recall	F1-score
Train set of non-spam	1.00	1.00	1.00
Train set of spam	1.00	0.98	0.99
Test set of non-spam	0.97	1.00	0.99
Test set of spam	1.00	0.83	0.91

6 Discussion

When such data have been collected, they can be used to either market their products or by hackers to overload the victims' servers or maybe infect their mail or the gadget they sign in with, which can result in the exposure of severe personal information [28]. This is the other reason spam filters exist, to prevent users from being coerced into opening these types of emails and clicking on suspicious links or trusting the content of the mail without thinking twice. Some other improved and advanced models are still being developed to overcome these types of activities since spammers are also getting and employing advanced methods to try to beat the present models. To prevent organizations and users from losing revenues, organizations like Google (in Gmail) direct the messages to a different folder than important mails, hence preventing network servers from being overwhelmed by unwanted alerts [29].

7 Conclusion

Spam filtering is a must and must be a primary need for email service providers since they cost people and organizations daily. This research discussed spam email filtering methods that apply advanced technologies based on Machine Learning algorithms. It tackled the issue of spam filtering based on Machine Learning using naïve Bayes. The model used text classification to analyse the corpus presented as a training set with an accuracy rate of 97.8% and turned out to be better than the Support Vector Model, a Machine Learning-based model with a slight difference of 0.2%. Due to the severe impact, spam messages can cause to victims, like, phishing attacks and other malicious acts, it is best to use Naive Bayes than Support Vector because the small margin can give fatal results if not taken seriously. Since every classification model has weaknesses, spammers can spoil it is excellent to use the model that applies different methods to handle the other model's flaws.

References

1. Abhila B, Delphin Periyanayagi M, Koushika M, Joseph MN, Dhanalakshmi R (2021) Spam detection system using supervised ML. In: 2021 International conference on system, computation, automation and networking, ICSCAN 2021. https://doi.org/10.1109/ICSCAN53069.2021.9526421
2. Hosseinalipour A, Ghanbarzadeh R (2022) A novel approach for spam detection using horse herd optimization algorithm. Neural Comput Appl 2022:1–15. https://doi.org/10.1007/S00521-022-07148-X
3. Mashaleh AS, Binti Ibrahim NF, Al-Betar MA, Mustafa HMJ, Yaseen QM (2022) Detecting spam email with machine learning optimized with Harris Hawks optimizer (HHO) algorithm. Procedia Comput Sci 201(C):659–664. https://doi.org/10.1016/J.PROCS.2022.03.087

4. Baral MM, Mukherjee S, Chittipaka V, Jana B (2023) Impact of blockchain technology adoption in performance of supply chain. In: Blockchain driven supply chains and enterprise information systems. pp 1–20. https://doi.org/10.1007/978-3-030-96154-1_1
5. Faris H et al (2019) An intelligent system for spam detection and identification of the most relevant features based on evolutionary Random Weight Networks. Inform Fusion 48:67–83. https://doi.org/10.1016/J.INFFUS.2018.08.002
6. Roy R, Chekuri K, Sandhya G, Pal SK, Mukherjee S, Marada N (2022) Exploring the blockchain for sustainable food supply chain. J Inf Optim Sci 43(7):1835–1847. https://doi.org/10.1080/02522667.2022.2128535
7. Ahmed N, Amin R, Aldabbas H, Koundal D, Alouffi B, Shah T (2022) Machine learning techniques for spam detection in email and IoT platforms: analysis and research challenges. Sec Commun Netw https://doi.org/10.1155/2022/1862888
8. Roy R, Babakerkhell MD, Mukherjee S, Pal D, Funilkul S (2022) Evaluating the intention for the adoption of artificial intelligence-based robots in the university to educate the students. IEEE Access 10:125666–125678. https://doi.org/10.1109/ACCESS.2022.3225555
9. Sharma VD, Yadav SK, Yadav SK, Singh KN, Sharma S (2021) An effective approach to protect social media account from spam mail—a machine learning approach. Mater Today Proc. https://doi.org/10.1016/J.MATPR.2020.12.377
10. Petersen LN (2018) The ageing body in Monty Python Live (Mostly). Eur J Cult Stud 21(3):382–394. https://doi.org/10.1177/1367549417708435
11. Bansal C, Sidhu B (2021) Machine learning based hybrid approach for email spam detection. In: 2021 9th international conference on reliability, infocom technologies and optimization (Trends and Future Directions), ICRITO 2021. https://doi.org/10.1109/ICRITO51393.2021.9596149
12. Mukherjee S, Baral MM, Pal SK, Chittipaka V, Roy R, Alam K (2022) Humanoid robot in healthcare: a systematic review and future research directions. In: 2022 International conference on machine learning, big data, cloud and parallel computing (COM-IT-CON), pp. 822–826. https://doi.org/10.1109/COM-IT-CON54601.2022.9850577
13. Raza M, Jayasinghe ND, Muslam MMA (2021) A comprehensive review on email spam classification using machine learning algorithms. In: International conference on information networking, vol. 2021, pp 327–332. https://doi.org/10.1109/ICOIN50884.2021.9334020
14. Roy R, Baral MM, Pal SK, Kumar S, Mukherjee S, Jana B (2022) Discussing the present, past, and future of Machine learning techniques in livestock farming: a systematic literature review. In: 2022 International conference on machine learning, big data, cloud and parallel computing (COM-IT-CON). pp. 179–183. https://doi.org/10.1109/COM-IT-CON54601.2022.9850749
15. Kontsewaya Y, Antonov E, Artamonov A (2021) Evaluating the effectiveness of machine learning methods for spam detection. Procedia Comput Sci 190:479–486. https://doi.org/10.1016/J.PROCS.2021.06.056
16. Srinivasan S, Ravi V, Alazab M, Ketha S, Al-Zoubi AM, Kotti Padannayil S (2021) Spam emails detection based on distributed word embedding with deep learning. Stud Comp Intell 919:161–189. https://doi.org/10.1007/978-3-030-57024-8_7/COVER
17. Kant Pal S, Mukherjee S, Baral MM, Aggarwal S (2021) Problems of big data adoption in the healthcare industries. Asia Pacific J Health Manag 16(4):282–287. https://doi.org/10.24083/apjhm.v16i4.1359
18. Blanzieri E, Bryl A (2008) A survey of learning-based techniques of email spam filtering. Artif Intell Rev 29(1):63–92. https://doi.org/10.1007/S10462-009-9109-6
19. Islam MK, al Amin M, Islam MR, Mahbub MNI, Showrov MIH, Kaushal C (2021) Spam-detection with comparative analysis and spamming words extractions. In: 2021 9th International conference on reliability, infocom technologies and optimization (Trends and Future Directions), ICRITO. https://doi.org/10.1109/ICRITO51393.2021.9596218
20. Trivedi SK (2016) A study of machine learning classifiers for spam detection. In: 2016 4th International symposium on computational and business intelligence ISCBI, pp 176–180. https://doi.org/10.1109/ISCBI.2016.7743279

21. Mukherjee S, Chittipaka V (2021) Analysing the adoption of intelligent agent technology in food supply chain management: an empirical evidence. FIIB Bus Rev 231971452110592. https://doi.org/10.1177/23197145211059243
22. Saleh AJ, et al (2019) An intelligent spam detection model based on artificial immune system. Information (Switzerland) 10(6). https://doi.org/10.3390/INFO10060209
23. Taloba AI, Ismail SSI (2019) An intelligent hybrid technique of decision tree and genetic algorithm for e-mail spam detection. In: Proceedings–2019 IEEE 9th international conference on intelligent computing and information systems, ICICIS 2019, pp 99–104. https://doi.org/10.1109/ICICIS46948.2019.9014756
24. bin Siddique Z, Khan MA, Din IU, Almogren A, Mohiuddin I, Nazir S (2021) Machine learning-based detection of spam emails. Sci Program 2021. https://doi.org/10.1155/2021/6508784
25. Hossain F, Uddin MN, Halder RK (2021) Analysis of optimized machine learning and deep learning techniques for spam detection. In: 2021 IEEE international IOT, electronics and mechatronics conference, IEMTRONICS 2021–Proceedings. https://doi.org/10.1109/IEMTRONICS52119.2021.9422508
26. Nayak R, Amirali Jiwani S, Rajitha B (2021) Spam email detection using machine learning algorithm. Mater Today Proc https://doi.org/10.1016/J.MATPR.2021.03.147
27. Nandhini S, Marseline DJ (2020) Performance evaluation of machine learning algorithms for email spam detection. In: International conference on emerging trends in information technology and engineering, ic-ETITE 2020. https://doi.org/10.1109/IC-ETITE47903.2020.312
28. Kumar N, Sonowal S (2020) Email spam detection using machine learning algorithms. In: Proceedings of the 2nd international conference on inventive research in computing applications, ICIRCA, pp. 108–113. https://doi.org/10.1109/ICIRCA48905.2020.9183098
29. Govil N, Agarwal K, Bansal A, Varshney A (2020) A machine learning based spam detection mechanism. In: Proceedings of the 4th international conference on computing methodologies and communication, ICCMC 2020, pp 954–957. https://doi.org/10.1109/ICCMC48092.2020.ICCMC-000177

IoT-Based Automated Hydroponic Cultivation System: A Step Toward Smart Agriculture for Sustainable Environment

Snehal V. Laddha

Abstract In recent years, plants have been grown without soil using a technique called hydroponics by directly feeding them the ideal amount of nutrients from the earth in water. Plants grown by using the hydroponic technique usually yield more, require less space, and conserve soil and water. The main aim of our work is to save lots of water, reduce pesticide use and factors that affect land quality, and improve crop quality. The automated hydroponics system based on the Internet of things (IoT) is developed to facilitate cultivation and the system can regulate and govern important environmental factors that affect plant growth. An efficient technique to grow plants is with an automated Internet of things-based hydroponics system for cultivating, regulating, and governing important environmental factors that influence plant growth. We have developed a system that can monitor and control complete hydroponic farming from anywhere using IoT via the web, therefore parameters like pH level, light intensity, electrical conductivity, water level, temperature, and room humidity parameters are often viewed in real-time.

Keywords IoT · Hydroponics · Smart agriculture · Cultivation system

1 Introduction

A hydroponics system is a modern plant system where the plants are grown without soil, however, makes use of water with Essentials Nutrients to save area planting and is not polluted with chemical substances in the soil. The modern approach of hydroponics does not confine itself to growing plants in water like nutrient film technology, deep flow technique, dynamic root floating technique, etc [1].

S. V. Laddha (✉)
Department of Electronics Engineering, Shri Ramdeobaba College of Engineering and Management, Nagpur, Maharashtra, India
e-mail: laddhasv2@rknec.edu

© The Author(s), under exclusive license to Springer Nature Singapore Pte Ltd. 2023
A. Mishra et al. (eds.), *Advances in IoT and Security with Computational Intelligence*, Lecture Notes in Networks and Systems 755,
https://doi.org/10.1007/978-981-99-5085-0_7

1.1 *Methods of Hydroponics*

1.1.1 A Wick System

This hydroponics is a type of hydroponic method where plants are grown in a growing medium, such as Rockwool coco coir, and nutrients are delivered to the plants via a wick. The wick is a piece of fabric, rope, or other material that is placed in contact with the nutrient solution and extends into the growing medium. The wick acts as a conduit, drawing the nutrient solution up to the roots of the plants by capillary action.

One of the main advantages of a wick system hydroponics is that it is relatively simple to set up and maintain. The system does not require any pumps or electricity to operate, making it a great option for off-grid or remote growing. The wick system also allows for a more consistent and steady flow of nutrients to the plants, which can result in healthier and more vigorous growth.

However, a wick system has also some limitations. It is not as efficient as other hydroponic methods, such as ebb and flow or drip systems, in delivering the nutrient solution to the plants. The plants' growth rate may be slower than other hydroponics systems, and the size of the plants is generally smaller. Additionally, the wick system requires more frequent attention, checking the water level, and refilling the nutrient solution.

1.1.2 Nutrient Film Technique

A structure similar to PVC pipes is implemented in this technique. A collection of plant life is grown in line by the usage of this shape of the structure [2]. This nutrient solution is forced into a pipe-like formation and then collected through another sensor back into the tank, where an outlet is created for the solution to flow out. By using this method, recirculation occurs while at the same time, water conservation is demonstrated.

1.1.3 The Ebb and Flow Method

It works by providing the roots of the plants with a consistent supply of oxygen and nutrients. When the solution is pumped in, the roots are flooded with the nutrient-rich solution, allowing the plants to absorb the necessary nutrients for growth. When the solution is drained out, the roots are exposed to air, which allows them to breathe and prevents them from becoming waterlogged.

One of the main benefits of ebb and flow systems is that they are relatively low maintenance and easy to set up. They are also well suited for plants that have a high water requirement. However, one of the main drawbacks of ebb and flow systems is that they can be prone to leaks and power outages can interrupt the pumping schedule, which can be detrimental to the plants.

1.1.4 Deep Water Culture

Deep Water Culture (DWC) hydroponics is a type of hydroponic method in which plants are grown in a floating platform, such as a styrofoam raft, that is suspended in a nutrient-rich solution. The roots of the plants are suspended directly into the solution, allowing them to absorb the nutrients they need to grow.

One of the main advantages of DWC hydroponics is that it can support very rapid plant growth. The roots are in constant contact with the nutrient solution and have access to oxygen, which allows them to absorb large amounts of water and nutrients. This results in healthier and more vigorous plant growth compared to traditional soil-based growing methods. Additionally, DWC systems are relatively simple to set up and maintain, and they are very space-efficient.

However, DWC systems also have some limitations. One of the main disadvantages is that the systems must be closely monitored to ensure that the nutrient solution is at the correct pH and temperature and that the solution is oxygenated. If not properly maintained, the plants can suffer from oxygen deprivation or nutrient imbalances, which can lead to stunted growth or death. Additionally, DWC systems can be expensive to set up, especially if large numbers of plants are grown in one container.

1.1.5 Drip System

A Drip System hydroponics is a type of hydroponic method where nutrient solution is delivered to the roots of the plants through a network of tubing and emitters. The nutrient solution is delivered in a controlled manner, typically through a timer-controlled pump, to each individual plant or group of plants.

One of the main advantages of a Drip System hydroponics is its efficiency and precision. By delivering the nutrient solution directly to the roots of the plants, it ensures that the plants have access to the right balance of nutrients at the right time. Additionally, Drip systems are relatively easy to set up and maintain. They can be customized to suit the needs of different types of plants and can be scaled up or down depending on the size of the operation.

However, Drip systems also have some limitations. One of the main disadvantages is that they require a consistent and reliable water supply and the systems must be closely monitored to ensure that the nutrient solution is at the correct pH and temperature. If not properly maintained, the plants can suffer from nutrient imbalances or water-logging, which can lead to stunted growth or death. Additionally, Drip systems can be expensive to set up, especially if large numbers of plants are grown in one container.

1.1.6 Aeroponics

This is much like the Nutrient Film Technique. Unlike recirculation, there is only one difference here: sprinklers are used instead of recirculation. These sprinklers do the task of misting the root zone with nutrient solution [3].

2 Related Work

The market for new sustainable agricultural techniques employing the Internet of Things and the rising global food demand are the issues that need to be tackled in this way. There have been several research studies and projects focused on IoT-based automated hydroponic cultivation systems in recent years [1, 4–7].

In the study, [8] presents a design and implementation of an IoT-based automated hydroponic system. Using NodeMcu, Node Red, MQTT, and sensors selected during component selection that supported the necessary parameters, this design was carried out. It was then sent to the cloud for monitoring and readiness. In the beginning, the prototype was built, coded, and tested in the same way that sensor data from two distinct environments were collected and tracked on a cloud-based website with a mobile application. A bot has also been deployed to control the supply chain and serve as a notification system.

A tool using Arduino Uno microcontroller and a loyal smartphone was developed [9]. It uses technology to continuously watch the rush of nutrients to the shop, device can also shoot data of temperature and fluid reach around the shop to the doper. Doper can view these data using smartphone. Nutritional water rush system by exploiting detector distance has been successfully done.

A small hydroponic farmer created a farming service that makes use of unused space [10]. A low-cost detector module and an agriculturist with an integrated remote-control system have been developed in order to provide freshmen in farming with a full-fledged accomplishment strategy for fruits and vegetables (Fig. 1).

The data from detectors and the control of a pump are covered by MQTT, a less important IoT protocol. TLS/SSL is used to encrypt data securely. A USB-connected camera-based still-image photography function with move discovery is also included, in addition to covering detector data.

One of the researchers used the innovative concept of creating a hydroponic smart farming system that can be tracked online via their internet platform was developed [11]. It was designed to keep track of the necessary agricultural system parameters, including conductivity, pH, temperature, humidity, and intensity level (EC). The prototype is designed to work with sensors like the DHT11 module, LDR, pH device module, and electrical conductivity device, all of which link straight to a Raspberry Pi 3. Additionally, a message bot that enables online sensor viewing via messages is established. The hydroponic system monitoring becomes much more flexible with the integration of the Physical System (Raspberry Pi, sensor) and Social System (Telegram Messenger) connected online via IoT.

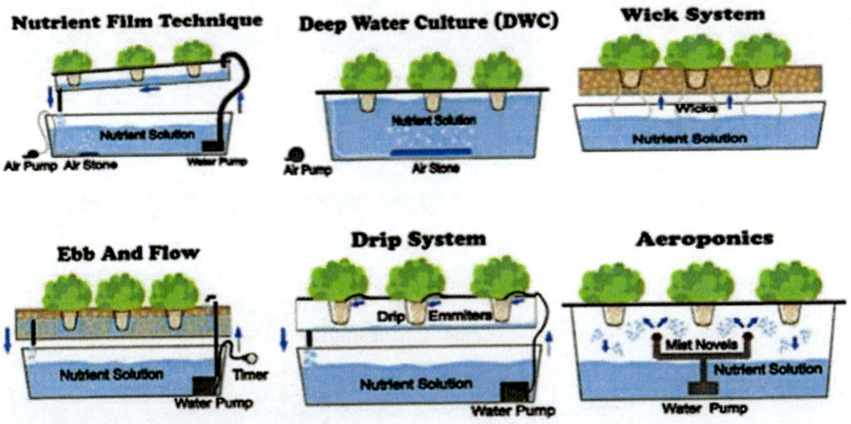

Fig. 1 Methods of hydroponics

Hariono developed a method to collect data using an ESP8266 on an IOT-based hydroponic plant automation system [12]. The DS18B20 sensor measures temperature, while the TDS Meter v1.0 sensor measures PPM, the Ph Meter v1.1 sensor measures pH, the TIP 42C sensor measures water level, and the ESP32Cam sensor continuously measures plant growth to be observed. Hydropower plants employ sonar panels as a further energy reserve. Each sensor's data are transferred to the Firebase real-time database, and any changes to the data are recorded in the MySQL server database. The outcomes of the data gathering are shown in real time on the dashboard page for user readability. The results of the function and effectiveness tests were good.

3 Materials and Method

The seeds are grown in cocopeat as shown in Fig. 2 to develop small saplings, which are then transferred into net pots where their roots are supported with pebbles and exposed to the nutrient-rich oxygenated water solution as shown in Fig. 3.

The plants cultivated in the container required parameters in ranges as follows:

- For the surrounding ambient temperature, the plants require 15–21 °C. Similarly,
- Humidity: 10–50%
- White Light of 300–800 lumens
- pH: 5.5–6.5 (Nutrient solution)
- ppm: 800–1500 ppm (Nutrient solution)

Fig. 2 Germination of spinach seed

Fig. 3 Transplanting of plants into net pot

3.1 ESP 8266 Node MCU

The node MCU is a Wi-Fi module integrated into the Arduino board that works with all the sensors present in the system and is used to send the information received from the sensors to the database over Wi-Fi. The MCU is the heart of the IoT system, the main part of which processes data at 1% and runs the software stack connected to the wireless device for connectivity [13].

3.2 Dht 11

Temperature and Humidity sensor: DHT11 performs this task. It provides ±0.1% accuracy and a good response time of 2 seconds at temperatures up to 80 degrees Celsius. A moisture measurement range of 0–100% is sufficient for this method. Plants are very sensitive to changes in temperature and humidity when both parameters are high. Plants lack the ability to germinate. Therefore, it has a negative impact on their growth. However, as the roots mature, they adapt to higher humidity ranges. An RTC (timer circuit) is implemented to track the time and the system can recognize whether it is day or night [3].

3.3 PH Sensor

The pH sensor is an electronic chemical splitting device that detects and measures the pH of water by measuring it on an acid basis, the pH suitable for measurement is 5.5–7 [13].

3.4 TDS Meter

The term TDS (Total Dissolved Solids) describes the amount of soluble solids that have dissolved in one liter of water. Working current is 3–6 mA, and the range varies from 0 to 1000 ppm. It has an input voltage of DC 3.3–5.5 volts and an output voltage of 0–2.3 volts.

For single bus data format, synchronization between DHT11 and NODE MCU sensor is used. This process takes 4 milliseconds. All the sensors are connected to our microcontroller Node MCU. As the supply starts the Node Mcu starts obtaining values from the sensors through the Wi-Fi module. After collecting the data from sensors, it then sends it to the Application through the Wi-Fi module. The values of Ph, temperature, humidity, and overall dissolves solids can be seen in the application [14–16].

As we can see in the block diagram, there are sensors like Ph sensor to measure the PH of the solution, TDS meter is used to indicate the Total Dissolved Solids in a solution, usually water by measuring its conductivity by the dissolved salts and minerals and then DHT 11 to measure the temperature and humidity of the environment. On the other hand, here, grow lights have been used so the plants will get ample light to grow efficiently. Relay and water pump are used to supply nutrient solution to the structure (Fig. 4).

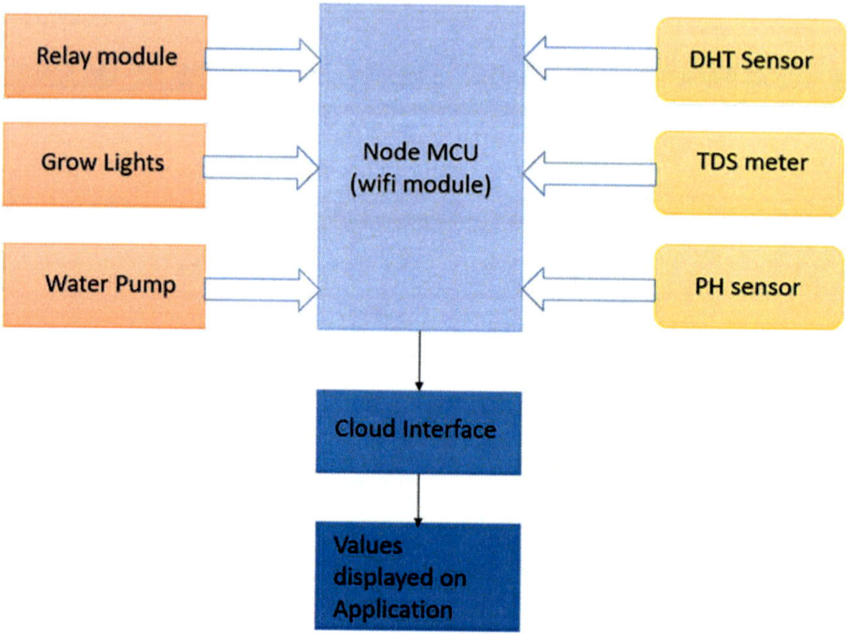

Fig. 4 Block diagram of automated hydroponics system

4 Results

This project is IoT-based as it includes connecting the hardware to an existing Wi-Fi network and communicating with a cloud-based real-time database to monitor after the physical setup, we interfaced the various sensor with node MCU (ESP8266) as shown in Fig. 5.

The data collected from various sensors were sent to ThingSpeak Cloud. As shown in Fig. 6, various parameters such as humidity, temperature, pH value, and TDS values are displayed.

We have accessed the data through ThingSpeak Cloud, and it was monitored and displayed through the mobile application, i.e. MIT A2 companion as shown in Fig. 7.

After the germination of seeds, we transferred the saplings into the physical structure and started the flow of nutrient solution. After 3 weeks, we observed that the saplings have grown as shown in Fig. 8. So we have used spinach in our project and the time needed for a spinach plant to grow fully is 5–6 weeks under the essential environmental factors.

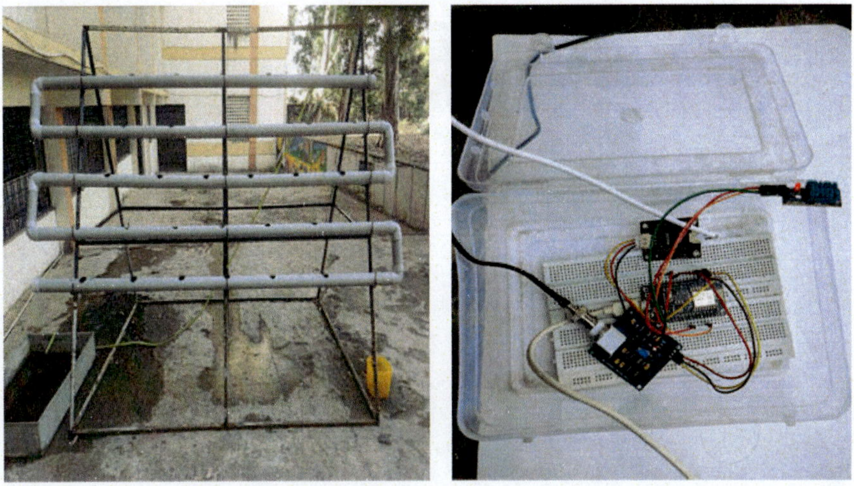

Fig. 5 Hardware setup of the system

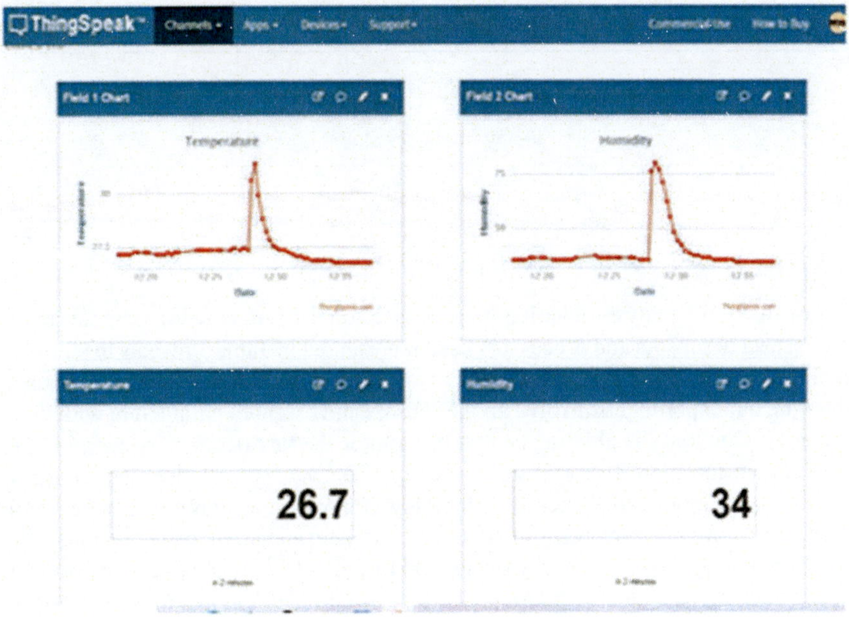

Fig. 6 Data values on the ThingSpeak Cloud

Fig.7 Results displayed on mobile application (MIT A2 Companion)

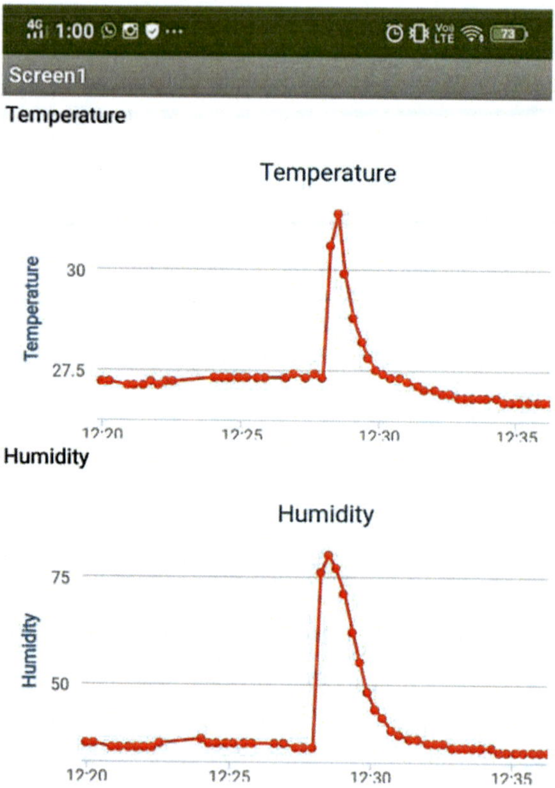

A comparative analysis of various papers based on IoT-based hydroponics systems with respect to sensors and processors used represented in Table 1 reveals that not all previous researchers used all the sensors required for monitoring various parameters. However, our experiments involve all the sensors and essential parameters with high efficiency. Our study is able to accurately monitor the key parameters such as pH, nutrient concentration, temperature, humidity, TDS values, and electrical conductivity using a combination of sensors. This allows for real-time monitoring and control of the hydroponics system, resulting in more consistent and predictable plant growth. By incorporating all the necessary sensors on IoT-based Things Speak platform, our study offers insights into the potential for further optimization and improvement of the system by controlling various essential parameters collected from the sensors so that the corrective actions can be taken at the right time to improve the yield

Fig. 8 Plant grown in a hydroponics system

and reduce the chances of plant diseases by continuously monitoring plant growth. Also, ESP8266 NodeMCU provides an excellent balance of affordability, size, power consumption, connectivity, programming language, and built-in support for sensors compared to Raspberry Pi and Arduino, making it a popular choice for IoT-based hydroponic systems. Further, artificial intelligence can also be incorporated for plant growth prediction and plant disease detection or classification.

Overall, IoT-based hydroponics is a significant and important development in the field of hydroponics. It has the potential to improve the efficiency and productivity of hydroponic systems, as well as to reduce costs and labor. This technology can help to make hydroponic growing more accessible and affordable for a wide range of growers, from hobbyists to commercial operations.

Table 1 Comparative analysis of various parameters essential for the hydroponics system

Reference	Paper title	Processor	Humidity sensor	pH sensor	EC sensor	Temperature sensor	TDS sensor
[17]	Automation of hydroponics greenhouse farming using IOT	Raspberry Pi	No	Yes	Yes	No	No
[18]	Hydroponics farming	Arduino UNO	No	Yes	Yes	No	No
[19]	Designing and implementing the Arduino-based nutrition feeding automation system of a prototype scaled nutrient film technique (NFT) hydroponics using total dissolved solids (TDS) sensor	Arduino UNO	Yes	No	No	No	Yes
[20]	IoT-based automation of hydroponics using node MCU interface	Node MCU	Yes	Yes	Yes	No	No
[21]	IoT-based hydroponics system using deep neural networks	Raspberry Pi	Yes	Yes	Yes	Yes	No
[22]	An AI-based system design to develop and monitor a hydroponic farm hydroponic farming for smart city	Raspberry Pi	Yes	Yes	Yes	Yes	No
Our model	Our model	ESP 8266 Node MCU	Yes	Yes	Yes	Yes	Yes

5 Conclusion

IoT-based automated hydroponic cultivation systems are innovative farming systems that use IoT technology to monitor and control the various parameters of a hydroponics system, such as temperature, humidity, light, and nutrient levels. These systems are designed to provide optimal growing conditions for plants, resulting in higher crop yields and better-quality crops. The use of IoT in hydroponics can provide several advantages that include real-time monitoring, automation, increased efficiency, remote management, increase in crop yields, data collection, and analysis. It can also help to reduce the environmental impact of hydroponics systems and increase the overall sustainability of the system.

6 Future Scope

The future scope of IoT-based hydroponics is quite promising, as it has the potential to revolutionize the way we grow crops and manage agriculture. IoT-based automated hydroponic cultivation systems can also be integrated with other technologies such as smart greenhouses, vertical farming, and blockchain to create highly efficient and sustainable growing environments. Furthermore, the integration of artificial intelligence and machine learning can optimize crop growth conditions, predict future crop yields, and also detect and diagnose issues in the system more efficiently. It can also play a key role in addressing global food security challenges in the future.

References

1. Griffiths MD (2014) The design and implementation of a hydroponics control system. Types of hydroponics https://hydrobuilder.com/learn/types-of-hydroponics-systems
2. Sardare M (2013) A review on plant without soil–hydroponics. Int J Res Eng Technol 02:299–304. https://doi.org/10.15623/ijret.2013.0203013
3. Pawar S, Tembe S, Acharekar R, Khan S, Yadav S (2019) Design of an IoT enabled automated hydroponics system using NodeMCU and Blynk. In: 2019 IEEE 5th international conference for convergence in technology (I2CT), 1–6. https://doi.org/10.1109/I2CT45611.2019.9033544
4. Stŏces M, Vaněk J, Masner J, Pavlík J (2016) Internet of Things (IoT) in agriculture—selected aspects. Agris online Pap Econ Inform 8:83–88
5. Statista Forecasted Market Value of Precision Farming Worldwide in 2018 and 2023. https://www.statista.com/statistics/721921/forecasted-market-value-of-precision-farming-worldwide/ Accessed 1 Oct 2019
6. Kite-Powell J (2018) Why precision agriculture will change how food is produced. https://www.forbes.com/sites/jenniferhicks/2018/04/30/why-precision-agriculture-will-change-how-food-is-produced/#1aa438ec6c65 Accessed 18 Nov 2018
7. Madoka S, Ramaswamy R, Tripathi S (2015) Internet of Things (IoT): a literature review. J Comput Commun 3:164–173
8. Lakshmanan R, Guedi M, Perumal S, Abdulla R (2020) Automated smart hydroponics system using internet of things. Int J Electr Comp Eng (IJECE) 10:6389

9. Sihombing P, Karina NA, Tarigan JT, Syarif MI (2018) Automated hydroponics nutrition plants systems using arduino uno microcontroller based on android. J Phys Conf Ser 978:012014. https://doi.org/10.1088/1742-6596/978/1/012014
10. Satoh A (2018) A hydroponic planter system to enable an urban agriculture service industry. In: 2018 IEEE 7th global conference on consumer electronics (GCCE), pp 281–284. https://doi.org/10.1109/GCCE.2018.8574661
11. Sisyanto REN, Kurniawan NB (2017) Hydroponic smart farming using cyber physical social system with telegram messenger. In: International conference on information technology systems and innovation (ICITSI), pp 239–245. https://doi.org/10.1109/ICITSI.2017.8267950
12. Hariono T, Putra C, Hasbullah KAW (2021) Data acquisition for monitoring IoT-based hydroponic automation system using ESP8266. 1(1):1–7
13. Kularbphettong K, Ampant U, Kongrodj N (2019) An automated hydroponics system based on mobile application. Int J Inform Educ Technol 9:548–552. https://doi.org/10.18178/ijiet.2019.9.8.12642
14. Zhao JC, Zhang JF, Feng Y, Guo JX (2010) The study and application of the IoT technology in agriculture. In: Proceedings of the IEEE 2010 3rd international conference on computer science and information technology, Chengdu, China, vol 2, pp 462–465
15. Verdouw C, Sundmaeker H, Tekinerdogan B, Conzon D, Montanaro T (2019) Architecture framework of IoT-based food and farm systems: a multiple case study. Comput Electron Agric 165:104939
16. Saraswathi D, Manibharathy P, Gokulnath R, Sureshkumar E, Karthikeyan K (2018) Automation of hydroponics green house farming using IOT. In: 2018 IEEE international conference on system, computation, automation and networking (ICSCA), pp 1–4. https://doi.org/10.1109/ICSCAN.2018.8541251
17. Nalwade R, Mote T (2017) Hydroponics farming. Int Conf Trends Electron Inform (ICEI) 2017:645–650. https://doi.org/10.1109/ICOEI.2017.8300782
18. Eridani D, Wardhani O, Widianto ED (2017) Designing and implementing the Arduino-based nutrition feeding automation system of a prototype scaled nutrient film technique (NFT) hydroponics using total dissolved solids (TDS) sensor. In: 2017 4th International conference on information technology, computer, and electrical engineering (ICITACEE), pp 170–175. https://doi.org/10.1109/ICITACEE.2017.8257697
19. Manohar G, Sundari VK, Pious AE, Beno A, Anand LDV, Ravikumar D (2021) IoT based automation of hydroponics using node MCU interface. In: 2021 Third international conference on inventive research in computing applications (ICIRCA), pp 32–36. https://doi.org/10.1109/ICIRCA51532.2021.9544637
20. Mehra M, Saxena S, Sankaranarayanan S, Tom RJ, Veeramanikandan M (2018) IoT based hydroponics system using deep neural networks. Comput Electron Agric 155:473–486
21. Lakshmanan R, Djama M, Selvaperumal SK, Abdulla R (2019) Automated smart hydroponics system using internet of things. Malaysia Int J Inform Educ Technol 9(8)
22. An AI Based System Design to Develop and Monitor a Hydroponic Farm Hydroponic Farming for Smart City Glenn Dbritto Computer Department St. Francis institute of Technology Mumbai, India

GCD Thresholding Function Applied on an Image with Global Thresholding

Hussain Kaide Johar Manasi and Jyoti Bharti

Abstract In this paper, we have developed and applied a new threshold function over an image globally and found the results to be quite promising. The method utilizes the feature of calculating the greatest common divisor (GCD) of the pixels within blocks formed in the image. The results obtained when compared with other standard thresholding techniques show us further insight with regard to the robust quality and performance of the newly devised thresholding function. In this paper, 3 progressive algorithms are presented with their particular challenges and shortcomings. Of these, the third algorithm is the main successful implementation of this thresholding technique.

Keywords GCD thresholding · Computer vision · Global thresholding · Digital image processing

1 Introduction

Up until now, many thresholding techniques have been devised and the main concept of thresholding has been preserved to this date and that is to separate the foreground of any image from its background. Also called segmentation it allows us to bring into focus the main objective of the image, and we can then go ahead and perform further descriptive analysis on the obtained part of the image. The obtained part of the image depending on the specific application can refer to written text, targets, defective materials, etc.

Otsu Thresholding has by far proven to be a reliable thresholding method albeit quite a computation-intensive process depending on the range in the image [10]. Others like the p-tile method, several entropic methods, and even before that we

H. K. J. Manasi (✉) · J. Bharti
Maulana Azad National Institute of Technology, Bhopal, India
e-mail: hussainjmanasi@gmail.com

J. Bharti
e-mail: jyotibharti@manit.ac.in

© The Author(s), under exclusive license to Springer Nature Singapore Pte Ltd. 2023
A. Mishra et al. (eds.), *Advances in IoT and Security with Computational Intelligence*, Lecture Notes in Networks and Systems 755,
https://doi.org/10.1007/978-981-99-5085-0_8

had seen the initial development of fuzzy clustering-based methods [12] try to make the thresholding more optimized. Also devised were the histogram transformations which took into consideration more than just the isolated pixel, but the majority and minority proportionalities of the available pixels in the overall image [1–3, 13].

The next step normally pursued after selecting and employing an appropriate thresholding function is to perform segmentation on the image. Here, global thresholding is the simplest as applying a single standard across all pixels of the image is pretty straightforward as depicted in Eq. 1. Conversely, there are also adaptive thresholding techniques wherein a threshold is recalculated and applied separately for different parts of the image. This is useful when the image has a high amount of variance from one part of the image to another but within its disparate regions, the frequency is quite low. Every image performs differently for different algorithms, and there is no perfect algorithm that suits all images perfectly yet [4–9]:

$$\text{Pixel},\ p = \begin{cases} 255, & p > T \\ 0, & p \leq T \end{cases} \ \text{where, } T \text{ is Threshold value} \tag{1}$$

The obtained part of the image in question can be anything from the foreground, i.e. a person in an image like in the lena.bmp standard. Or in several other situations, users may want to perform edge detection which is possible by the Sobel, Prewitts, and Roberts operators. Contour-based line detection is also a well-in-demand problem statement achievable by Hough transforms and Hough lines. When it comes to whole objects, however, the 2 main techniques highlighted are region growing and region splitting and merging. More derivative segmentation methods that provide nuance to the obtained part as well are defined in the watershed method and gradient transform [11].

In lieu of this search for unique thresholding techniques, this paper aims to propose its own approach to thresholding and attempts to build upon a quite well-known concept in simple mathematics. The Greatest Common Divisor (GCD) is an effective function that gives good insight and understanding with regards to a couple or group of numbers, and utilizing this relationship of factors among pixel values poses an appropriate approach to building good threshold values that can target a wide range of images. The rest of the paper is structured to first deliver on the initial hindrances that are encountered when developing a novel thresholding technique in Sect. 2, after which all the possible implementations and their evolutions are explained in Sect. 3. Section 4 contains all the output tables and images, with brief descriptions of all of them. Observations and insights are made in Sect. 5 and finally, the paper is wrapped up in Sect. 6 with hints of future possible work on this topic.

2 Implementation Challenges

The steps applied to gain the threshold value of images were first done in a general fashion to calculate the GCD after which several modifications were performed to fix, enhance, and test the following parameters:

- **Runtime of the algorithm**: At first, the runtime of the algorithm was observed to be acceptable for smaller images, particularly of size 256 × 256. But the developed algorithm did not perform well with respect to larger images in the range of 3068 × 2457 and such. For this reason, the usage of a larger block size within the image was proposed and implemented, but with that came the challenge of inflated and adulterated results. Additionally, the core problem of the algorithm having a high running time complexity did not seem to be tackled. However, fortunately, this problem was tackled using appropriate data structures and that solved other troubles fairly easily as well.
- **Robustness against block size**: It was imperative that the threshold value not vary with respect to block size as it may not be feasible to reiterate the image, again and again, using different thresholding block sizes. And even if it was possible, there seemed to be no appropriate method to decide the most correct block size for that particular image without active human involvement and selection. The varying results would also have been a fair consideration should the implementation have been similar to adaptive thresholding; however, that is not the case and such results are treated as undesirable when global thresholding is applied. This solution was achieved with a fairly simple implementation as we will see later in the paper.
- **Failure of the algorithm with extremely dark or bright images**: In the first few major constructions of the algorithms, it was observed that the algorithm performed significantly terribly with darker images which had a slightly less discernible foreground. This was later rectified by considering only the unique values in the image as at first glance it is easily identified that repeating values in an image can produce outliers to the calculation of the threshold value. However, for uniformly bright images, no working algorithm could be devised as the output received would be a completely black image.
- **Distinguishing factor**: Certain math applied in the upcoming pages can prove to look quite redundant, and the initial challenge of working on this was to find an algorithm that not only performs in a manner different from currently established standards but also delivers results that are equally promising. Whether this delivers convincingly innovative solutions is a decision best left to the reader.

Each of the above challenges was tackled in consequent iterations of the working program, and the solutions will be elaborated on in the following sections. However, there still does remain a certain problem that even standard thresholding techniques have failed to solve, and that is that this algorithm performs quite poorly with images that have a regularly high brightness throughout the image. That would be the recommended focus and target going forward if this approach of implementing GCD proves to truly be an effective approach.

Additionally, modern digital image processing has entered the age of computer vision. Incorporating this method of thresholding with some degree of convolutional neural network layer processing during the segmentation and object detection stage would be truly insightful as to the future usability of the algorithms.

3 Implementations

3.1 Algorithm 1

The first approach was fairly simple with limited considerations taken, for example, whether the pixel value was even or odd, and whether the cumulative sum of the GCDs of the pixel pairs exceeded the mean value. This approach was admittedly naive as in almost all realistic scenarios the cumulative sum would almost always exceed the mean value of the pixels and hence, the threshold would be brought down to the mean value and the results obtained would be no different than the traditional average of the image. This approach was subject to the fourth challenge stated above, as the results were never particularly unique or useful in any way than if simple averaging were employed. In fact, certainly in certain situations, the simple averaging would have actually delivered better results. The steps of the algorithm are as below.

1. Calculate the mean of the pixels of the image and initialize the current running GCD threshold value ($currgcd$) of the image as the first pixel in the image.
2. Loop through the pixel values in the image and perform operations depending on which case is encountered
   (a) **Case 1**: The pixel value is even; then calculate GCD like normal between pixel value and $currgcd$ and store in $pgcd$.
   (b) **Case 2**: The pixel value is odd, then add 1 to pixel value and then calculate GCD like normal between pixel value and $currgcd$ and store in $pgcd$.
3. For each $pgcd$, add it to $currgcd$ and check if it is greater than the mean, and if it is then $currgcd$ is brought down to equal the mean value.

3.2 Algorithm 2

The second approach is significantly more valuable for consideration. Here, a standard initial block size is selected as 4, and pixels are taken in blocks of 16 values. For 16 values, they are all first considered in pairs. That corresponds to $\frac{16*15}{2} = 120$ pairs. This calculation is important as it shows a degree of how long the algorithm has to run when considering block sizes.

$Block\_size = 3 \rightarrow 9 \rightarrow 36$ pairs
$Block\_size = 5 \rightarrow 25 \rightarrow 300$ pairs
$Block\_size = 6 \rightarrow 36 \rightarrow 630$ pairs

The pairs are increasing in a quite steep fashion, and this means the complexity of the algorithm increases proportionally too. However, the larger the block size the faster the entire image gets processed, so having a larger block size also corresponds to the image being processed quickly. Hence, one can safely devise Eq. 2:

$$Time\_Complexity(O) \propto \frac{Image\_size}{Block\_size} \tag{2}$$

Without further digressions, the steps performed in Algorithm 2:

1. First take all pixel values in that block and reshape them into a 1-D array of just normal values.
2. Take the GCD of all the pairs of values in the formed 1-D array.
3. Take the summation of the elements in the blocks and store them in another array that stores the sums of the different blocks.
4. Find the mean (%255) of all the elements in the list of sums of the blocks. That is the threshold value that will be applied.

The main concept while developing this algorithm was to harness the regional differences in the image while combining them at the same time. The summation of each block gives us an idea of the variations and GCDs of the pixels within that block, while averaging all the blocks gives the image a chance to balance out the regions of high brightness and deep darkness. We still, in this method, encounter a number of difficulties particularly.

1. **Higher thresholds**: This wouldn't normally be a problem if it actually reflected the characteristics of the image. The results obtained here are significantly more generous, as anything that isn't matching the gradient of a bright background gets caught in the threshold value and is labeled as foreground.
2. **High execution time**: This corresponds to the first challenge described earlier. Running time for larger images of sizes of up to 3068 × 2457 can take upwards of 15 min to completely execute. This is simply too high and had to be worked on as the time complexity of this method also is $O(n^2)$.
3. **Increasing block sizes failure**: An attempt to increase block size can on paper seem like a great option to speed up execution but, in reality, it didn't change execution time by much. Moreover, the acquired threshold values were simply too high to be appropriate. This corresponds to the second challenge described above. The thresholding function is sensitive to changing block sizes, and this is not a good phenomenon as this means certain block sizes would work better for different images, and this as of now has not yet been clearly discerned.

These challenges were however overcome in the next and final iteration of building this algorithm.

3.3 Algorithm 3

Finally, this devised algorithm is the working success of this paper. The primary difference here is that the 1-D array holding all the pixel values gets substituted for a dictionary holding a count of all the pixel values appearing in the block. Initially, all the pixel values and their GCDs were considered but that had to be modified to consider the fact that the threshold values seemed to be stagnating at ≈127. Considering only unique values resulted in more dynamic threshold values being obtained due to lesser repeating outliers of GCD calculation like having multiple repeating

255 pixels in a single image. Further, the block size that performed best was size 6. This contrast to the earlier selected value of 4 performing best can be explained by Eq. 3

$$Block\_size = 6 \rightarrow 36 \rightarrow 630\%255 = 120\, pairs \tag{3}$$

Regardless, the steps and algorithm for this process are elaborated below

1. Initialize a dictionary of 256 keys and initialize their values to 0.
2. Loop over each pixel for each block and increment value of corresponding keys in the dictionary.
3. Then iterate over the dictionary to calculate the GCDs of all the key pairs with non-zero values. For pixel values that repeat more than once, their GCDs among the other pixels are simply calculated by factoring in their dictionary values.
4. Do this for each block while accommodating the leftover pixels that wouldn't fit perfectly into a block.
5. Finally find the mean of the obtained resulting list of sums of all blocks.

Algorithm 1: Best results for GCD thresholding by below algorithm

```
Initialize: 1. A dictionary with 256 keys all with value 0; dict
2. An empty list that stores cumulative calculated values of each block; blockgcd
for i = h0, h0+1, ..., h0 + block_size do
    end
    for j = w0, w0+1, ..., w0 + block_size do
    end
    dict[img[i,j]] += 1
for k1,v1 in dict.items() do
    end
    for k2, v2 in dict.items() do
    end
    if k1 == k2 or v1 == 0 or v2 == 0 then
    end
    exit()
GCDsum = GCDsum + GCD(k1,k2) + v1*v2
return GCDsum                                   ▷ Calculated threshold value
```

4 Results

The results obtained are first tabulated as a progression of threshold values obtained throughout the process of development by different algorithms in Table 1.

Table 2 then provides insights with respect to comparisons between our algorithm and results obtained from Otsu's thresholding.

Table 1 Progression of thresholds in our algorithms

Image	Algorithm 1	Algorithm 2	Algorithm 3
Lena	72.0	84.4	100.2
Shapes1	183.8	219.7	88
Shapes2	216.5	149.0	21.2
Scan1	88	166.6	126.1
Scan2	58.8	58.4	105.0
Dark	26.3	36.8	53.8

Table 2 Comparing our algorithm with that of Otsu's

Image Name	Algorithm 3	Otsu Method
Lena	100.2	71.0
Shapes1	88	173.0
Shapes2	21.2	115.0
Scan1	126.1	168.0
Scan2	105.0	87.0
Dark	53.8	46.0

Fig. 1 Shapes1 through Algorithm 1

Fig. 2 Shapes1 through Algorithm 2

Figures 1, 2, and 5 are all different outputs of the same image consisting of different shapes with different outlines and colors. Figures 4 and 7 are results obtained from implementing our algorithms on medical scans. Standard image processing images like the lena.bmp are used in Figs. 3 and 6. Lastly, Fig. 5 is an example of the 3rd algorithm's performance on darker images (Fig. 8).

Fig. 3 Lena through Algorithm 2

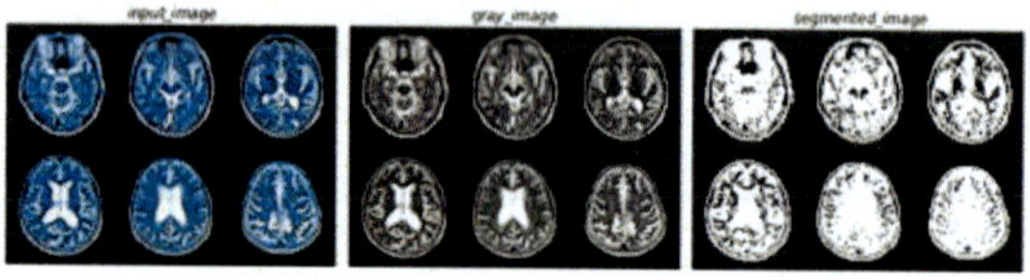

Fig. 4 Medical scans through Algorithm 2

Fig. 5 Shapes1 through Algorithm 3

Fig. 6 Lena through Algorithm 3

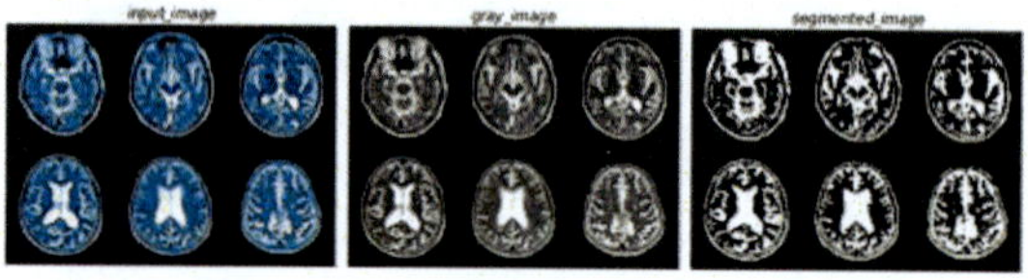

Fig. 7 Medical scans through Algorithm 3

Fig. 8 Dark images through Algorithm 3

5 Discussions

The purpose of discussing is not just the final observations but the process of developing the algorithm as well as to highlight the exact complications encountered when developing a thresholding function. Modifying the program bit by bit brings us to a conclusion that looks completely unrecognizable from the original intent and implementation. Regardless, an obvious observation from the finally devised algorithm is that it manages to perform quite well even in low lighting conditions but fails terribly when uniformly bright images are considered, and that too unexpectedly the problem seems to be that the threshold comes out too high for the regularly bright images with no clear background and foreground.

Hence, it is safe to conclude that GCD thresholding is a lower bound thresholding method. The value for the threshold will under very specific unrealistic circumstances reach a value higher than 130. This is because the GCD of any values at the 200 range would never have a factor greater than 130. The taken and considered GCD of any real image without changing the image itself can never be higher than 130. Hence, brighter images miss out on this thresholding method.

6 Conclusion

In this paper, we have developed an algorithm that accommodates the GCD of the pixel values in the image. We first begin from a naive understanding of the problem statement and then evolve toward a fully developed solution tackling each of the encountered problems one by one. The performance of the algorithm improves as well and is documented throughout development and is then compared with results from performing Otsu thresholding on the images. We find that it is a promising method of thresholding that can further be developed with reasonable interest, especially with the new establishment of computer vision and CNNs processing images and videos at extremely high speeds.

References

1. Kaur N, Kaur R (2011) A review on various methods of image thresholding. Int J Comput Sci Eng 3(10):3441
2. Sankur B, Sezgin M (2001) Image thresholding techniques: a survey over categories. Pattern Recogn 34(2):1573–1607
3. Guruprasad P (2020) Overview of different thresholding methods in image processing. In: TEQIP sponsored 3rd national conference on ETACC
4. Sahoo P, Soltani S, Wong A, Chen Y (1988) A survey of thresholding techniques. Comput. Vision Graph Image Process 41(2)
5. Cuevas E et al (2009) A novel multi-threshold segmentation approach based on artificial immune system optimization. In: Advances in computational intelligence. Springer, Berlin, Heidelberg, pp 309–317
6. Dalal N, Triggs B (2005) Histograms of oriented gradients for human detection. In: 2005 IEEE computer society conference on computer vision and pattern recognition (CVPR'05), vol 1. IEEE
7. Singh TR et al (2012) A new local adaptive thresholding technique in binarization. arXiv:1201.5227
8. Dhanachandra N, Manglem K, Chanu YJ (2015) Image segmentation using K-means clustering algorithm and subtractive clustering algorithm. Procedia Comput Sci 54:764–771
9. Wang Q, Chi Z, Zhao R (2002) Image thresholding by maximizing the index of nonfuzziness of the 2-D grayscale histogram. Comput Vis Image Underst 85(2):100–116

10. Bangare S, Dubal A, Bangare P, Patil S (2015) Reviewing Otsu's method for image thresholding. Int J Appl Eng Res. https://doi.org/10.37622/IJAER/10.9.2015.21777-21783
11. Priyadharshini KS, Singh T (2012) Research and analysis on segmentation and thresholding techniques. Int J Eng Res Technol (IJERT) 1(10):1–8
12. Henila M, Chithra P (2020) Segmentation using fuzzy cluster-based thresholding method for apple fruit sorting. IET Image Proc 14(16):4178–4187
13. Tan KS, Isa NAM (2011) Color image segmentation using histogram thresholding-Fuzzy C-means hybrid approach. Pattern Recogn 44(1):1–15

Image Semantic Segmentation: Scene-Aware System for Autonomous Vehicle

Neha Tomar, Mihir Mendse, Rupal Paliwal, Vivek Wanve, and Gouri Morankar

Abstract Understanding urban street scenes is an essential part of the perceptual job involved in the use of autonomous vehicles. The interpretation of scenes through semantic segmentation has been employed widely, and this helps with the task of autonomous driving including planning of path, object identification, and motion control. However, precise semantic image segmentation in computer vision is a difficult job. The development and study of image segmentation are the main topics of this paper. On the Cityscapes dataset, we analyzed our methodology. In the proposed work, the U-Net model is utilized since it is a well-known semantic segmentation network for image semantic segmentation, and it is a compact, elegant architecture and has much fewer convolutions. For recognizing the object, YOLOv3 algorithm is used. The result of the proposed work is the image segmented along with bounding boxes, labels, and detecting accuracy.

Keywords Artificial intelligence · Machine learning · Deep learning · Image semantic segmentation · Autonomous vehicles · Scene aware system · U-Net architecture · YOLOv3 · Cityscape dataset · Intersection over union (IoU)

N. Tomar (✉) · M. Mendse · R. Paliwal · V. Wanve · G. Morankar
Department of Electronics Engineering, Shri Ramdeobaba College of Engineering and Management, Nagpur 440013, India
e-mail: tomarns@rknec.edu

M. Mendse
e-mail: mendsemn@rknec.edu

R. Paliwal
e-mail: paliwalrs_1@rknec.edu

V. Wanve
e-mail: wanvevv@rknec.edu

G. Morankar
e-mail: morankarg@rknec.edu

© The Author(s), under exclusive license to Springer Nature Singapore Pte Ltd. 2023

A. Mishra et al. (eds.), *Advances in IoT and Security with Computational Intelligence*, Lecture Notes in Networks and Systems 755,
https://doi.org/10.1007/978-981-99-5085-0_9

1 Introduction

Machine learning is a subset of artificial intelligence technology that focuses on training computers to learn more rapidly and intelligently. The goal of machine learning is to make AI solutions smarter and quicker, enabling them to produce better outcomes for any task they are given. Machine learning has gained popularity in recent years due to advances in technology and increased accessibility to large-scale datasets. The growth in computing power and the availability of high-speed Internet have made it possible to quickly and accurately evaluate massive and complex datasets. Machine learning has a wide range of applications, including suggesting goods and services, identifying cyber security breaches, and enabling self-driving automobiles. Its purpose is to reduce costs and risks, and enhance the quality of life. As data and computing power become more accessible, machine learning is becoming increasingly common and will soon be integrated into many aspects of daily life.

Machine learning involves training a machine to perform a specific task using a set of training data. One of the techniques of machine learning is image semantic segmentation. Also known as pixel-level categorization, image semantic segmentation groups portions of an image that correspond to the same object class. This process is considered pixel-level prediction because it assigns each pixel to a category. The goal of semantic segmentation is to provide pixel-level classification of images, identify the scene unit and its location, and mark semantic information about the scene unit. Both global information (information about the farthest pixel from the point) and local information (information about the closest pixel to the point) should be taken into account when determining the category of a pixel. While object detection and semantic segmentation may seem similar, as both involve locating and classifying items in an image, they are distinct methods. Object detection requires locating all items in an image, putting them inside a bounding box, and giving them a name to be identified as an object. Semantic segmentation, on the other hand, labels each pixel with a class at the individual pixel level, providing labels to each area of the image. Image semantic segmentation has various applications, such as recognizing road signs, analyzing geographic pictures, segmenting different land types, and mapping land. It is widely used in the medical industry for detecting tumors and brains, and identifying and tracking surgical equipment. In self-driving cars, semantic segmentation is a crucial method of perception, as more and more autonomous vehicles are being developed. For self-driving cars, the AI-based software supporting the function requires perfect training. The semantic segmentation technique helps increase the accuracy of lane and traffic sign detection, as well as recognize and classify items that the car encounters. To prevent passengers from being in hazardous situations, real-time monitoring algorithms must be predictive and able to react immediately to new information.

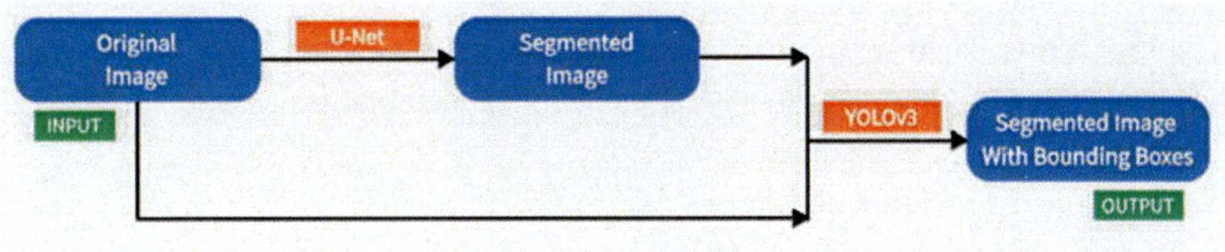

Fig. 1 Flowchart

2 Scene-Aware System

Autonomous driving has become increasingly popular in recent years, with vehicles that can navigate roads without human aid. Scene understanding is essential for autonomous driving as it lays the foundation for key functions such as path planning, simultaneous localization and mapping, and motion control. To achieve this, semantic segmentation of images is used, which involves pixel-level categorization of each pixel in an image into classes such as a car, pedestrian, building, lane, traffic signal, sky, and tree. In this work, the objective is to develop a machine learning model for semantic segmentation that can be used by autonomous vehicles to understand their surroundings. The model is built using an encoder-decoder architecture, U-Net architecture, and YOLOv3. To train the model, a suitable dataset is collected and pre-processed, including image cropping, rotation, brightness and contrast adjustments, re-mapping class labels, etc. After the model is trained, it can produce a segmented image of its surroundings, which provides the perception required for autonomous vehicles to understand their surroundings and safely integrate onto roads. The goal is to identify each pixel in an image with the matching class of what it represents, allowing autonomous vehicles to make informed decisions and navigate roads safely (Fig. 1).

The proposed work involves the implementation of U-Net and YOLOv3 models to achieve image semantic segmentation for autonomous vehicles. The original image is used as input for both models, with U-Net being used to generate the segmented image, while YOLOv3 is used for object recognition. The resulting output image displays the segmented image with bounding boxes and labels, along with the accuracy of the model's predictions. This work aims to provide autonomous vehicles with the necessary perception to understand their surroundings and ensure their safe integration onto existing roads.

3 Review of Literature

Due to CNN's capacity to extract features and provide judgments by processing the pertinent feature data, for image processing tasks several researchers have adapted models based on CNN. To execute convolution operations on the image or extracting feature matrix, convolution layers employ a series of weight matrices referred to as kernels or filters. There are numerous designs based on convolution neural networks that have been created for the semantic picture segmentation process.

1. In this study, the team of Torres et al. investigates the use of Deep Learning Neural Networks (DL-NNs) for semantic segmentation of multispectral remote sensing images. Using images captured by the European Space Agency's Sentinel-2 satellite system, the researchers divide each pixel into five classes: vegetation, soil, water, clouds, and cloud shadows. They find that the recommended AI approach offers higher accuracy in image segmentation compared to current state-of-the-art algorithms and is done in a reasonable amount of time [1].
2. In a study by Sugirtha and Sridevi, the U-Net model's encoder component was replaced with a Convolutional Neural Network (CNN) architecture to improve accuracy. The effectiveness of VGG-16 and ResNet-50 CNN architectures was evaluated. Results showed that the U-Net model with a VGG-16 encoder performed better than the ResNet-50 encoder and improved the mean Intersection over Union by 2% compared to SegNet and Fully Convolutional Network, which were used for semantic segmentation [2].
3. The paper "ScleraSegNet: An Attention-Assisted U-Net Model for Accurate Sclera Segmentation" by Wang et al. proposes a new sclera segmentation technique using an attention-assisted U-Net model. The original U-Net has increased attention modules in its core bottleneck or skip connection, enabling the model to highlight important features for sclera segmentation and suppress unimportant parts in an input image, eliminating the need for an external ROI localization model [3].
4. In this paper, Liu et al. divide traditional and contemporary DNN approaches for semantic image segmentation into two categories and provide a brief overview of the conventional approach and datasets. They then extensively examine the latest DNN-based techniques, which include supervised, loosely supervised, and unsupervised methods. The paper concludes with a summary of their findings [4].
5. The objective of the study by Yi et al. was to train a semantic image segmentation neural network in various scenarios. They separated the semantic segmentation dataset into three groups and trained the network in three different scenarios to yield three models. The evaluation of the methods showed that scene-aware semantic segmentation outperforms traditional semantic segmentation by incorporating category information and leads to more accurate models for image analysis [5].
6. The study by Dong optimizes the ENet network for image semantic segmentation by using pruning and convolution optimization methods. The network's structure is altered to improve the segmentation results and enhance the bottleneck convolution structure. The squeeze-and-excitation module is added to the upgraded ENet network, enabling automatic recognition of feature channel relevance and improving small-target segmentation accuracy. The open dataset was used to verify the results, and this method outperforms existing techniques in terms of mean intersection over union and mean pixel accuracy values while providing efficient and correct segmentation in a short operating time [6].

7. The ENet architecture is a deep neural network developed by Paszke et al. for low-latency applications. ENet achieves equivalent or better accuracy compared to previous models and is up to 18× faster, requiring 75× fewer FLOPs and having 79× fewer parameters. It was tested on CamVid, Cityscapes, and SUN datasets and outperformed cutting-edge methods. The authors also discussed the trade-off between accuracy and processing performance and provided performance metrics and recommendations for software enhancements to speed up ENet further [7].

4 Proposed Methodology

In this, we made use of the deep learning framework TensorFlow to construct the model and imported the Keras framework as well as the TensorFlow library. After importing the necessary libraries, we proceed with developing the model. We apply a loss function. For the purpose of calculating the overall error from residuals for every training batch, neural network models use loss functions. The choice of loss function directly affects model performance since it influences how they modify their internal weights during backpropagation. Dice Coefficient or Intersection over Union (IoU) is used to assess the performance of models. A popular statistic for pixel segmentation called the Dice coefficient may be changed to operate as a loss function. Typically, these loss functions take the form 1−f(x), where f(x) is the metric in the issue. To monitor the loss function, call-back functions are used:

$$\text{Dice coefficient(DC)} = \frac{2|X \cap Y|}{|X| + |Y|} \tag{1}$$

The suggested method carries out semantic segmentation of images using the representative deep learning model of U-net [2]. The encoder-decoder network in a U-shape was created for semantic segmentation. The expanding path is another name for the decoder path, and is employed to deduce the class labels, whereas the encoder section is used to extract features and is sometimes referred to as a contracting path. In addition to end-to-end segmentation, the encoder and decoder are connected by skip links. Expanding route recognizes pixel positions and the object's identity is recognized by the contracting path. Two convolutions followed by ReLU, batch normalization and one max pooling, are used in each step of the encoder's standard convolution network architecture. In the decoder one convolution is transposed, and there are two convolutions that upsample the feature map before performing the convolution followed by a ReLU and batch normalization in each step [8]. The decoder component's purpose is to increase resolution as a result. In order to infer the class labels, upsampled features acquired in the decoder using transposed convolution and high-resolution features obtained in the encoder component using skip connections were combined. Then the model was trained to see its performance on the data using Adam optimizer and sparse_categorical_crossentropy loss till 20 epochs. Then we plot and analyze the val_accuracy and val_loss with respect to the number of epochs. We visualize a few outputs from the dataset (Fig. 2).

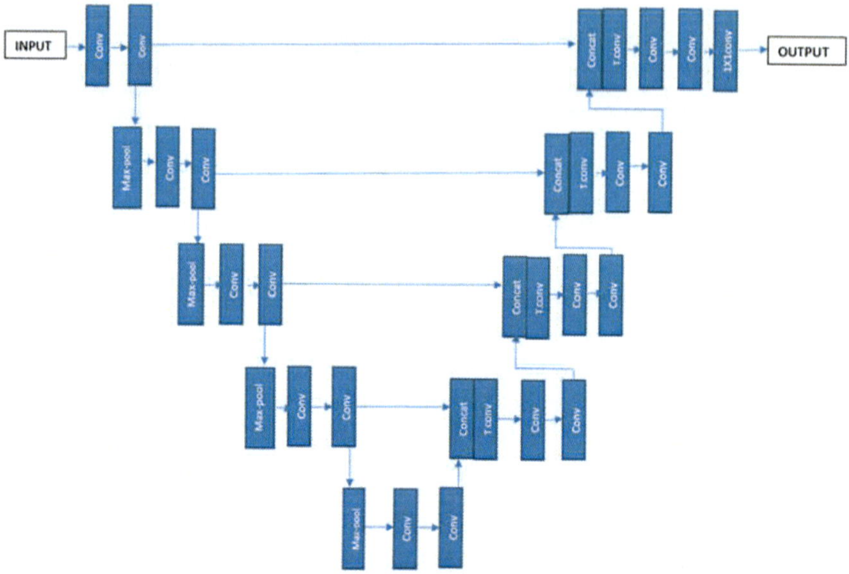

Fig. 2 Architecture of U-Net

For object detection, we have used the YOLOv3 algorithm that makes use of a convolution neural network. The input image of size 416 × 416 is sent to a convolutional neural network by YOLOv3. A more sophisticated feature extractor named Darknet-53 is used in YOLOv3, which originally has a 53-layer network and 53 more layers are added to it for the purpose of detection, giving YOLOv3 an overall underlying architecture of 106 layers that is fully convolutional. The architecture of YOLOv3 is shown in Fig. 3.

The architecture consists of three different layer forms: Residual Layer, Detection Layer, and the Upsampling Layer. In the residual layer, the activation is easily forwarded to the deep layers of the neural network. The detection layer performs the task of detection at different scales by increasing the size of grids. Three different-sized feature maps with strides of 32, 16, and 8 are detected using the detection layer. This indicates that with an input of 416 × 416, detections are obtained on scales of 13 × 13, 26 × 26, and 52 × 52. The detections are made at the 82nd, 94th, and 106th layers. With the use of the three detection feature maps, YOLOv3 can detect at three different sizes. The first 13 × 13 map is used to detect large objects, the second 26 × 26 map for medium-sized objects, and the third 52 × 52 map for small objects. The upsampling layer which is the third layer increases the spatial resolution of the image. Before scaling the image is upsampled. The outputs of the preceding layer are combined with those of the current layer using the concatenation operation. In the figure, the pink blocks show the residual layer, the orange blocks are for the detection layer, and the green blocks are the upsampling layers.

The shape of the detection kernel is calculated using **1 x 1 x (B x (5 + C)), where B is the number of bounding boxes and C is the number of classes.** Darknet-53 is

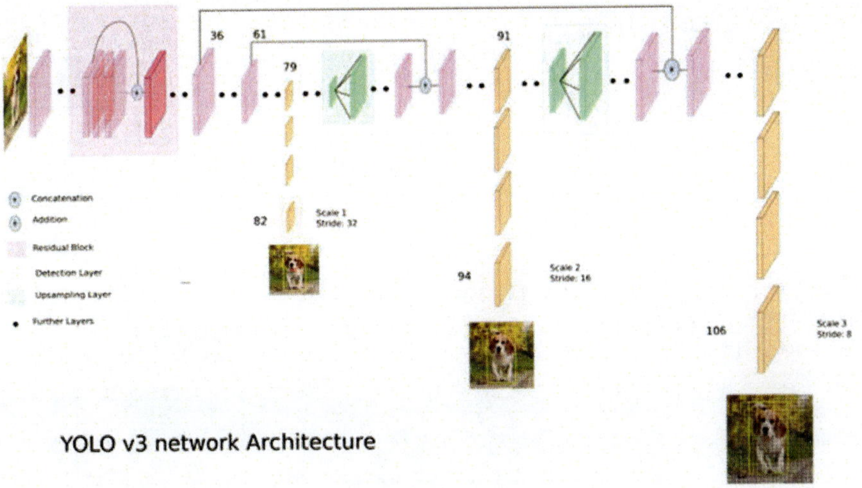

Fig. 3 YOLO v3 network architecture [10]

mostly made up of 1 × 1 and 3 × 3 filters with skip connections. Each of the 53 convolutional layers in YOLOv3 has been followed by a batch normalization layer and a Leaky ReLU activation layer. The threshold value we used is 0.6. The boxes will be suppressed based on the class threshold and non-max suppression, and we will then have the box that accurately detects the object. The classification loss is calculated for each label using binary cross-entropy, and logistic regression is used to predict object confidence and class predictions. A list of bounding boxes and the identified classes along with the accuracy is the result. Finally, we avoid selecting overlapping boxes by using the IoU (Intersection over Union) and Non-Max Suppression. Non-Max Suppression (NMS) ensures that the used algorithm detects the objects only once and it discards all the false detections. The YOLOv3 Object Detection provides high-quality image in the output with bounding boxes along with the labels of the respective objects.

5 Dataset and Evaluation Metrics

Dataset:The most well-known large-scale autonomous driving dataset is the Cityscape dataset, which contains images of urban street scenes. It is mainly dedicated to the semantic interpretation of urban street scenes. It has 30 classes that have been divided into eight categories (plane surfaces, vehicles, humans, constructions, nature, objects, sky, and void). There are about 5000 finely annotated images in the dataset, and there are about 20,000 coarsely annotated images. The 5,000 photos are divided into 1,525 test images, 500 validation images, and 2,975 training images [4].

The Evaluation Metrics that we used to evaluate our model are as follows.

Fig. 4 Formula of IoU [11]

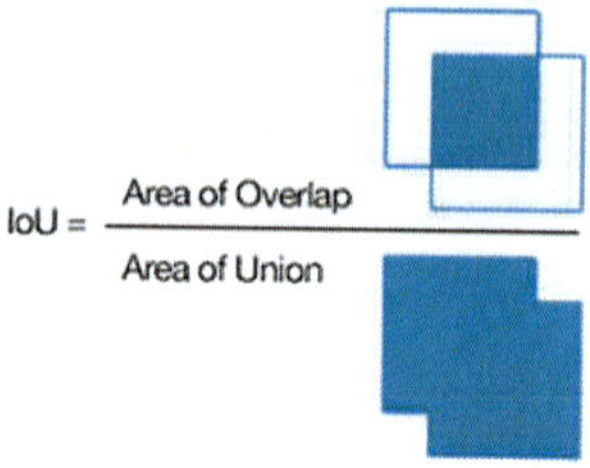

Fig. 5 Formula of Dice coefficient [12]

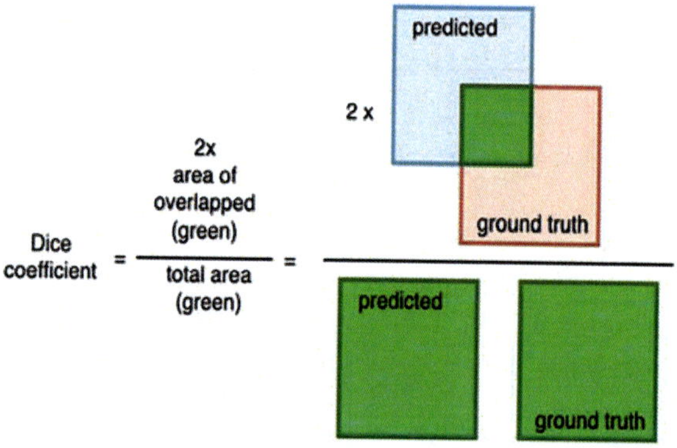

Intersection Over Union (IoU): IoU is calculated by dividing the ground truth's area of overlap by the anticipated segmentation's area of union. It gauges how precisely semantic segmentation is done.

Dice Coefficient: The F1-score is another name for the Dice coefficient. It is twice the Area of Overlap divided by the total number of pixels in both images. The Dice coefficient and the IoU are quite close [9] (Figs. 4 and 5).

IoU of the output is 0.67.

Dice Coefficient of the Output is 0.49.

6 Result

We have performed semantic segmentation using the U-Net model on the Cityscapes dataset, in which 20 parameters: [unlabelled, person, rider, vehicle, truck, bus, train, motorcycle, bicycle, sidewalk, building, wall, fence, pole, traffic light, traffic sign, plant, landscape, sky] are segmented and given a specific color and implemented over an image, which is given as an input image to the model. The accuracy of the segmentation model touches 80%. By implementing YOLOv3 Object Detection using Keras on a segmented image, it gives a high-quality output image with bounding boxes attached to object classes of our interest ["person", "bicycle", "car", "bus",

Fig. 6 Output image after performing image semantic segmentation on input image 1

Fig. 7 Output image after performing image semantic segmentation on input image 2

"truck", "traffic light", "train"]. In the output image, each bounding box has two attributes attached to it, namely label and accuracy. The U-Net architecture and YOLOv3 performed best for image semantic segmentation and labeling in this model and are able to give the desired output.

The work we proposed is different from the work proposed by other researchers, as the segmented output image contains the bounding box for each object and its accuracy. The output also contains the graphical analysis of the image processed, which makes it extremely useful for various future applications like smart traffic management, etc. (Figs. 6 and 7).

7 Graphs and Analysis

In Fig. 8, the first graph illustrates the total number of objects present in the segmented output image. The X-axis shows the name of the objects, and the Y-axis shows the number of objects. The second graph shows the objects present in the segmented image and their accuracy. The X-axis shows the name of each object, and the Y-axis shows the corresponding accuracy.

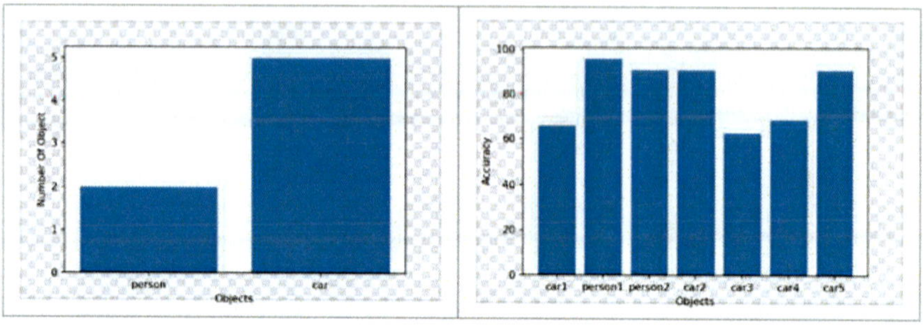

Fig. 8 Bar graphs representing number of objects and accuracy

In Fig. 9, the first graph shows the Loss with respect to Epoch. Y-axis represents Loss and X-axis represents Epoch. The graph shows that the Loss of the model decreases as the Epoch increases. Loss = 0.7255. The second graph shows the Accuracy with respect to Epoch. Y-axis represents Accuracy and X-axis represents Epoch. The graph shows that the Accuracy of the model increases as the Epoch increases. Accuracy = 78.18%.

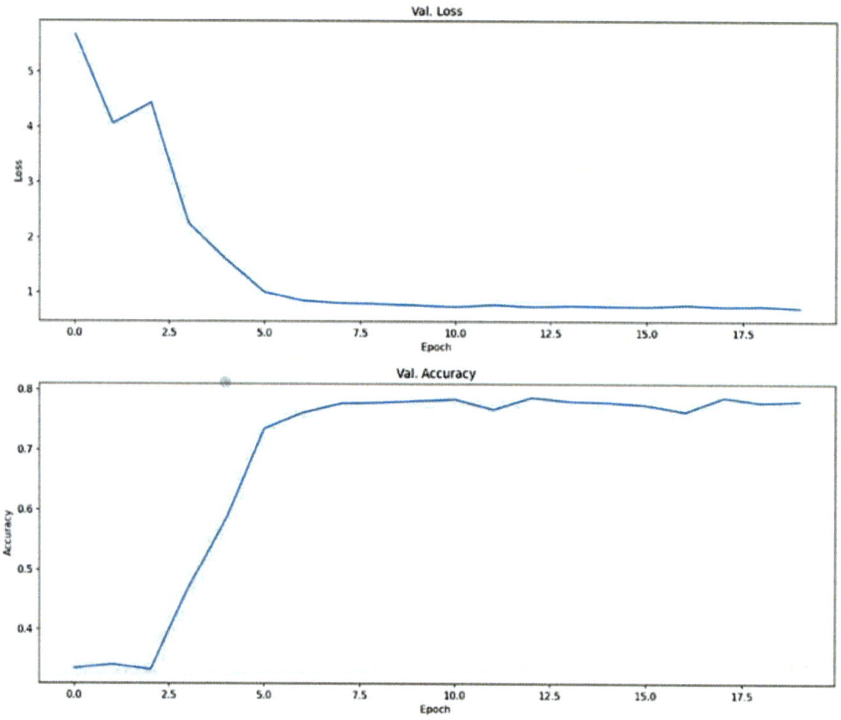

Fig. 9 Graphs representing loss and accuracy

8 Conclusion

Due to the ongoing progress in automated driving and security monitoring, higher standards have been set for image semantic segmentation in terms of model size, computation cost, and segmentation accuracy. During the past ten years, numerous CNN models were suggested for semantic segmentation as deep learning models improved. This research proposes the U-Net architecture-based model for image semantic segmentation and YOLOv3 architecture for object detection. YOLOv3 is faster and very accurate in detecting small objects. The YOLO algorithm can be used in a variety of fields to address a wide range of real-time issues, including security, traffic rule monitoring, and other issues. For testing and training, the Cityscapes dataset is used. The result is a high-quality image with bounding boxes and labels classifying each pixel into a certain class. In this model, the U-Net architecture showed the best performance for the semantic segmentation of images. In the future, the performance of such a model will be further enhanced to increase target boundary segmentation accuracy and the ability to segment tiny targets successfully, and to address the problem of segmenting discontinuous targets.

References

1. Lopez J, Santos S, Atzberger C, Torres D (2018) Convolutional neural networks for semantic segmentation of multispectral remote sensing images. In 2018 IEEE 10th Latin-American conference on communications (LATINCOM), pp 1–5. https://doi.org/10.1109/LATINCOM.2018.8613216
2. Sugirtha T, Sridevi M (2022) Semantic segmentation using modified U-Net for autonomous driving
3. Wang C, Wang Y, Liu Y, He Z, He R, Sun Z (Senior Member, IEEE) (2019) Sclera SegNet: an attention assisted U-Net model for accurate sclera segmentation
4. Liu X, Deng Z, Yang Y (2018) Recent progress in semantic image segmentation. Springer
5. Yi Z, Chang T, Li S, Liu R, Zhang J, Hao A (2019) Scene-aware deep networks for semantic segmentation of images. IEEE 7:2169–3536
6. Dong C (2021) Image semantic segmentation using improved ENet network
7. Paszke A, Chaurasia A, Kim S, Culurciello E (2016) ENet: a deep neural network architecture for real-time semantic segmentation
8. Ronneberger O, Fischer P, Brox T(2015) Unet: convolutional networks for biomedical image segmentation, CoRR, vol. abs/1505.04597
9. Nezla NA, MithunHaridas TP, Supriya MH (2021) Semantic segmentation of underwater images using UNet architecture based deep convolutional encoder decoder model
10. DEV, https://dev.to/afrozchakure/all-you-need-to-know-about-yolo-v3-you-only-look-once-e4m
11. Medium (Towards Data Science), https://towardsdatascience.com/metrics-to-evaluate-your-semantic-segmentation-model-6bcb99639aa2
12. StackExchange (Data Science), https://datascience.stackexchange.com/questions/75708/neural-network-probability-output-and-loss-function-example-dice-loss

Estimation of Reliability in Multicomponent Stress–Strength Model for Generalized Half-Logistic Distribution

Taruna Kumari and Anupam Pathak

Abstract In the literature, so many authors have discussed about the applicability of the stress–strength reliability models having one component in the various fields of engineering. Substantial work for the estimation of the reliability function in a single-component set-up is already carried out by the researchers. Presently, reliability in multicomponent stress–strength set-up has attracted a number of researchers from different extremes because of its ability of application in various fields such as industrial communication systems, military systems, civil engineering, logistic systems, safety-security analysis, maintenance and many more. Herein, we have considered a stress–strength model having k strength components, which are subjected to a common stress. We have considered both classical and Bayesian methods to estimate the reliability of the considered multicomponent system. Monte Carlo simulation technique is exercised to investigate the potential of the obtained estimators. The applicability of these estimators is also discussed for an engineering dataset.

Keywords Generalized half-logistic distribution · Reliability in multicomponent set-up · Maximum likelihood method of estimation · Bayesian method of estimation · Monte Carlo simulation technique

1 Introduction

In a single-component stress–strength reliability ambience, the reliability of a system is given by the probability P(X > Y), where X is the random strength of the system, which is exposed to a random stress Y. In a single-component set-up, a system fails iff the operative stress transcends the strength. In the last decade, single-component

T. Kumari
Discipline of Statistics, School of Sciences, Indira Gandhi National Open University, New Delhi 110068, India

A. Pathak (✉)
Department of Statistics, Ramjas College, University of Delhi, Delhi 110007, India
e-mail: pathakanupam24@gmail.com

© The Author(s), under exclusive license to Springer Nature Singapore Pte Ltd. 2023
A. Mishra et al. (eds.), *Advances in IoT and Security with Computational Intelligence*, Lecture Notes in Networks and Systems 755,
https://doi.org/10.1007/978-981-99-5085-0_10

stress–strength models have got great attention due to its applicability in various fields like engineering, psychology, genetics, and in numerous other fields. Enormous amount of work is drained in the literary works for Bayesian and non-Bayesian estimation of the reliability function P(X > Y). For deep understanding, one may go through Tyagi and Bhattacharya [11], Chaturvedi and Pathak [3], Chaturvedi et al. [4] and Kumari et al. [6].

Advanced engineering systems are usually prepared with the highest distinction. A redundant edifice is used to enhance the system reliability. For instance, a commercial airplane has an extra engine to deal with unexpected breakdowns. In such type of cases, the reliability of the system is generally measured by multicomponent stress–strength (MSS) models. For more understating of these type of models, one can consider the example of suspension bridges, where the floor of the bridge is sustained by k vertical cables and these cables are suspended from the pillars. Moreover, the bridge will only sustain if at least s (s < k) vertical cables are not broken when subjected to a common stress. From this example, it is also clear that the multicomponent reliability (MR) model can also be utilized in the system safety-security analysis and maintenance.

Bhattacharya and Johnson [1] developed the concept of the reliability in a MSS model. Rao [9] obtained the estimators of the reliability function in MSS set-up for generalized exponential distribution. Kızılaslan and Nadar [5] obtained the reliability estimates in MSS model for Weibull distribution.

In this article, we assume that the random variables (rvs) $X_1, X_2, ..., X_k \overset{identically}{\sim} F(x)$ and $Y \sim G(y)$, such that rvs $Y, X_1, X_2, ..., X_k$ are independent. Under these considerations, the reliability function in a MSS set-up comes out to be {Bhattacharya and Johnson [1]}.

$$R_{s,k} = \sum_{i=s}^{k} \binom{k}{i} \int_{-\infty}^{\infty} [1 - F(y)]^i [F(y)]^{k-i} dG(y), s \leq k. \tag{1}$$

A rv X is said to be follow the generalized half-logistic distribution (GHLD), if

$$F(x; \lambda, \theta) = 1 - \left(2e^{-x/\theta}\left(1 + e^{-x/\theta}\right)^{-1}\right)^{\lambda}; \ x > 0, \ \theta, \ \lambda > 0. \tag{2}$$

$$f(x; \lambda, \theta) = \frac{\lambda}{\theta}\left(2e^{-x/\theta}\left(1 + e^{-x/\theta}\right)^{-1}\right)^{\lambda}\left(1 + e^{-x/\theta}\right)^{-1}; \ x > 0, \ \lambda, \ \theta > 0, \tag{3}$$

where $F(x; \lambda, \theta)$ and $f(x; \lambda, \theta)$ represent the cumulative distribution function (cdf) and probability density function (pdf) of X, respectively. Throughout this paper, we denote GHLD with the parameters $(\lambda, \ \theta)$ by GHLD$(\lambda, \ \theta)$.

The layout of this chapter is fivefold. The maximum likelihood (ML) and Bayes estimators of $R_{s,k}$ are derived in Sects. 2 and 3, respectively. In order to obtain the Bayes estimators of $R_{s,k}$, two different techniques, namely; Lindley's and MCMC approximation techniques are adopted and considerations are given to squared error loss function (SELF). In Sect. 4, Monte Carlo simulation technique is exercised along with the real data example. Lastly, in Sect. 5, conclusions on this chapter are provided.

2 ML Estimation of $R_{s,k}$

Here, we consider GHLD(λ_1, θ) as the common pdf of $X_1, X_2, ..., X_k$ and GHLD(λ_2, θ) as the pdf of Y. Then, using (1) and (2), $R_{s,k}$ comes out to be

$$R_{s,k} = \Phi\lambda_2[\lambda_1(i+j)+\lambda_2]^{-1}, \text{ where } \Phi = \sum_{i=s}^{k}\sum_{j=0}^{k-i}\binom{k}{i}\binom{k-i}{j}(-1)^j. \quad (4)$$

Now, assume that n systems are placed on testing assay from each strength universe and m systems are placed on testing assay from stress universe. Suppose in this case, the following observations are recorded: $X_{i1}, X_{i2}, ..., X_{ik}$ and Y_l, $i = 1, 2, ..., n$, $l = 1, 2, ..., m$. Therefore, the log-likelihood function based on the above data can be written as

$$\ln(l) = nk\ln\lambda_1 + m\ln\lambda_2 - (nk+m)\ln\theta + \lambda_1\sum_{i=1}^{n}\sum_{j=1}^{k}\ln\left(2e^{-x_{ij}/\theta}\left(1+e^{-x_{ij}/\theta}\right)^{-1}\right)$$
$$-\sum_{i=1}^{n}\sum_{j=1}^{k}\ln\left(1+e^{-x_{ij}/\theta}\right) + \lambda_2\sum_{l=1}^{m}\ln\left(2e^{-y_l/\theta}\left(1+e^{-y_l/\theta}\right)^{-1}\right) - \sum_{l=1}^{m}\ln\left(1+e^{-y_l/\theta}\right). \quad (5)$$

From (5), the normal equations for λ_1, λ_2 and θ are given by

$$0 = \frac{\partial\ln(l)}{\partial\lambda_1} = \frac{nk}{\lambda_1} + \sum_{i=1}^{n}\sum_{j=1}^{k}\ln\left(2e^{-x_{ij}/\theta}\left(1+e^{-x_{ij}/\theta}\right)^{-1}\right), \quad (6)$$

$$0 = \frac{\partial\ln(l)}{\partial\lambda_2} = \frac{m}{\lambda_2} + \sum_{l=1}^{m}\ln\left(2e^{-y_l/\theta}\left(1+e^{-y_l/\theta}\right)^{-1}\right) \quad (7)$$

and

$$0 = \frac{\partial\ln(l)}{\partial\theta} = -\frac{(nk+m)}{\theta} + \frac{\lambda_1}{\theta^2}\sum_{i=1}^{n}\sum_{j=1}^{k}\left(x_{ij}\left(1+e^{-x_{ij}/\theta}\right)^{-1}\right)$$
$$-\frac{1}{\theta^2}\sum_{i=1}^{n}\sum_{j=1}^{k}\left(x_{ij}e^{-x_{ij}/\theta}\left(1+e^{-x_{ij}/\theta}\right)^{-1}\right) + \frac{\lambda_2}{\theta^2}\sum_{l=1}^{m}\left(y_l\left(1+e^{-y_l/\theta}\right)^{-1}\right)$$

$$-\frac{1}{\theta^2}\sum_{l=1}^{m}\left(y_l e^{-y_l/\theta}\left(1+e^{-y_l/\theta}\right)^{-1}\right). \tag{8}$$

If, $\tilde{\lambda}_1$, $\tilde{\lambda}_2$ and $\tilde{\theta}$ denote the ML estimators of λ_1, λ_2 and θ, respectively, then $\tilde{\lambda}_1$, $\tilde{\lambda}_2$ and $\tilde{\theta}$ can be obtained by using (6), (7) and (8). Now, using (4), ML estimators of $\lambda_1, \lambda_2, \theta$ and invariance property of the ML estimators one can easily obtain the ML estimator of $R_{s,k}$, i.e.,

$$\tilde{R}_{s,k} = \Phi\tilde{\lambda}_2\left[\tilde{\lambda}_1(i+j)+\tilde{\lambda}_2\right]^{-1}. \tag{9}$$

3 Bayesian Estimation of $R_{s,k}$

In order to obtain the Bayes estimators of $R_{s,k}$, three non-informative independent priors, $\pi(\lambda_1)$, $\pi(\lambda_2)$ and $\pi(\theta)$ are considered as follows:

$$\pi(\lambda_1) = \tfrac{1}{\lambda_1},\ \lambda_1 > 0,\ \pi(\lambda_2) = \tfrac{1}{\lambda_2},\ \lambda_2 > 0 \text{ and } \pi(\theta) = \tfrac{1}{\theta},\ \theta > 0. \tag{10}$$

Then, the joint posterior density can be obtained as

$$\pi(\lambda_1, \lambda_2, \theta|X, \underline{y}) \propto \frac{\lambda_1^{nk-1}\lambda_2^{m-1}}{\theta^{nk+m+1}}\prod_{i=1}^{n}\prod_{j=1}^{k}\left(2e^{-x_{ij}/\theta}\left(1+e^{-x_{ij}/\theta}\right)^{-1}\right)^{\lambda_1}\prod_{i=1}^{n}\prod_{j=1}^{k}\left(1+e^{-x_{ij}/\theta}\right)^{-1}$$
$$\cdot\prod_{l=1}^{m}\left(2e^{-y_l/\theta}\left(1+e^{-y_l/\theta}\right)^{-1}\right)^{\lambda_2}\prod_{l=1}^{m}\left(1+e^{-y_l/\theta}\right)^{-1}. \tag{11}$$

Thus, the Bayes estimator of $R_{s,k}$ under SELF is

$$\breve{R}_{s,k} = \int_0^\infty\int_0^\infty\int_0^\infty R_{s,k}\pi(\lambda_1, \lambda_2, \theta|X, \underline{y})d\lambda_1 d\lambda_2 d\theta. \tag{12}$$

Since the analytical solution of (12) is not attainable therefore, we utilize two approaches namely; Lindley's approximation and MCMC techniques to obtain the same.

3.1 Lindley's Approximation

To derive the Bayes estimator of $R_{s,k}$, $\breve{R}_{s,k}^{Lin}$ say, by using Lindley's approximation technique, suppose that the posterior expectation $I(\varphi)$, where $\varphi = R_{s,k}$, is representable in the following form

$$I(\varphi) = E(\varphi|\underline{x}) = \frac{\int\limits_{(\lambda_1,\lambda_2,\theta)} \varphi e^{l(\lambda_1,\lambda_2,\theta)+\rho(\lambda_1,\lambda_2,\theta)} d(\lambda_1,\lambda_2,\theta)}{\int\limits_{(\lambda_1,\lambda_2,\theta)} e^{l(\lambda_1,\lambda_2,\theta)+\rho(\lambda_1,\lambda_2,\theta)} d(\lambda_1,\lambda_2,\theta)}, \tag{13}$$

where $\rho(\lambda_1, \lambda_2, \theta) = -\ln\lambda_1 - \ln\lambda_2 - \ln\theta$.

For sufficiently large m and n, we get {see Lindley [7]}

$$\begin{aligned} E(\varphi|\underline{x}) = {} & \varphi + \left(\varphi_{\lambda_1} a_{\lambda_1} + \varphi_{\lambda_2} a_{\lambda_2} + \varphi_\theta a_\theta\right) + \left(\varphi_{\lambda_1\lambda_2}\sigma_{\lambda_1\lambda_2} + \varphi_{\lambda_1\theta}\sigma_{\lambda_1\theta} + \varphi_{\lambda_2\theta}\sigma_{\lambda_2\theta}\right) \\ & + \frac{1}{2}\Big[\varphi_{\lambda_1\lambda_1}\sigma_{\lambda_1\lambda_1} + \varphi_{\lambda_2\lambda_2}\sigma_{\lambda_2\lambda_2} + \varphi_{\theta\theta}\sigma_{\theta\theta} + A\left(\varphi_{\lambda_1}\sigma_{\lambda_1\lambda_1} + \varphi_{\lambda_2}\sigma_{\lambda_1\lambda_2} + \varphi_\theta\sigma_{\lambda_1\theta}\right) \\ & + B\left(\varphi_{\lambda_1}\sigma_{\lambda_2\lambda_1} + \varphi_{\lambda_2}\sigma_{\lambda_2\lambda_2} + \varphi_\theta\sigma_{\lambda_2\theta}\right) + C\left(\varphi_{\lambda_1}\sigma_{\theta\lambda_1} + \varphi_{\lambda_2}\sigma_{\theta\lambda_2} + \varphi_\theta\sigma_{\theta\theta}\right)\Big]\Big|_{(\lambda_1,\lambda_2,\theta)=(\tilde{\lambda}_1,\tilde{\lambda}_2,\tilde{\theta})}, \end{aligned}$$

where,

$$\begin{aligned} a_{\lambda_1} &= \rho_{\lambda_1}\sigma_{\lambda_1\lambda_1} + \rho_{\lambda_2}\sigma_{\lambda_1\lambda_2} + \rho_\theta\sigma_{\lambda_1\theta}, \; a_{\lambda_2} = \rho_{\lambda_1}\sigma_{\lambda_2\lambda_1} + \rho_{\lambda_2}\sigma_{\lambda_2\lambda_2} + \rho_\theta\sigma_{\lambda_2\theta}, \\ a_\theta &= \rho_{\lambda_1}\sigma_{\theta\lambda_1} + \rho_{\lambda_2}\sigma_{\theta\lambda_2} + \rho_\theta\sigma_{\theta\theta}, \\ A &= \sigma_{\lambda_1\lambda_1}L_{\lambda_1\lambda_1\lambda_1} + 2\sigma_{\lambda_1\lambda_2}L_{\lambda_1\lambda_2\lambda_1} + 2\sigma_{\lambda_1\theta}L_{\lambda_1\theta\lambda_1} + 2\sigma_{\lambda_2\theta}L_{\lambda_2\theta\lambda_1} + \sigma_{\lambda_2\lambda_2}L_{\lambda_2\lambda_2\lambda_1} + \sigma_{\theta\theta}L_{\theta\theta\lambda_1}, \\ B &= \sigma_{\lambda_1\lambda_1}L_{\lambda_1\lambda_1\lambda_2} + 2\sigma_{\lambda_1\lambda_2}L_{\lambda_1\lambda_2\lambda_2} + 2\sigma_{\lambda_1\theta}L_{\lambda_1\theta\lambda_2} + 2\sigma_{\lambda_2\theta}L_{\lambda_2\theta\lambda_2} + \sigma_{\lambda_2\lambda_2}L_{\lambda_2\lambda_2\lambda_2} + \sigma_{\theta\theta}L_{\theta\theta\lambda_2}, \\ C &= \sigma_{\lambda_1\lambda_1}L_{\lambda_1\lambda_1\theta} + 2\sigma_{\lambda_1\lambda_2}L_{\lambda_1\lambda_2\theta} + 2\sigma_{\lambda_1\theta}L_{\lambda_1\theta\theta} + 2\sigma_{\lambda_2\theta}L_{\lambda_2\theta\theta} + \sigma_{\lambda_2\lambda_2}L_{\lambda_2\lambda_2\theta} + \sigma_{\theta\theta}L_{\theta\theta\theta}. \end{aligned}$$

The terms involved in the above expression are defined as follows: $\rho_{\lambda_1} = \frac{\partial\rho}{\partial\lambda_1}$, $\varphi_{\lambda_1} = \frac{\partial\varphi}{\partial\lambda_1}$, $\varphi_{\lambda_1\lambda_1} = \frac{\partial^2\varphi}{\partial\lambda_1\partial\lambda_1}$, $\varphi_{\lambda_2\lambda_2} = \frac{\partial^2\varphi}{\partial\lambda_2\partial\lambda_2}$, $\varphi_{\lambda_1\lambda_2} = \frac{\partial^2\varphi}{\partial\lambda_1\partial\lambda_2}$, $L_{\lambda_1\lambda_2\theta} = \frac{\partial^3 \ln l}{\partial\lambda_1\partial\lambda_2\partial\theta}$. Other terms can be defined in the similar manner.

Moreover, σ_{ij} denotes the jth element present in the ith row of the inverse of the matrix $-L$, where,

$$\begin{aligned} -L &= \{-L_{ij}\}, i, j = \lambda_1, \lambda_2, \theta \\ L_{\lambda_1\theta} &= L_{\theta\lambda_1} = \frac{1}{\theta^2}\sum_{i=1}^{n}\sum_{j=1}^{k} x_{ij}\left(1+e^{-x_{ij}/\theta}\right)^{-1}, \; L_{\lambda_2\lambda_2} = \frac{-m}{\lambda_2^2}, \; L_{\theta\lambda_2} = L_{\lambda_2\theta} = \frac{1}{\theta^2}\sum_{l=1}^{m} y_l\left(1+e^{-y_l/\theta}\right)^{-1}, \\ L_{\theta\theta} &= \frac{(nk+m)}{\theta^2} - \frac{(\lambda_1+1)}{\theta^4}\sum_{i=1}^{n}\sum_{j=1}^{k} x_{ij}^2 e^{-x_{ij}/\theta}(1+e^{-x_{ij}/\theta})^{-2} - \frac{(\lambda_2+1)}{\theta^4}\sum_{l=1}^{m} y_l^2 e^{-y_l/\theta}(1+e^{-y_l/\theta})^{-2} \\ & - \frac{2(\lambda_1+1)}{\theta^3}\sum_{i=1}^{n}\sum_{j=1}^{k} x_{ij}(1+e^{-x_{ij}/\theta})^{-1} - \frac{2(\lambda_2+1)}{\theta^3}\sum_{l=1}^{m} y_l(1+e^{-y_l/\theta})^{-1} \\ & + \frac{2}{\theta^3}\left(\sum_{i=1}^{n}\sum_{j=1}^{k} x_{ij} + \sum_{l=1}^{m} y_l\right), \\ \rho_{\lambda_1} &= -\frac{1}{\lambda_1}, \; \rho_{\lambda_2} = -\frac{1}{\lambda_2}, \; \rho_\theta = -\frac{1}{\theta}, \; L_{\lambda_1\lambda_1} = \frac{-nk}{\lambda_1^2}, \\ L_{\lambda_1\lambda_2} &= L_{\lambda_2\lambda_1} = 0, \; L_{\lambda_1\lambda_1\lambda_1} = \frac{2nk}{\lambda_1^3}, \; L_{\lambda_2\lambda_2\lambda_2} = \frac{2m}{\lambda_2^3}, \\ L_{\lambda_2\lambda_2\lambda_1} &= L_{\lambda_1\lambda_2\lambda_1} = 0, \; L_{\lambda_1\lambda_1\lambda_2} = L_{\lambda_1\lambda_2\lambda_2} = 0, \end{aligned}$$

$$
\begin{aligned}
L_{\theta\theta\lambda_1} &= L_{\lambda_1\theta\theta} = -\sum_{i=1}^{n}\sum_{j=1}^{k}\frac{1}{\theta^4}x_{ij}\left(x_{ij}e^{-x_{ij}/\theta} + 2\theta\left(1+e^{-x_{ij}/\theta}\right)\right)\left(1+e^{-x_{ij}/\theta}\right)^{-2},\\
L_{\lambda_1\theta\lambda_1} &= L_{\lambda_2\theta\lambda_1} = 0,\ L_{\lambda_2\lambda_2\theta} = L_{\lambda_2\lambda_1\theta} = 0,\\
L_{\theta\theta\lambda_2} &= L_{\lambda_2\theta\theta} = -\sum_{l=1}^{m}\frac{1}{\theta^4}y_l\left(y_l e^{-y_l/\theta} + 2\theta\left(1+e^{-y_l/\theta}\right)\right)\left(1+e^{-y_l/\theta}\right)^{-2},\\
L_{\lambda_1\theta\lambda_2} &= L_{\lambda_2\theta\lambda_2} = 0,\\
L_{\theta\theta\theta} &= -\frac{2(nk+m)}{\theta^3} - (\lambda_1+1)\sum_{i=1}^{n}\sum_{j=1}^{k}\frac{1}{\theta^6}x_{ij}^2 e^{-x_{ij}/\theta}\left(x_{ij}(1-e^{-x_{ij}/\theta}) - 4\theta\left(1+e^{-x_{ij}/\theta}\right)\right)\left(1+e^{-x_{ij}/\theta}\right)^{-3}\\
&\quad - (\lambda_2+1)\sum_{l=1}^{m}\frac{1}{\theta^6}y_l^2 e^{-y_l/\theta}\left(y_l(1-e^{-y_l/\theta}) - 4\theta\left(1+e^{-y_l/\theta}\right)\right)\left(1+e^{-y_l/\theta}\right)^{-3}\\
&\quad + 2(\lambda_1+1)\sum_{i=1}^{n}\sum_{j=1}^{k}\frac{1}{\theta^5}x_{ij}\left(x_{ij}e^{-x_{ij}/\theta} + 3\theta\left(1+e^{-x_{ij}/\theta}\right)\right)\left(1+e^{-x_{ij}/\theta}\right)^{-2}\\
&\quad + 2(\lambda_2+1)\sum_{l=1}^{m}\frac{1}{\theta^5}y_l\left(y_l e^{-y_l/\theta} + 3\theta\left(1+e^{-y_l/\theta}\right)\right)\left(1+e^{-y_l/\theta}\right)^{-2} - \frac{6}{\theta^4}\left(\sum_{i=1}^{n}\sum_{j=1}^{k}x_{ij} + \sum_{l=1}^{m}y_l\right),\\
L_{\lambda_1\lambda_1\theta} &= L_{\lambda_1\lambda_2\theta} = 0,\\
\varphi_\theta &= \varphi_{\lambda_1\theta} = \varphi_{\lambda_2\theta} = \varphi_{\theta\lambda_1} = 0\\
\varphi_{\lambda_1} &= -\Phi\lambda_2(i+j)[\lambda_1(i+j)+\lambda_2]^{-2},\ \varphi_{\lambda_2} = \Phi\lambda_1(i+j)[\lambda_1(i+j)+\lambda_2]^{-2},\\
\varphi_{\theta\lambda_2} &= \varphi_{\theta\theta} = 0,\ \varphi_{\lambda_1\lambda_1} = \Phi 2\lambda_2(i+j)^2[\lambda_1(i+j)+\lambda_2]^{-3},\\
\varphi_{\lambda_2\lambda_2} &= -\Phi 2\lambda_1(i+j)[\lambda_1(i+j)+\lambda_2]^{-3},\\
\varphi_{\lambda_1\lambda_2} &= \varphi_{\lambda_2\lambda_1} = \Phi(i+j)(\lambda_2-\lambda_1(i+j))[\lambda_1(i+j)+\lambda_2]^{-3}.
\end{aligned}
$$

These terms are evaluated at ML estimators of the parameters. Also, using above we get

$A = \sigma_{\lambda_1\lambda_1}L_{\lambda_1\lambda_1\lambda_1} + \sigma_{\theta\theta}L_{\theta\theta\lambda_1}$, $B = \sigma_{\lambda_2\lambda_2}L_{\lambda_2\lambda_2\lambda_2} + \sigma_{\theta\theta}L_{\theta\theta\lambda_2}$, $C = 2\sigma_{\lambda_1\theta}L_{\lambda_1\theta\theta} + 2\sigma_{\lambda_2\theta}L_{\lambda_2\theta\theta} + \sigma_{\theta\theta}L_{\theta\theta\theta}$.Thus, the Lindley estimator of $R_{s,k}$ is given by the following expression:

$$
\breve{R}^{L}_{s,k} = \varphi + \left(\varphi_{\lambda_1}a_{\lambda_1} + \varphi_{\lambda_2}a_{\lambda_2}\right) + \left(\varphi_{\lambda_1\lambda_2}\sigma_{\lambda_1\lambda_2}\right) + \frac{1}{2}[\varphi_{\lambda_1\lambda_1}\sigma_{\lambda_1\lambda_1} + \varphi_{\lambda_2\lambda_2}\sigma_{\lambda_2\lambda_2} + A(\varphi_{\lambda_1}\sigma_{\lambda_1\lambda_1} + \varphi_{\lambda_2}\sigma_{\lambda_1\lambda_2}) + B(\varphi_{\lambda_1}\sigma_{\lambda_2\lambda_1} + \varphi_{\lambda_2}\sigma_{\lambda_2\lambda_2}) + C(\varphi_{\lambda_1}\sigma_{\theta\lambda_1} + \varphi_{\lambda_2}\sigma_{\theta\lambda_2})]_{|(\lambda_1,\lambda_2,\theta)=(\tilde{\lambda}_1,\tilde{\lambda}_2,\tilde{\theta})}. \tag{14}
$$

3.2 MCMC Method

From (11), one can easily obtain the marginal posterior densities of λ_1, λ_2 and θ as

$$\lambda_1|\theta, X, \underline{y} \sim \gamma\left(nk, -\sum_{i=1}^{n}\sum_{j=1}^{k}\ln\left(2e^{-x_{ij}/\theta}\left(1+e^{-x_{ij}/\theta}\right)^{-1}\right)\right), \tag{15}$$

$$\lambda_2|\theta, X, \underline{y} \sim \gamma\left(m, -\sum_{l=1}^{m}\ln\left(2e^{-y_l/\theta}\left(1+e^{-y_l/\theta}\right)^{-1}\right)\right) \tag{16}$$

and

$$\pi(\theta|\lambda_1, \lambda_2, X, \underline{y}) \propto \frac{1}{\theta^{nk+m+1}}\prod_{i=1}^{n}\prod_{j=1}^{k}\left(2e^{-x_{ij}/\theta}\left(1+e^{-x_{ij}/\theta}\right)^{-1}\right)^{\lambda_1}\prod_{i=1}^{n}\prod_{j=1}^{k}\left(1+e^{-x_{ij}/\theta}\right)^{-1}$$
$$\cdot\prod_{l=1}^{m}\left(2e^{-y_l/\theta}\left(1+e^{-y_l/\theta}\right)^{-1}\right)^{\lambda_2}\prod_{l=1}^{m}\left(1+e^{-y_l/\theta}\right)^{-1}, \tag{17}$$

respectively.

Let us assume the proposal such that

$$q(\theta) = \frac{\delta^{\xi}\theta^{\xi-1}e^{-\delta\theta}}{\Gamma(\xi)}, \ \xi, \delta, \theta > 0. \tag{18}$$

The algorithm to compute $R_{s,k}^{MCMC}$ is as follows:

1. Start with initial guess $\theta^{(0)}$.
2. Set i = 1.
3. Generate $\lambda_1^{(i)}$ from (15).
4. Generate $\lambda_2^{(i)}$ from (16).
5. Generate $\theta^{(i)}$ from (17) by using (18) and the MH algorithm.

   5.1 Let $u = \theta^{(i-1)}$
   5.2 Invoke v from (18).
   5.3 Let $p(u, v) = min\left\{1, \frac{\pi(v|\lambda_1^{(i)},\lambda_2^{(i)},X,\underline{y})q(u)}{\pi(u|\lambda_1^{(i)},\lambda_2^{(i)},X,\underline{y})q(v)}\right\}$.
   5.4 Draw $r \sim Uniform(0, 1)$ and set $\theta^{(i)} = \begin{cases} v; & \text{if } r \le p(u, v), \\ u; & \text{otherwise.} \end{cases}$

6. Compute $R_{s,k}^{(i)}$ at $\lambda_1^{(i)}$, $\lambda_2^{(i)}$ and $\theta^{(i)}$.
7. Increase i by 1 and repeat the previous steps from 2 to 6 by N number of times and obtain the MCMC estimate under SELF of $R_{s,k}$ by using

$$\breve{R}_{s,k}^{MCMC} = \frac{1}{N-N_1} \sum_{i=N_1+1}^{N-N_1} R_{s,k}^{(i)},$$

where N_1 represents the burn in period.

4 Simulation Study and Real Data Analysis

In this section, we first investigate the potentials of the developed estimators, namely; ML, MCMC and Lindley's of $R_{s,k}$. To do so, 1000 random samples are generated from (3) corresponding to each of the strength variables and one stress variable of sizes (n, m), where, n = 5, 10, 15 and m = 5, 10, 20. Thus, in each case, 1000 estimates of $R_{s,k}$ are evaluated. Also, the MCMC estimates are computed with proposal distribution parameter $(\xi, \delta) = (1.2, 0.5)$ and $(N, N_1) = (5000, 2000)$. Average estimates of $R_{s,k}$ along with their corresponding mean squared errors (in parenthesis) are obtained for different values of the parameters $(\lambda_1, \lambda_2, \theta)$ as (2, 2, 2) and (2, 1, 2). For (s, k) = (1, 3) and (2, 3), the obtained results are shown in Table 1.

Next, we take different set of parameter values of the distribution, $(\lambda_1,\lambda_2,\theta)$ as (0.5, 0.2, 0.3) and (0.5, 0.5, 0.3). For (s, k) = (1, 3) and (2, 3), we have generated 1000 random samples from (3), corresponding to each of the strength variables and one stress variable. The MCMC estimates are computed with proposal distribution parameter $(\xi, \delta) = (1.2, 0.5)$. The obtained average ML, MCMC and Lindley's estimates of $R_{s,k}$ with their corresponding mean squared errors (in parenthesis) are shown in Table 2.

To perform the real data analysis, we considered the following data of the failure log times to breakdown of an insulating fluid experiment {see Nelson [8]}:

Table 1 Estimates of $R_{s,k}$

$R_{s,k}$	θ	λ_1	λ_2	m	n	$\tilde{R}_{s,k}$	$R_{s,k}^{MCMC}$	$\breve{R}_{s,k}^{L}$
$R_{1,3} = 0.75$	2	2	2	5	5	0.741(0.0182)	0.71(0.0202)	0.7139(0.0173)
				10	10	0.7495(0.0092)	0.7343(0.0085)	0.7328(0.0093)
				15	20	0.7426(0.0042)	0.736(0.004)	0.7345(0.0042)
$R_{1,3} = 0.543$		2	1	5	5	0.5802(0.0236)	0.5664(0.0182)	0.5574(0.0195)
				10	10	0.5637(0.0126)	0.5572(0.0111)	0.5524(0.0114)
				15	20	0.5471(0.0057)	0.5475(0.0053)	0.5426(0.0054)
$R_{2,3} = 0.5$		2	2	5	5	0.5333(0.0296)	0.514(0.0229)	0.5171(0.0256)
				10	10	0.5139(0.0079)	0.505(0.0066)	0.5064(0.0072)
				15	20	0.498(0.0066)	0.4961(0.006)	0.495(0.0064)
$R_{2,3} = 0.3143$		2	1	5	5	0.3438(0.0127)	0.3471(0.0112)	0.3344(0.0121)
				10	10	0.3312(0.006)	0.3355(0.0056)	0.3292(0.0056)
				15	20	0.326(0.0035)	0.3286(0.0034)	0.3243(0.0034)

Table 2 Estimates of $R_{s,k}$

$R_{s,k}$	θ	λ_1	λ_2	m	n	$\tilde{R}_{s,k}$	$R_{s,k}^{MCMC}$	$\breve{R}_{s,k}^{L}$
$R_{1,3}$ = 0.4747899	0.3	0.5	0.2	5	5	0.4946(0.0239)	0.4885(0.0186)	0.4563(0.0329)
				10	10	0.508(0.0122)	0.5063(0.011)	0.4987(0.0109)
				15	20	0.4784(0.0073)	0.4789(0.0068)	0.4745(0.007)
$R_{1,3}$ = 0.75		0.5	0.5	5	5	0.739(0.0195)	0.7106(0.018)	0.7086(0.0211)
				10	10	0.7505(0.008)	0.7351(0.0077)	0.7343(0.0081)
				15	20	0.748(0.0051)	0.7405(0.0045)	0.74(0.0051)
$R_{2,3}$ = 0.2647059		0.5	0.2	5	5	0.281(0.0115)	0.2882(0.0107)	0.2768(0.0106)
				10	10	0.28(0.0062)	0.2821(0.006)	0.2777(0.0059)
				15	20	0.2706(0.0024)	0.2747(0.0024)	0.2706(0.0024)
$R_{2,3}$ = 0.5		0.5	0.5	5	5	0.5155(0.0209)	0.4994(0.0159)	0.499(0.0187)
				10	10	0.5006(0.0089)	0.4968(0.0081)	0.4933(0.0084)
				15	20	0.4977(0.004)	0.4946(0.0036)	0.4949(0.0039)

0.270027, 1.02245, 1.15057, 1.42311, 1.54116, 1.57898, 1.8718, 1.9947, 2.08069, 2.11263, 2.48989, 3.45789, 3.48186, 3.52371, 3.60305, 4.28895.

Seo et al. [10] showed that these data are given good fit to the GHLD distribution by applying Kolmogorov-Smirnov test. Later Chaturvedi et al. [4] used these data for the analysis purpose.

The above data can also be arranged in matrix form corresponding to the triplet (X_1, X_2, X_3) and Y such that n = 4 and m = 4, as follows:

$$\begin{array}{c} X_1\ X_2\ X_3 \end{array}$$

$$X = \begin{pmatrix} 0.270027 & 1.02245 & 1.15057 \\ 1.423110 & 1.54116 & 1.57898 \\ 1.871800 & 1.99470 & 2.08069 \\ 2.112630 & 2.48989 & 3.45789 \end{pmatrix} \quad Y = \begin{pmatrix} 3.48186 \\ 3.52371 \\ 3.60305 \\ 4.28895 \end{pmatrix}$$

If we set the data in the above matrix form then we have k = 3. For (ξ, δ) = (1.5, 0.5) and s = 1 the various estimates of $R_{s,k}$ come out to be $\tilde{R}_{s,k} = 0.48146$, $R_{s,k}^{MCMC} = 0.4799627$ and $\breve{R}_{s,k}^{L} = 0.4609595$.

5 Conclusion

From Tables 1 and 2, we conclude that the Bayes estimator for the different values of (n, m) performs better than the ML estimators as their corresponding MSEs are lower than that of ML estimators. Moreover, MCMC estimates are better than that Lindley estimates.

References

1. Bhattacharyya GK, Johnson RA (1974) Estimation of reliability in a multicomponent stress-strength model. J Am Stat Assoc 69(348):966–970
3. Chaturvedi A, Pathak A (2013) Bayesian estimation procedures for three parameter exponentiated Weibull distribution under entropy loss function and type II censoring. InterStat. interstat.statjournals.net/YEAR/2013/ abstracts/1306001.php
4. Chaturvedi A, Kang SB, Pathak A (2016) Estimation and testing procedures for the reliability functions of generalized half logistic distribution. J Korean Stat Soc 45(2):314–328
5. Kızılaslan F, Nadar M (2015) Classical and bayesian estimation of reliability in multicomponent stress-strength model based on weibull distribution. Revista Colombiana de Estadistica 2:467–484
6. Kumari T, Chaturvedi A, Pathak A (2019) Estimation and testing procedures for the reliability functions of Kumaraswamy-G distributions and a characterization based on records. J Stat Theory Practice 13(22). https://doi.org/10.1007/s42519-018-0014-7
7. Lindley DV (1980) Approximate Bayes Method. Trabajos de Estadistica 3:281–288
8. Nelson WB (1982) Applied life data analysis. John Wiley and Sons, New York
9. Rao GS (2012) Estimation of reliability in multicomponent stress-strength model based on generalized exponential distribution. Colombian J Stat 35(1):67–76
10. Seo JI, Kim Y, Kang SB (2013) Estimation on the generalized half logistic distribution under Type-II hybrid censoring. Commun Stat Appl Methods 20:63–75
11. Tyagi RK, Bhattacharya SK (1989) A note on the MVU estimation of reliability for the Maxwell failure distribution. Estadistica 41:73–79

PNA-DCN: A Deep Convolution Network to Detect the Pneumonia Disease

Rishikesh Bhupendra Trivedi, Anuj Sahani, and Somya Goyal

Abstract Pneumonia (PNA) refers to a viral or bacterial infection that causes inflammation of the lungs. Pneumonia is known to cause long-term damage to the lungs even after recovery. Timely detection of pneumonia can prevent long-term damage to the lungs and save the patient's condition from worsening. Diagnosis of X-ray images of the lungs is being employed by the researchers to detect pneumonia. The X-ray images of a pneumonia-ridden lung tend to show abnormal opacification of the lung. This paper proposes a deep convolution network to detect the pneumonia disease named PNA-DCN. It deploys VGG19 architecture over the publicly available dataset, namely, Labeled Optical Coherence Tomography (OCT) and Chest X-ray images for classification. The proposed model is found experimentally significant with an accuracy of 91.2%. After an empirical comparison with baseline models from the literature, it is concluded that the PNA-DCN model is significantly accurate to effectively predicting pneumonia in humans.

Keywords Chest X-rays · Pneumonia disease · VGG19 · Deep learning

1 Introduction

Lung inflammation due to viral, bacterial, or any other infection is the condition of pneumonia disease. Pneumonia may result in irreversible damage to the lungs. A survey suggests that pneumonia accounts for 14% of all deaths of children under 5 years of age around the world [1]. The early detection of pneumonia can be a potential help in saving the lives of thousands [2]. Chest X-rays have become an important tool to identify pneumonia [3]. Interpretation of X-ray images requires experience and time. The radio experts and some technical supports are always needed. A wide range of Deep Learning (DL) techniques is available for the detection of pneumonia. The X-ray of the chest and lungs contains numerous indicators that

R. B. Trivedi · A. Sahani · S. Goyal (✉)
Manipal University Jaipur, Jaipur, Rajasthan 303007, India
e-mail: somyagoyal1988@gmail.com

© The Author(s), under exclusive license to Springer Nature Singapore Pte Ltd. 2023

A. Mishra et al. (eds.), *Advances in IoT and Security with Computational Intelligence*, Lecture Notes in Networks and Systems 755,
https://doi.org/10.1007/978-981-99-5085-0_11

effectively help us to identify pneumonia. Deep Learning Techniques contributed a lot to the detection of pneumonia by making the process faster and more efficient. Convolution Neural Network (CNN) has extensively been used to detect pneumonia [4].

This paper contributes a novel PNA-DCN model with transfer learning using VGG-19 architecture to detect Pneumonia (PNA) disease in its early condition. The model classifies the X-ray images into healthy or unhealthy categories. The organization of the paper is as follows. The next section discusses the related literature work. Section 3 describes the methodology used. Section 4 discusses the experimental results obtained. The work is concluded in Sect. 5 with remarks on future scope.

2 Related Works

This section highlights the state-of-the-art in the domain of pneumonia detection (see Table 1).

ML techniques including SVM, ANN, Decision Trees [15, 16], and Ensembles [17] are being deployed for the PNA classification problem. Data pre-processing methods for selecting features help to improve the accuracy of the models [18–20].

Table 1 Current trends in PNA detection

Serial number	Study	Dataset used	Technique used	Evaluation criteria
1	Zech et al. [5]	MSH-NIH dataset	CNN	AUC
2	Ge et al. [6]	Hospital EHR	RNN	AUC
3	Dey et al. [4]	OCT dataset	CNN	Accuracy
4	Muhammad et al. [7]	UCI Kaggle dataset	CNN	AUC, precision, recall, accuracy
5	Asnaoui [8]	Covid dataset	ResNet	F1-measure
6	Bermejo-Peláez et al. [9]	Covid dataset	Deep networks	AUC
7	Goyal and Singh [10]	C19RD and CXIP datasets	RNN + LSTM	Accuracy, F1-measure
8	Hasan et al. [11]	Kaggle dataset	RNN	Accuracy
9	Hashmi et al. [12]	OCT dataset	ResNet, DenseNet	AUC, accuracy
10	Hammoudi et al. [13]	Kaggle dataset	CNN	Accuracy
11	Asnaoui et al. [14]	OCT dataset	ResNet, DenseNet	AUC, accuracy

Nowadays, Deep Learning (DL) [21, 22]-based classification models are prevailing in the domain of PNA detection.

3 Proposed Research Methodology

This section briefs about the adopted research methodology including the experimental set-up, dataset, and evaluation criteria. The dataset used for the experiments is OCT Classification dataset with chest X-rays [23]. The dataset has two varieties of classes, Normal and Pneumonia. PNA is a two-class classification problem. The Deep Convolutional Network (DCM)-based classification model using VGG19 architecture is devised in this work.

3.1 Methodology

The proposed deep convolution network (DCN) model for pneumonia (PNA) detection utilizes VGG19 architecture that is 19 layers deep. It deploys a pretrained model over the ImageNet Database as shown in Fig. 1. The forward and backward propagation iterate through the training samples in the network till the time the network determines the optimal weights, and the most predictive neurons are activated so that the network can make the predictions. The cost function determines the average loss across all the samples, which indicates the overall performance. The description of the layers is given in Table 2.

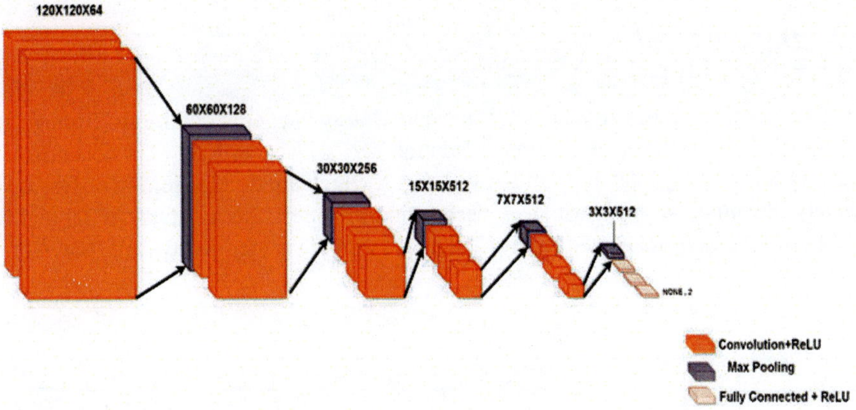

Fig. 1 Proposed PNA-DCN model

Table 2 Layers in PNA-DCN model

Layer number	Layer name	Layer size
1	Input layer	224 X 224 X 3
2	Convolution2D	224 X 224 X 64
3	Convolution 2D	224 X 224 X 64
4	Max pooling 2D	112 X 112 X 64
5	Convolution 2D	112 X 112 X 128
6	Convolution 2D	112 X 112 X 128
7	Max pooling 2D	56 X 56 X 128
8	Convolution 2D	56 X 56 X 256
9	Convolution 2D	56 X 56 X 256
10	Convolution 2D	56 X 56 X 256
11	Convolution 2D	56 X 56 X 256
12	Max pooling 2D	28 X 28 X 256
13	Convolution 2D	28 X 28 X 512
14	Convolution 2D	28 X 28 X 512
15	Convolution 2D	28 X 28 X 512
16	Convolution 2D	28 X 28 X 512
17	Max pooling 2D	14 X 14 X 512
18	Max pooling 2D	14 X 14 X 512
19	Convolution 2D	14 X 14 X 512
20	Convolution 2D	14 X 14 X 512
21	Convolution 2D	14 X 14 X 512
22	Max pooling 2D	7 X 7 X 512

3.2 Dataset Used

The Dataset consists of validated OCT and chest X-ray images [23]. The OCT images are split into training, testing, and validation set, which consists of independent patients. There are a total of 5856 X-ray images, which are divided into two classes, namely, Pneumonia and Normal. A description of the dataset is given in Table 3. The normal X-rays are shown in Fig. 2. The Pneumonia ridden X-rays are shown in Fig. 3.

Table 3 Description to dataset

Serial number	Dataset (Hold-out)	Healthy	Pneumonic
1	Training	1341	3875
2	Testing	234	390
3	Validation	8	8

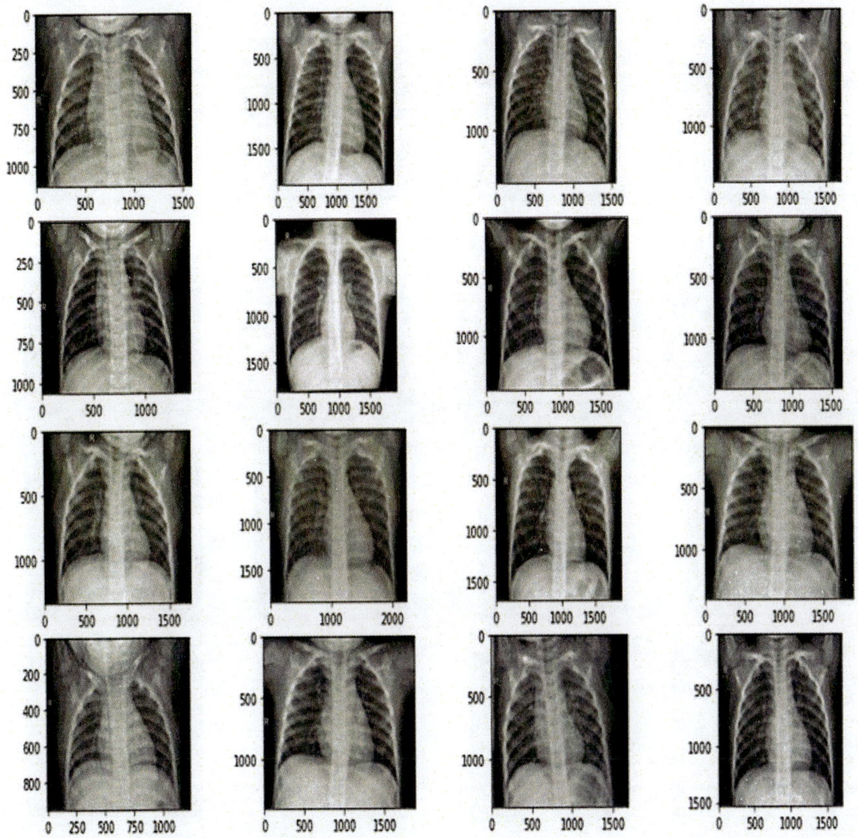

Fig. 2 X-ray images for a normal person

3.3 Performance Metrics

The criteria for performance evaluation of the proposed model (i.e., PNA-DCN) based on deep convolution network (DCN) model for Pneumonia (PNA) detection are accuracy, AUC, and ROC curve [21].

4 Results

This section reports the results of the work and discusses the inferences drawn from the analysis. The PNA-DCN models show the accuracy of 91.2%, and AUC of 84.06%. The ROC curve is plotted in Fig. 4. From the results, it is evident that the proposed model has significance to detect PNA.

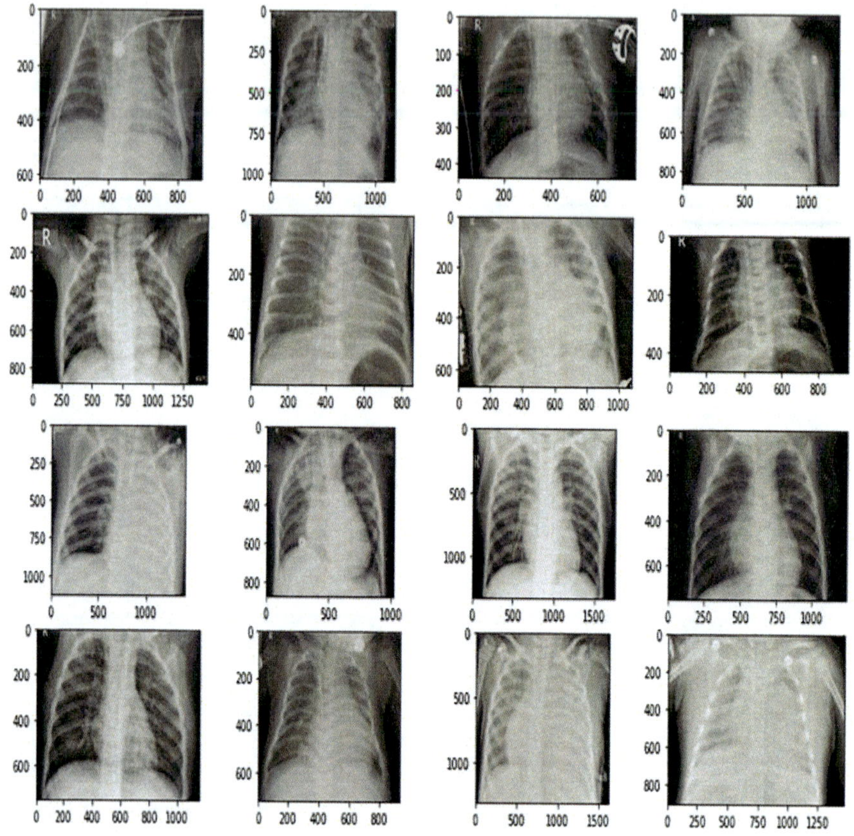

Fig. 3 X-ray images for Pneumonia ridden

Further, two studies from the literature are selected for carrying out the experimental comparison. The selected studies are Goyal and Singh [10] and Asnaoui et al. [14]. The comparison is drawn over the Accuracy measure and AUC measure. The results are shown in Table 4. The best value is highlighted in boldface. The ROC curves for Goyal and Singh [10] and Asnaoui et al. [14] are given in Figs. 5 and 6, respectively.

From the analysis, it is evident that the PNA-DCN has the highest accuracy among all the candidate models.

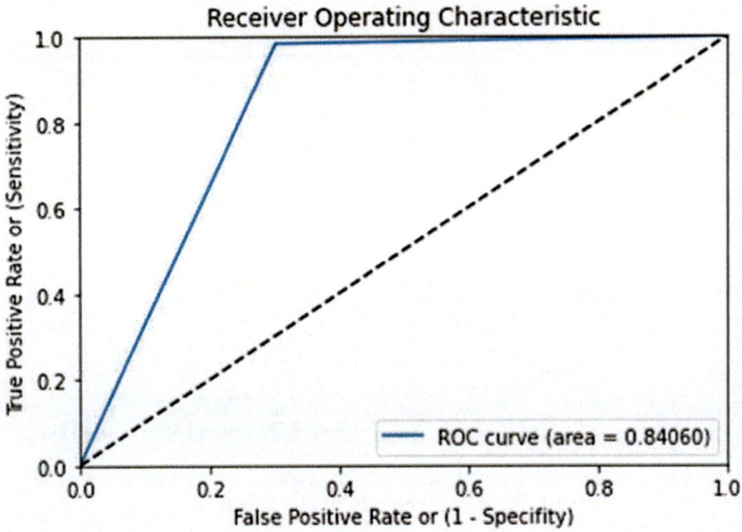

Fig. 4 ROC curve for the proposed PNA-DCN model

Table 4 PNA-DCN versus state-of-the-art models

Serial number	Study	Deep architecture	Accuracy (in %)	AUC (in %)
1	Goyal and Singh [10]	RNN	87.4	79.1%
2	Asnaoui et al. [14]	ResNet	88.2	83.1
3	Proposed PNA-DCN model	VGG19	**91.2**	**84.0**

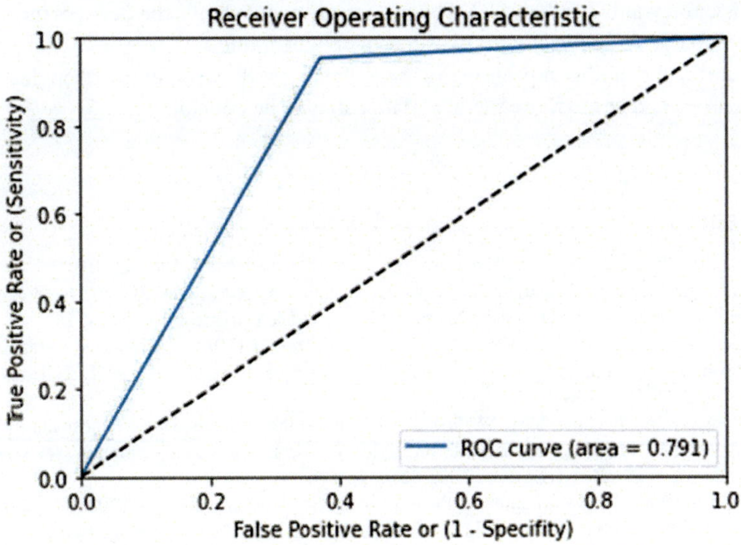

Fig. 5 ROC curve for Goyal and Singh [10]

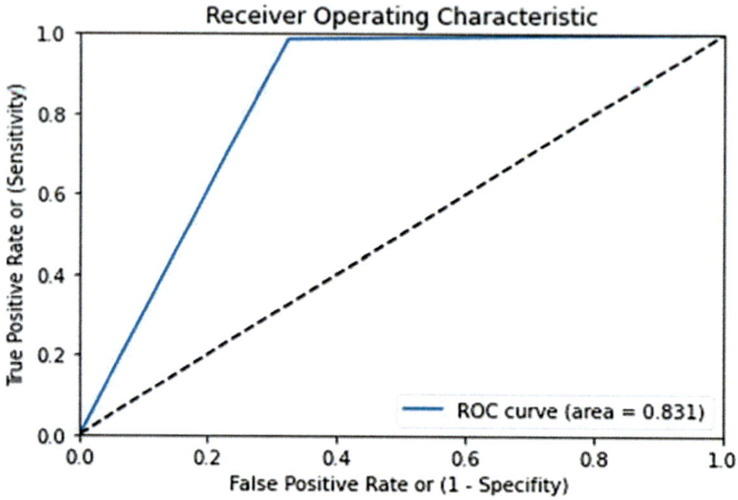

Fig. 6 ROC curve for Asnaoui et al. [14]

5 Conclusion and Future Scope of the Study

This paper contributes a novel approach to predict pneumonia in its early stages so that the proper care and medications can be done. In this way, hundreds of lives can be saved by early detection PNA disease specifically in young children and old people. The proposed model utilizes VGG19-based deep learning architecture and reports 91.2% accuracy and 84.0% AUC. The proposed model shows the best performance in the comparative analysis with the state-of-the-art models.

In future, the authors propose to extend the work with more deep learning architectures, and some more medical data can also be considered.

References

1. Jaiswal AK, Tiwari P, Kumar S, Gupta D, Khanna A, Rodrigues JJ (2019) Identifying pneumonia in chest X-rays: a deep learning approach. Measurement 145:511–518
2. Avni U, Greenspan H, Konen E, Sharon M, Goldberger J (2011) X-ray categorization and retrieval on the organ and pathology level, using patch-based visual words. IEEE Trans Med Imaging 30(3):733–746
3. Pattrapisetwong P, Chiracharit W (2016) Automatic lung segmentation in chest radiographs using shadow filter and multilevel thresholding. In: International computer science and engineering conference (ICSEC). IEEE 2016, pp 1–6
4. Dey N, Zhang YD, Rajinikanth V, Pugalenthi R, Sri Madhava Raja N (2021) Customized VGG19 architecture for Pneumonia detection in chest X-Rays. Pattern Recogn Lett 143:67–74, ISSN 0167-8655

5. Zech JR, Badgeley MA, Liu M, Costa AB, Titano JJ, Oermann EK (2018) Variable generalization performance of a deep learning model to detect pneumonia in chest radiographs: a cross-sectional study. PLoS Med 15(11):e1002683
6. Ge Y, Wang Q, Wang L, Wu H, Peng C, Wang J, Xu Y, Xiong G, Zhang Y, Yi Y (2019) Predicting post-stroke pneumonia using deep neural network approaches. Int J Med Inf 132:103986
7. Muhammad Y, Alshehri MD, Alenazy WM, Vinh Hoang T, Alturki R (2021) Identification of pneumonia disease applying an intelligent computational framework based on deep learning and machine learning techniques. Mobile Inf Syst
8. El Asnaoui K (2021) Design ensemble deep learning model for pneumonia disease classification. Int J Multimedia Inf Retrieval 10(1):55–68
9. Bermejo-Peláez D, San José Estépar R, Fernández-Velilla M, Palacios Miras C, Gallardo Madueño G, Benegas M, Gotera Rivera C, Cuerpo S, Luengo-Oroz M, Sellarés J, Sánchez M (2022) Deep learning-based lesion subtyping and prediction of clinical outcomes in COVID-19 pneumonia using chest CT. Sci Rep 12(1):1–11
10. Goyal S, Singh R (2021) Detection and classification of lung diseases for pneumonia and Covid-19 using machine and deep learning techniques. J Ambient Intell Human Comput:1–21
11. Hasan MD, Ahmed S, Abdullah ZM, Monirujjaman Khan M, Anand D, Singh A, AlZain M, Masud M (2021) Deep learning approaches for detecting pneumonia in COVID-19 patients by analyzing chest X-ray images. Math Probl Eng
12. Hashmi MF, Katiyar S, Keskar AG, Bokde ND, Geem ZW (2020) Efficient Pneumonia detection in chest xray images using deep transfer learning. Diagnostics 10(6):417
13. Hammoudi K, Benhabiles H, Melkemi M, Dornaika F, Arganda-Carreras I, Collard D, Scherpereel A (2021) Deep learning on chest X-ray images to detect and evaluate pneumonia cases at the era of COVID-19. J Med Syst 45(7):1–10
14. El Asnaoui K, Chawki Y, Idri A (2021) Automated methods for detection and classification pneumonia based on x-ray images using deep learning. In: Artificial intelligence and blockchain for future cybersecurity applications. Springer, Cham, pp 257–284
15. Goyal S (2021) Effective software defect prediction using support vector machines (SVMs). Int J Syst Assur Eng Manag. https://doi.org/10.1007/s13198-021-01326-1
16. Goyal S (2022) Comparative analysis of machine learning techniques for software effort estimation. intelligent computing techniques for smart energy systems. Lecture Notes in Electrical Engineering, vol 862. Springer, Singapore, pp 63–73. https://doi.org/10.1007/978-981-19-0252-9_7
17. Goyal S (2022) Effective software effort estimation using heterogenous stacked ensemble. In: IEEE International conference on signal processing, informatics, communication and energy systems (SPICES), pp 584–588. https://doi.org/10.1109/SPICES52834.2022.9774231
18. Goyal S (2022) 3PcGE: 3-parent child-based genetic evolution for software defect prediction. Innovations Syst Softw Eng. https://doi.org/10.1007/s11334-021-00427-1
19. Goyal S (2022) Genetic evolution-based feature selection for software defect prediction using SVMs. J Circuits Syst Comput:2250161
20. Goyal S (2022) Software fault prediction using evolving populations with mathematical diversification. Soft Comput. https://doi.org/10.1007/s00500-022-07445-6
21. Goyal S (2022) Static code metrics-based deep learning architecture for software fault prediction. Soft Comput. https://doi.org/10.1007/s00500-022-07365-5
22. Soni K, Kirtivasan A, Ranjan R, Goyal S (2022) Analysis of fifteen approaches to automated covid-19 detection using radiography images. In: Advanced machine intelligence and signal processing. Lecture Notes in Electrical Engineering, vol 858. Springer, Singapore. https://doi.org/10.1007/978-981-19-0840-8_2
23. Kermany D, Zhang K, Goldbaum M (2018) Labeled optical coherence tomography (oct) and chest x-ray images for classification. Mendeley Data 2(2)

Performance Analysis of Selection and Migration for Virtual Machines in Cloud Computing

Rashmi Sindhu, Vikas Siwach, and Harkesh Sehrawat

Abstract A cloud computing environment is composed of numerous heterogeneous resources and for efficient utilization of resources endorses virtualization. In virtualization, a cloud provider creates multiple instances of virtual machine which can be configured on a single host, hence increases the resource utilization. Though this can be further enhanced through the consolidation process, it requires selecting appropriate VM allocation and selection policy that work together to complete the process. To achieve the consolidation process with the physical data center is quite challenging therefore a virtual data center can be created with the help of simulators where by tasks can be simulated with varied configurations in different simulation environments. The simulation gives output corresponding to the simulation environment and is based on the performance metrics. In this work, a study is carried out on one such simulator named as CloudSim and analysis of its VM allocation along with selection policies. These policies will be implemented using PlanetLab workload to conclude about their performance based on few metrics.

Keywords SLA · VM · IaaS · PaaS · SaaS · ISA · API · ABI

1 Introduction

With the growing demand of data, need of quality resources is also increasing which can be achieved by usage of high power and high-end equipment which itself often becomes quite expensive and unattainable. In order to carry out these prerequisites, organisations and the users are actively deploying cloud computing framework which is affordable to users, offers promptness, flexibility as well as operative use of the resources with dilute cost. In cloud computing, a shared pool of resources

R. Sindhu (✉) · V. Siwach · H. Sehrawat
Maharishi Dayanand University, Rohtak, Haryana, India
e-mail: rashmisindhu88@gmail.com

V. Siwach
e-mail: vikas.siwach.uiet@mdurohtak.ac.in

© The Author(s), under exclusive license to Springer Nature Singapore Pte Ltd. 2023

A. Mishra et al. (eds.), *Advances in IoT and Security with Computational Intelligence*, Lecture Notes in Networks and Systems 755,
https://doi.org/10.1007/978-981-99-5085-0_13

is retrieved by the users depending upon their requirements. These resources can be designated as networks, services and storage space etc. which are managed by the service provider. According to National Institute of Standards and Technology (NIST), cloud computing has defined three basic service delivery models for which it has two different stakeholders namely cloud vendor or service provider and the other is client or user who is consuming the cloud services. Each service model defines its responsibility area for each stakeholder. Software as a Service (SaaS) is a complete operating environment with application management and user interface. Platform as a Service (PaaS) provides virtual machines (VMs), application services, operating system, development framework, transactions and control structures. Infrastructure as a service (IaaS) provides VMs, virtual storage, virtual infrastructure, and other hardware assets as resources that clients can provision.

1.1 Virtualization

Cloud computing framework consists of large data centres which enclose an extensive number of computing devices that consume plenteous electric energy primarily due to incompetent usage of cloud computing resources. In order to manage these resources efficiently, a technology called virtualization is embraced in the cloud computing environment.

Virtualization is a broad technology which can provide virtual environment at the operating system level, programming level as well as at the application level. Here, focus is towards the hardware virtualization where the abstract layer of virtualization provides virtual resources such as VMs, virtual storage, virtual memory and virtual network to the users. The hardware virtualization is enabled using a special system software known as Hypervisor or Virtual Machine Manager (VMM) as shown below in Fig. 1.

A traditional or non-virtualized host is a tightly coupled system in which carrying out a required modification is a challenging task due to its tightly coupled environment. At the bottom is the hardware; over the above is the operating system (OS)

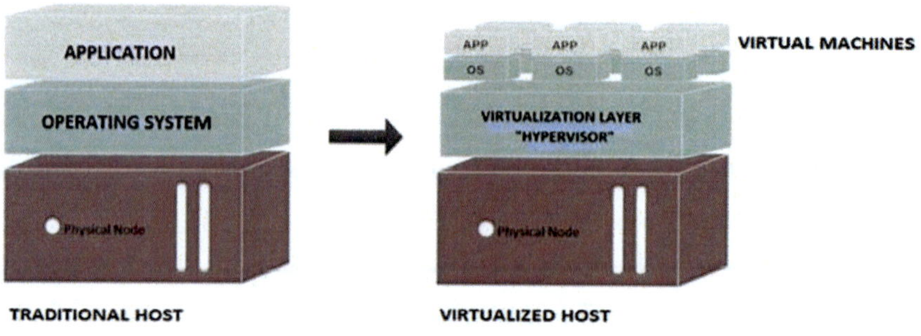

Fig. 1 Traditional host versus virtualized host

which is installed on that particular hardware. OS communicates with the hardware using instruction set architecture (ISA). The framework contains certain set of libraries above the OS which support certain kind of deployment environment. Libraries communicate with OS using arbitrary binary interfaces (ABI).

At the top applications are installed which interact with the libraries using application programming interface (API). As the traditional system is a tightly coupled system, the virtualized system overcomes this system up to an extent. In the virtualized systems, an extra abstraction layer is added between the hardware and the OS in order to share a single hardware across multiple users based on the requirement and also for the better utilization of the resources. At the bottom of the system is the hardware layer and above the system is the abstraction layer i.e. hypervisor or the VMM layer. This hypervisor allows creating VMs which allow the user to create a wrapper environment for the OS installation. In the virtualized system, there are two ISA's between the VMs and the hardware. VMs are sending ISA instructions to the hypervisor which in turn is further communicating to the hardware. This system is called loosely coupled because of its flexibility of consolidation of VMs among different hosts.

1.2 CloudSim

There exist several simulators that provide the same framework as that of a real data centre with minimum cost on which repetitive tests that can be performed with the flexibility of changing the core execution environment in real time. One such simulation which is being widely used by the cloud researchers is Cloudsim. Cloudsim is a simulation framework comprising an asset of class libraries written in java that support both modelling and simulation of cloud computing based systems as well as application provisioning environment. Cloudsim also supports the implementation of energy efficient policies with support for aggregation and isolation of virtual resources. When the cloud researchers are working on power aware algorithms of resource scheduling or task scheduling or service level agreement (SLA) algorithms then CloudSim fulfils all the parameters which are required by their implementation. Cloudsim consists of a data centre, data centre broker, cloud information service (CIS) which are termed as entities in the simulation environment.

CIS contains the information about the resources available on the cloud. **DATA CENTER** need to be registered through CIS and each data center has some characteristics and these characteristics are associated with the host. Each host has some hardware configuration and this can be a number of processing elements, RAM and bandwidth. In CloudSim, it is supposed to have a **BROKER** which will submit task to the data centre. This broker entity is created by the CloudSim which at the initial state talks to the CIS and retrieves the resource information which is registered with the CIS.

The whole framework works on different policies such as VM allocation policy which are used by the data centre, VM scheduler policies which are used by the host and the cloudlet scheduler policies which are used for the processing of cloudlets on VMs. All these policies are either time shared or space shared, if time shared then a specific time will be allotted to each VM and if space shared then all VMs share the space.

1.3 VM Allocation and Selection

This paper emphasizes on the power aware VM allocation and selection policies of CloudSim embedded in the power package namely org. cloudbus. CloudSim. power. This power package contains two set of classes which are relevant to the VM allocation and selection policies. These set of classes support the real time VM consolidation or migration based on real time circumstances. Power aware VM allocation classes are responsible for the identification of overutilized or underutilized host which is selected on the criterion of threshold. In power VM selection policy, a VM is selected from the overutilized host so that it can be migrated to an underutilized host and it may also be selected from an underutilized host to make the host as unoccupied host or can be turned off to reduce power consumption.

Cloudsim allows application of host overload detection mechanism for VM allocation and for this analysis an auto adjustable upper and lower threshold value is maintained which depends on the previous data collected from the lifetime of VMs. These mechanisms are categorized as following:

(a) Interquartile Range (IQR): It uses an adaptive utilization threshold IQR = $Q_3 - Q_1$ where Q_1 and Q_3 are the lower and upper middle values of the data.
(b) Middle Absolute Deviation (MAD): MAD is headway over the standard deviation as it is less affected by the outliers and thus, the change in the magnitude of outlier's distance is not irrelevant [1].
(c) Local Regression (LR): A curve is built in LR where approximation of original data is done by setting up the sample data models to localized subset of data [1].
(d) Local Regression Robust (LRR): This mechanism is the advancement over the LR and is transformed by adding bi square which is a robust estimation method because of the vulnerability of the LR to outliers.

After identifying all the underutilized or overutilized hosts, selection of VMs is done among other VMs for migration. Cloudsim provides four techniques for selecting a VM for migration, namely:

(a) Minimum Migration Time (MMT): The VM having minimum migration time of transferring from one host to the other host will be selected for the migration.
(b) Maximum Correlation (MC): Those VMs which have the highest correlation of CPU utilization with other VMs will be selected for the migration [1].
(c) Random Choice (RC): A random variable X which is a discrete random variable and is uniformly distributed i.e. $X =^{d} U(0,|V_j|)$ whose value index is a set of VMs V_j allocated to a host j [2].

(d) Minimum Utilization (MU): In this technique the selection of VM is based upon the dependency of VM utilization on the CPU utilization and the VM with minimum dependency is selected to migrate.

1.4 Workload

In this paper, simulation of a heterogeneous power data centre is carried that applies varied combinations of VM allocation and VM selection policies. These policies will be applied on the data of PlanetLab's monitoring infrastructure CoMon project. This data had been collected from the CPU utilization of thousands of VMs imposed on servers which were located at 500 different places of the world. The interval of utilization measurements is 5 min [2]. From March 2011 to April, in one month period workload traces were collected for 10 random days. These workloads are being used by the CloudSim and are stored in workload PlanetLab package.

The PlanetLab data runs on a particular host which consumes power. To determine the exact power utilization of the host CloudSim consider the Standard performance Evaluation Corporation (SPEC) benchmark is considered which publishes the power consumption and information considering the servers. Cloudsim has implemented two sets of power consumption models namely:

(a) Mathematical or Generic mathematical model for power consumption.
(b) Real life server based models which are from two different companies i.e. HP and IBM.

2 Related Work

A vast number of papers were published describing several different VM placement algorithms [3]. Most of the computational approaches utilize heuristic algorithms; with the effectiveness of these algorithms, a few researchers had applied theoretical analysis and a few had applied empirical evaluation [4]. Few of the researchers evaluated the VM placement algorithms through real time systems [5]. As maximum researchers do not have access to the large data centre, therefore, it is not possible to test the fidelity of algorithm there. Hence, to check the scalability of their algorithms, most of the researchers used the simulators. Few of them had built their own simulators which is a quite challenging task [6]. Therefore, to check the feasibility of the VM placement and migration algorithms, majority of the researchers used the CloudSim simulator. In [2] focus was on the technique of resource management that minimizes operating costs and provides quality-of-service constraint. They also concluded that consolidating VMs based on their resource utilization results in the less power consumption. In [7] scheduling algorithms for the allocation of VMs was proposed to minimize the consumption of power during task execution in the cloud data centre environment. Researchers in [8] designed an algorithm which reduces the allocation load on servers needs to achieve less power consumption.

3 Result and Analysis

In this paper, comparison of the VM allocation and selection policies is done using the PlanetLab workload 20,110,303, 2 host type i.e. HP ProLiant ML110G4 and HP ProLiant ML110G5. The comparison is based on few performance metrics such as energy consumption, number of VM migrated, SLA (Service Level agreement), SLA per active host, overall SLA violation, average SLA violation, and the number of hosts shutdown (power off or suspend the VM). Based on these metrics best allocation and selection policy combination is selected which gives better throughput. Figures 2, 3, 4, 5, 6 and 7 illustrate the comparison between the policies based on above stated metrics.

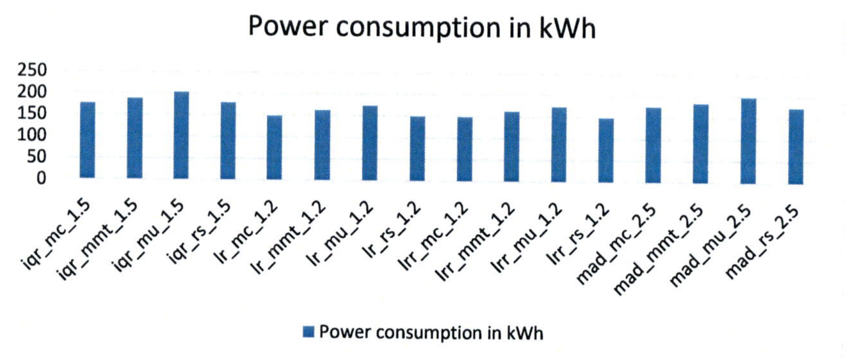

Fig. 2 Power consumption by different policies

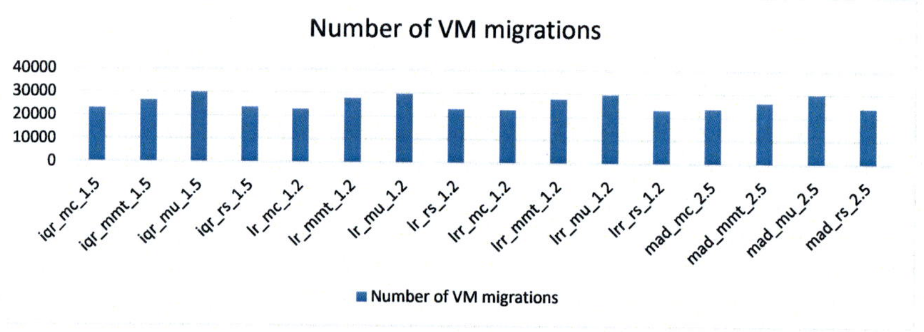

Fig. 3 Number of VMs migrated by different policies

Fig. 4 SLA violation (%) during the consolidation process

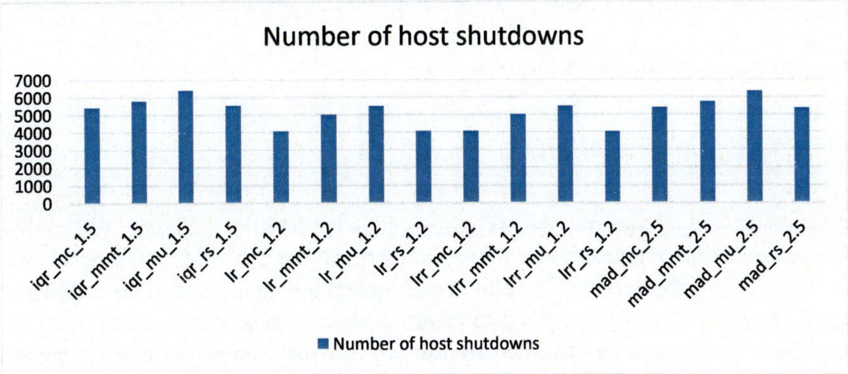

Fig. 5 Total number of host shutdown in process

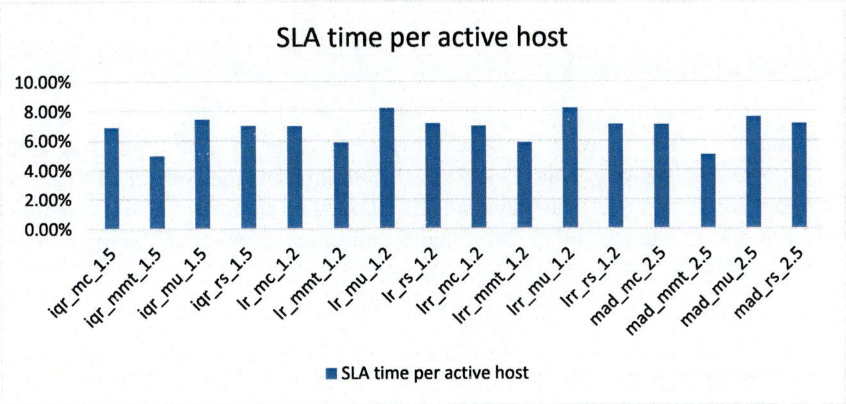

Fig. 6 SLA violation time per active host (%)

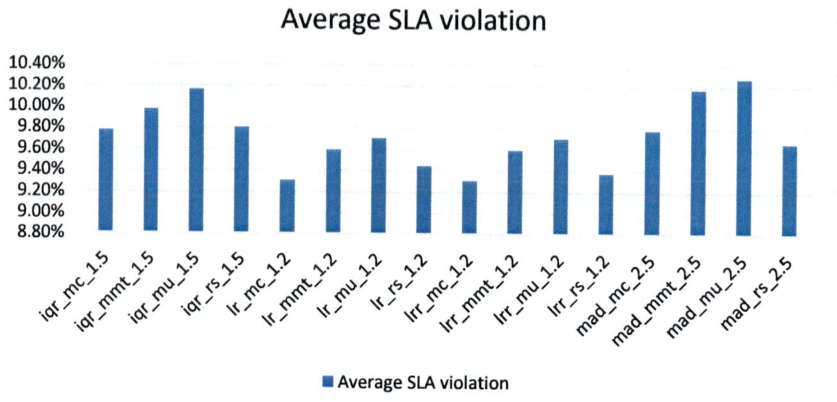

Fig. 7 Average SLA violation (%) in the process

3.1 *Power Consumption*

Power consumed by the data centre is measured in kWh (Kilowatt hour). VM placement and selection algorithms are merged to perform the consolidation process and the merged algorithms in the CloudSim environment are mentioned in the following figures. Power consumption by a data centre is always considered to be minimum. Therefore, this criterion is fulfilled by iqr_mu_1.5 that shows the highest power consumption in kWh whereas lr_mc_1.2 and lrr_mc_1.2 show the minimum power consumption as shown in Fig. 2.

3.2 *VM Migrations*

Migration of VMs is a costly process, thus, the number of VMs to be migrated should not be too high. Focus is always towards minimizing the number of VMs for migration. Algorithm mad_mu_2.5 results in maximum number of VM migrations whereas lr_mc_1.2 and lrr_mc_1.2 result in the minimum number of VM migrations as shown in Fig. 3.

3.3 *SLA Violation*

The SLA violation factor consists of all elements and service metrics like power, temperature, network availability, bandwidth etc. The overall violation should be minimum which is satisfied by iqr_mmt_1.5 algorithm, whereas mad_rs_2.5 shows the maximum SLA violation as shown in the Fig. 4.

3.4 Number of Host Shutdown

This metrics is concerned with the number of host machines that are turned off in the migration process in order to minimize the power consumption. The focus should be towards reducing the number of active machines in the data centre but this should not interfere the overall functioning of the data centre. Algorithm iqr_mu_1.5 shuts down the maximum number of host in the process whereas lrr_rs_1.2 shuts down the maximum number of host in the process as shown in Fig. 5.

3.5 SLA Time Per Active Host

SLA violation factor is also considered for a particular active host in the data centre. An active host should satisfy the SLA factor that is fulfilled by the iqr_mmt_1.5 which shows the minimum SLA violation per active host whereas lr_mu_1.2 and lrr_mu_1.2 show the maximum SLA violation per active host as shown in Fig. 6.

3.6 Average SLA Violation

The overall SLA violation of the system should be minimum. Algorithm mad_mu_1.2 shows the maximum average violation during the entire consolidation process whereas lr_mc_1.2 shows the maximum average violation during the entire consolidation process.

4 Conclusion and Future Scope

In the summary it is concluded that for power consumption, number of VM migrations, number of host shutdown and for average SLA violation **lr_mc_1.2, lr_rs_1.2** and the robust version of lr i.e.**lrr_mc_1.2, lrr_rs_1.2** contrived beneficial results, but for SLA violation and SLA violation per active host **iqr_mmt_1.5, mad_mmt_2.5, lr_mmt_1.2** and **lrr_mmt_1.2** made better results. In future, this conclusion may be followed by associating some task scheduling algorithm with the cloudlets and research can be done towards merging the best allocation and selection algorithm.

References

1. Chowdhury MR, Mahmud MR, Rahman RM (2015) Implementation and performance analysis of various VM placement strategies in CloudSim. J Cloud Comp 4:20
2. Beloglazov A, Buyya R (2011) Optimal online deterministic algorithms and adaptive heuristics for energy and performance efficient dynamic consolidation of virtual machines in cloud data centers. Concurrency Comput Pract Exp 24(13):1397–1420
3. Mann ZÁ (2015) Allocation of virtual machines in cloud data centers—a survey of problem models and optimization algorithms. ACM Comput Surv. https://doi.org/10.1145/2797211
4. Mann ZÁ (2015) Rigorous results on the effectiveness of some heuristics for the consolidation of virtual machines in a cloud data center. Future Gener Comput Syst. https://doi.org/10.1016/j.future.2015.04.004
5. Feller E, Rilling L, Morin C (2012) Snooze: a scalable and autonomic virtual machine management framework for private clouds. In: Proceedings of the 2012 12th IEEE/ACM international symposium on cluster, cloud and grid computing
6. Guazzone M, Anglano C, Canonico M (2012) Exploiting VM migration for the automated power and performance management of green cloud computing systems. In: 1st International workshop on energy efficient data centers
7. Beloglazov A, Buyya R (2010) Energy efficient allocation of virtual machines in cloud data centers. In: 10th IEEE/ACM International conference on cluster, cloud and grid computing (CCGrid)
8. Binder W, Suri N (2009) Green computing energy consumption optimized service hosting. In: Nielsen M, Kučera A, Miltersen PB, Palamidessi C, Tůma P, Valencia F (eds) SOFSEM 2009. LNCS. Springer, Heidelberg

A Multi-objective Path–Planning Based on Firefly Algorithm for Mobile Robots

Prachi Bhanaria, Maneesha, and Praveen Kant Pandey

Abstract Currently there are lots of Swarm Intelligence algorithms available that can be used in optimization problems. The modified Multi-objective Firefly algorithm proposed in the present work is used to find an optimal path for mobile robots in a static environment. The algorithm is based on the flashing behavior of the fireflies in nature, hence called the Firefly Algorithm (FA). The proposed approach takes into account three objectives to obtain efficient and optimal solutions and is used to plan a path for mobile robots in a static environment. The three objectives are as follows: path length, path safety and path smoothness. The results were obtained after optimizing for all three parameter values over several Iterations (generations) for different population sizes and different sets of obstacles. The performance of the proposed multi-objective Firefly algorithm was compared to the results obtained using the classical NSGA-II approach.

Keywords Multi-objective path planning · Path length · Path safety · Path smoothness

1 Introduction

The path-planning algorithm is the heart of every robot which allows it to navigate through a complex environment. The main objective of path planning algorithms is to calculate an optimum path for the mobile robot from the start to the destination without collisions. For decades, many researchers have been working to develop

P. Bhanaria (✉)
Department of Electronic Science, University of Delhi, South Campus, Delhi, India
e-mail: prachibhanariya30@gmail.com

Maneesha · P. K. Pandey
Maharaja Agrasen College, University of Delhi, Delhi, India
e-mail: maneesha@mac.du.ac.in

P. K. Pandey
e-mail: pkpandey.du@gmail.com

© The Author(s), under exclusive license to Springer Nature Singapore Pte Ltd. 2023

A. Mishra et al. (eds.), *Advances in IoT and Security with Computational Intelligence*, Lecture Notes in Networks and Systems 755,
https://doi.org/10.1007/978-981-99-5085-0_14

more powerful algorithms to achieve the best possible solution [1]. There are many traditional path planning algorithms which include, Artificial Potential field, Simulated Annealing, Fuzzy logic and Recursive methods. Meanwhile, more advanced algorithms like Swarm Intelligence algorithms and Genetic Algorithms were developed. The need for path planning is a result of several factors, including the need for optimization and exploration capabilities [2]. An artificial intelligence problem may require robots to navigate every corner of an environment to ensure that all areas are checked for targets or obstacles [3, 4]. A design optimization technique for a specific task like Engineering, Robot Path planning, and Industrial automation is the most researched topic for mobile robots [5, 6].

Some of the approaches only focus on a single objective, normally minimizing the path length, avoiding collision (or obstacle detection and avoidance), or path smoothness. The path length objective is to find the path as short as possible. The Path safety concept is to provide a safe path for mobile robot and path smoothness provide the path with a smaller number of turns. The elitist non-dominated sorting genetic algorithm (NSGA II) [7] was proposed to minimize two objectives, namely, path length and path vulnerability.

Hiba et al. [8] developed an offline hybrid algorithm in which Tabu Search is first applied followed by the Firefly Algorithm to get the optimal path between two given points. Peng et al. [9] provide the optimal solutions using Path length and Path smoothness are considered two important objectives for optimization. Patle et al. [1] developed a FA based controller which gives the optimal path with effective obstacle avoidance in a short time at a low computational cost. Hidalgo et al. [10] fulfill the task of navigation in the presence of static obstacles. Liu et al. [11] provided application of FA for underwater robot navigation in the presence of static obstacles. The proficiency in self-planning, self-organization and self-adaption of the firefly algorithm is used by many researchers for various optimization problems such as program optimization, constrained optimization problem, space and time optimization, cooperative networking problem, combinatorial optimization problem and many more [8, 12].

Multi-objective Path planning is a method for tracking down a practical course for a portable robot between two points: Source and Goal. As opposed to a solitary objective improvement, multi-objective path planning is troublesome and complex. In the Multi-objective Firefly algorithm (MOFA), the fascination of one firefly towards the other firefly because of the variety of their splendor is the critical idea of the proposed study [13]. The calculation intends to accomplish three sorts of streamlining goals for the desired path: Path length, Path smoothness, and Path safety by presenting a multi-objective Firefly calculation [14]. The results show that the proposed MOFA is a better choice to tackle the path planning issue. It has been seen that the path generated by the FA is short and gives the most secure path [15, 16].

The paper is organized as follows; Sect. 2 discusses the Multi-Objective Path-Planning problem. In Sect. 3, the methodology used to test MOFAs, that is, the parameters configuration and the metrics used are presented. The simulation results and comparisons have been shown in Sect. 4 and finally, the conclusion is presented in Sect. 5.

2 Multi-objective Path-Planning

MOPP is the way of finding a feasible path for a mobile robot that requires traveling from a starting point A to a goal point B. The most common objectives in this path-planning method are: reaching the target through the shortest path, avoiding obstacles, limiting energy consumption, optimizing the difficulty of the path, etc. [17]. There are various optimization techniques that are available. In [18], the authors used A* algorithm to minimize difficulty, danger, elevation, and length of the path from a start to a goal point. Duan et al. [19] used particle swarm optimization (PSO) to find smooth and shorter paths, with parameter tuning, using a 2D map. The first and most important parameter for multi-objective optimization is to find the shortest path. The second one is to find and avoid obstacles in the path. Mobile robot energy consumption is an important issue directly related to smooth path planning [20–22].

Path planning (PP) is an NP-hard optimization problem, for this reason, it can be tackled by using MOFA. This paper presents an efficient multi-objective version of the firefly algorithm (FA), a swarm intelligence algorithm based on the flashing behavior of fireflies in nature [23]. In the present work, the PP problem solved in a multi-objective manner, using an efficient Multi-objective Firefly algorithm MOFA with the three essential objectives, that is, the path safety, the path length, and the path smoothness.

Firefly Algorithm

The standard Firefly algorithm was developed by Xin-she Yang for the continuous optimization problem [16]. FA was based on the flashing behaviour of the firefly. The brightness of fireflies is described by the objective function [24]. Firefly algorithm uses the following three rules:

1. Fireflies are unisex, so attracted to each other regardless of their sex.
2. The attraction of the firefly is proportional to its brightness, so they decrease with distance.
3. The brightness of the firefly is calculated using the objective function.

Let, β_0 = Attractiveness of firefly at r = 0, Coefficient of light absorption = γ, x_i = Position of first firefly, x_j = Position of second firefly, i = first firefly, j = second firefly, x_i^k = current value of i firefly at kth dimension, r_{ij} = distance between xi and xj, d = dimension, t = iteration, α = randomization parameter.

The attractiveness of the firefly is defined by Eq. 1.

$$\beta= \beta_0 e^{-\gamma r^2} \tag{1}$$

Thus, it will vary with the distance r_{ij} between firefly i and firefly j. For any given two fireflies x_i and x_j, the movement of the firefly i is attracted to another firefly j as determined by using Eq. 2.

$$x_i^{t+1} = x_i^t + \beta_0 e^{-\gamma r_{ij}^2}\left(x_j - x_i\right) + \alpha_t * \delta_t * (rand - 0.5)(UL - LL) \tag{2}$$

3 Multi-objective Firefly Algorithm

To extend the firefly algorithm for a single objective to multi-objective optimization, the following three objectives are combined in the single objective function without compromising each other (Figs. 1 and 2).

1. Path Length.
2. Path Safety.
3. Path Smoothness.

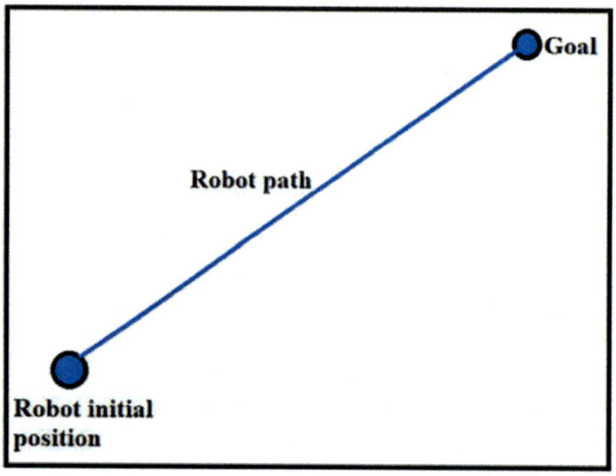

Fig.1 Navigation of robot in free environment

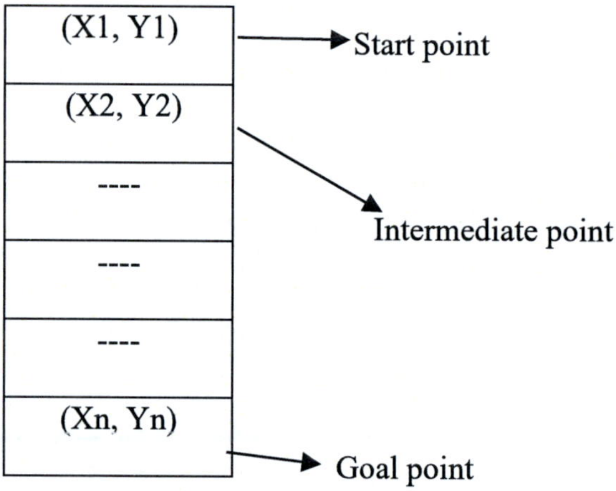

Fig. 2 Example of path encoding

3.1 Path Length

The aim is to find the paths as short as possible. The length is based on the Euclidean distance between the points (coordinates). The Euclidean distance between two points P1(x1, y1) and P2(x2, y2) is calculated using Eq. 3.

$$d(P1, P2) = \sqrt{(x1 - x2)^2 + (y1 - y2)^2} \tag{3}$$

The path is encoded as the list of n coordinates. The number of intermediate coordinates in the list is variable. To reduce the Path length the operator deletes a path coordinate when it finds the goal in the middle of the segment.

3.2 Path Smoothness

The path smoothness operator tries to obtain a path with minimum number of turns and as straight as possible. The algorithm calculates the number of turns in different paths and tries to select the path with smaller number of turns (Fig. 3).

3.3 Path-Safety

The Path-safety function aims to minimize the objects (obstacles) existing in the path and reduce the collisions with the obstacles.

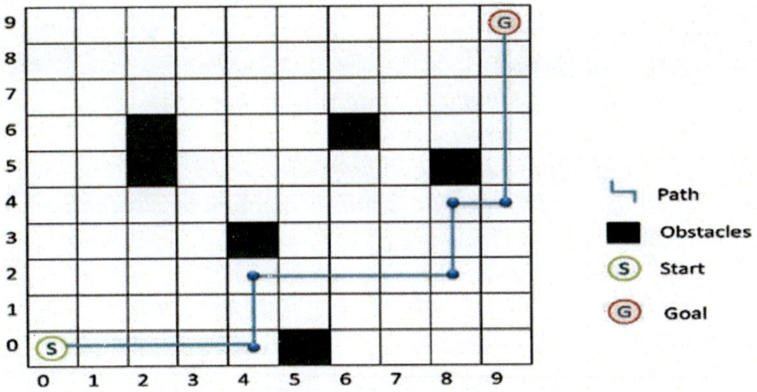

Fig. 3 Path smoothness (number of turns Calculation)

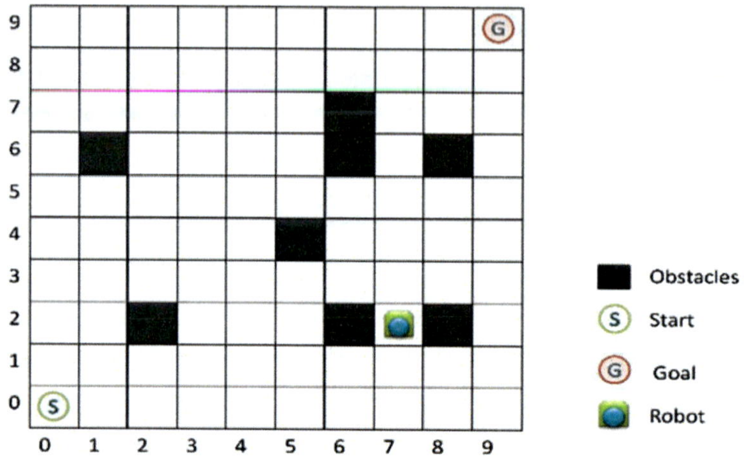

Fig. 4 Path safety (penalty calculation)

To avoid the objects, the operator determines the calculated path and if the obstacle is in between the two coordinate routes, it eliminates the complete path. To accomplish this goal, the operator takes as input a segment of the path and then tries to reduce the possible collisions presented in it. Fig. 4 shows that when the robot passes between two obstacles it will count as a penalty for the robot.

Pseudocode MO-FA

```
Initialize n fireflies to random positions
Calculate the value of objective function (intensity) for each firefly
for i in range (n-1):
    for j in range (n-1):
        if intensity of firefly (i) < intensity of firefly (j)
            compute attractiveness
            move firefly (i) toward firefly (j)
            update firefly (i) if it results in higher intensity for firefly (i)
        else
            Firefly (i) makes a random move
    end for
end for
Sort fireflies according to objective function
Calculate path length, path safety and path smoothness
Return the best firefly with minimum value of objective function
```

Multi-objective Firefly algorithm is described in the pseudocode given above. All the fireflies are first initialized with random positions using a random number generator. The algorithm then calculates the value of the objective function for each firefly. The value of the objective function is analogous to the brightness (intensity) of the firefly. The lower the value of the objective function, the higher the brightness of the firefly. The intensity of the ith firefly is compared with the rest of the fireflies. If the intensity of the ith firefly is less than that of the jth firefly, then the ith firefly moves towards the jth firefly, otherwise, the ith firefly makes a random move. The ith firefly is updated only if the intensity of the resultant firefly (i), after moving towards firefly (j), is higher that its original value. Once the loop is finished, the final fireflies are sorted according to their brightness (intensity) and the firefly with the highest intensity is returned as the solution.

4 Simulation Results

Firefly Algorithm is very popular among all metaheuristic techniques. The reason for its popularity is that it can deal with highly non-linear multi-model optimization problems. The convergence speed of the Firefly algorithm is very high compared to other techniques. FA can outperform other conventional algorithms based on statistical performance. The Firefly algorithm works based on global communications among the fireflies. Hence it can find global and local optimal simultaneously. There are several optimization areas where FA is used such as solving traveling salesman problems, Antenna design, Scheduling, Dynamic problems, etc. [16]. Based on the theoretical analysis and the problem-solving ability of firefly algorithms, we can summarize that MOFA is suitable for a path-planning optimization problem.

The MOFA algorithm was implemented using Python programing language. It was tested on a grid of size 50×50 with obstacles of different shapes and sizes placed randomly in the grid. The MOFA parameters (Table 1) were optimized for the current simulation environment. The starting position for the robot was (0,0) and the goal was (49,49). The results obtained after 20, 50, 80 and 100 iterations are shown in Fig. 5. Metaheuristic algorithms are non-deterministic algorithms and provide different simulation results after each iteration. It can be seen that the Simulation results for each objective i.e. path length, path safety (ascertain the punishment when passes close to the obstacles), and path smoothness (number of turns) improves themselves at the cost of the other two objectives. As the number of iterations increases, a mutually optimal solution is reached.

The results of the current work are compared with the results obtained by Duan [9] in which the NSGA-II algorithm is used to find the best solution as shown in Table 3. In this study, two reference points were used. The first one is the ideal reference point (the best values of the problem). On the other hand, nadir reference value is formed by the length of the theoretical path of the simulation. The path length obtained in Peng Duan's work was at least 29% more than the length of the ideal theoretical path of the simulation. In the present work, 4 different scenarios were implemented

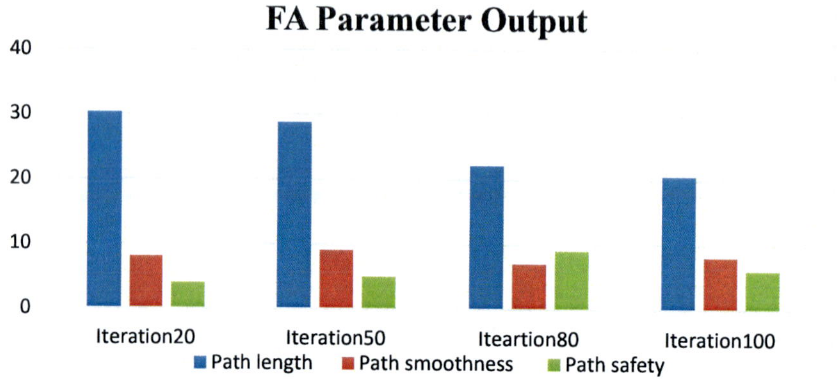

Fig. 5 Multi-objective firefly algorithm parameters

Table1 Parameters of FA

S.N	Parameters	Values in range
1	No. of fireflies	5–20
2	No. of generation(t)	20–50
3	Randomization parameter(α)	0.1–1
4	Light absorption coefficient (γ)	0.1–1
5	Attractiveness (β)	0.1–1
6	Random function [rand()]	0–1
7	Fitting parameter (δ)	0–1

with different start and goal positions along with the different number of obstacles in each scenario. The results obtained are shown in Table 2. The average increase in the path length for the present MOFA solution was 6.94%. Hence, the present model generates better results. From Tables 2 and 3, we could conclude that the modified firefly algorithm achieves a better performance, than the NSGA-II algorithm.

Table 2 Path length variation in the proposed MOFA algorithm

Path	Ideal path length	MOFA results	% increase
P1'	318.16	334.88	4.99
P2'	213.05	228.67	6.83
P3'	305.05	334.31	8.75
P4'	228.42	213.05	7.21

Table 3 Path length variation in the NSGA-II algorithm

Ideal value	NSGA-II Nadir reference	% increase
1079.0	1526.5	29.28

5 Conclusion

In the present work, a multi-objective Firefly algorithm is presented. Path length, path safety and Path smoothness were incorporated in the objective function to optimize the solution for all three parameters simultaneously. The simulation results are presented for 20, 50, 80 and 100 iterations. It can be verified from the results that if the value of any one parameter is optimized then it leads to a less optimal solution for the other two parameters. Hence, as the number of iterations increases, there is a trade-off between the three parameters and the best solution is achieved when all three parameters are collectively optimized.

The results were compared with those obtained with the NSGA-II algorithm also. The increase in path length obtained using the present MOFA when compared with the ideal theoretical path in the simulation was found to be 6.94%, whereas the increase in path length for the NSGA-II algorithm was more than 29%.

Hence, it may be concluded that the present MOFA method provides the best results with all the three parameters i.e. path length, path smoothness and path safety at their optimum values.

References

1. Patle BK, Pandey A, Jagadeesh A, Parhi DR (2018) Path planning in uncertain environment by using firefly algorithm. Defence Technol
2. Zhou J, Chen P, Liu H, Gu J, Zhang H, Chen H, Zhou H (2019) Improved path planning for mobile robot based on firefly algorithm. In: Proceeding of the IEEE international conference on robotics and biomimetics dali, China
3. Brand M, Yu X (2013) Autonomous robot path optimization using firefly algorithm. In: Proceedings of the 2013 international conference on machine learning and cybernetics. Tianjin. https://doi.org/10.1109/ICMLC.2013.6890747
4. Brand M, Yu XH (2013) Autonomous robot path optimization using firefly algorithm. In: Proceedings of the 2013 International conference on machine learning and cybernetics, Tianjin
5. Castillo O, Trujillo L (2005) Multiple objective optimization genetic algorithms for path planning in autonomous mobile robots. Int J Comput Syst Signals 6(1):48–63
6. Davoodi M, Panahi F, Mohades A, Hashemi SN (2013) Multi-objective path planning in discrete space. Appl Soft Comput 13(1):709–720
7. Deb K, Pratap A, Agarwal S, Meyarivan T (2002) A fast and elitist multiobjective genetic algorithm: NSGA-II. IEEE Trans Evol Comput 6(2)
8. Fister I, Yang XS, Brest J (2013) A comprehensive review of firefly algorithms. Swarm Evol Comput 13:34–46
9. Duan P, Li J, Sang H, Han Y, Sun Q (2018) A developed firefly algorithm for multi-objective path planning optimization problem. In: Proceedings of 2018 IEEE 8th annual international conference on cyber technology in automation, control, and intelligent systems

10. Hidalgo-Paniagua A, Vega-Rodríguez MA, Ferruz J, Pavón N (2015) Solving the multi-objective path planning problem in mobile robotics with a firefly-based approach. Methodol Appl
11. Liu C, Zhao Y, Gao F, Liu L (2015) Three-dimensional path planning method for autonomous underwater vehicle based on modified firefly algorithm. Math Probl Eng. https://doi.org/10.1155/2015/561394
12. Hliwa H, Daoud M, Abdulrahman N, Atieh B (2018) Optimal path planning of mobile robot using hybrid tabu search- firefly algorithm. Int J Comput Sci Trends Technol (IJCST) 6(6)
13. Tighzert L, Fonlupt C, Mendil B (2018) A set of new compact firefly algorithms. Swarm Evol Comput Base Data. https://doi.org/10.1016/j.swevo.2017.12.006
14. Zhang TW, Xu GH, Zhan XS, Han T (2021) A new hybrid algorithm for path planning of mobile robot. J Supercomput
15. Yang XS (2010) Firefly algorithm, stochastic test functions and design optimisation. Int J Bio-Inspired Comput 2:78–84
16. Yang XS (2014) Nature-inspired optimization algorithms, 1st edn. Elsevier
17. Sadhu AK, Konar A, Bhattacharjee T, Das S (2018) Synergism of firefly algorithm and Q-learning for robot arm path planning. Swarm Evol Comput. https://doi.org/10.1016/j.swevo.2018.03.014
18. Jeddisaravi K, Alitappeh RJ, Guimarães FG (2016) Multi-objective mobile robot path planning based on A* search. In: Proceedings of the 2016 6th International conference on computer and knowledge engineering (ICCKE), Iran, pp 7–12
19. Duan YQ, Zhang Y, Zhang B, Wang Y (2020) Path planning based on improved multi-objective particle swarm algorithm. In: IEEE 5th Information technology and mechatronics engineering conference
20. Kim H, Kim J, Ji Y, Park J (2013) Path planning of swarm mobile robots using firefly algorithm. J Instit Contr Robot Syst 19(5):435e41. https://doi.org/10.5302/J.ICROS.2013.13.9008
21. Masehian E, Sedighizadeh D (2010) Multi-objective PSO-and NPSO-based algorithms for robotpath planning. Adv Electr Comput Eng 10:69–76
22. Patle BK, Parhi DRK, Jagadeesh A, Kashyap SK (2017) On firefly algorithm: optimization and application in mobile robot navigation. World J Econ Eng 14(1):65e76
23. Lucas C, Hernandez-Sosa D, Caldeira R (2018) Multi-objective four-dimensional glider path planning using NSGA-II. In: Proceedings of the 2018 IEEE/OES autonomous underwater vehicle workshop (AUV), pp 1–5, Porto, Portugal
24. Zhou L, Ding L, Qiang X et al (2015) An improved discrete firefly algorithm for the traveling salesman problem. J Comput Theor Nanosci 12(7):1184–1189

Detection of Alzheimer Disease Using MRI Images and Deep Networks—A Review

Narotam Singh, D. Patteshwari, Neha Soni, and Amita Kapoor

Abstract Alzheimer's disease (AD) is the most common cause of dementia worldwide; it is a progressive degenerative neurological disorder; due to it, the brain cells die slowly. Early detection of the disease is crucial for deploying interventions and slowing its progression. In the past decade, many machine learning and deep learning algorithms have been explored to build automated detection for Alzheimer's. Advancements in data augmentation techniques and deep learning architectures have opened up new frontiers in this field, and research is moving rapidly. Hence, this survey aims to provide an overview of recent research on deep learning models for Alzheimer's disease diagnosis. In addition to categorizing the numerous data sources, neural network architectures, and commonly used assessment measures, we also classify implementation and reproducibility. Our objective is to assist interested researchers in keeping up with the newest developments and reproducing earlier investigations as benchmarks. In addition, we also indicate future research directions for this topic.

Keywords Alzheimer's detection · Deep neural networks · Hippocampus volume · MRI

N. Singh · D. Patteshwari
Department of Cognitive Neurosciences, School of Life Sciences, JSS Academy of Higher Education and Research, Mysuru, Karnataka, India

N. Soni
University of Delhi South Campus, Delhi, India

A. Kapoor (✉)
NePeur, Delhi, India
e-mail: amita.kapoor@nepeur.com

© The Author(s), under exclusive license to Springer Nature Singapore Pte Ltd. 2023

A. Mishra et al. (eds.), *Advances in IoT and Security with Computational Intelligence*, Lecture Notes in Networks and Systems 755,
https://doi.org/10.1007/978-981-99-5085-0_15

1 Introduction

Alzheimer's disease (AD) is the most common cause of dementia worldwide; it is a progressive degenerative neurological disorder; due to it, the brain cells die slowly. Forgetfulness and memory loss of recent events are the early symptoms of Alzheimer's disease. And as the disease progresses, it affects all cognitive functions, eventually making the patient completely dependent for even the essential functions of daily life. It is estimated that worldwide there are as many as 55 million people suffering from Alzheimer [1], and the number is expected to increase by 10 million every year. If unmanaged, this results in an expensive public health burden in the years to come.

There is no cure for Alzheimer's; if detected early, you can lead an everyday life with lifestyle changes and manage it for a considerable length of time. Deep learning models for Alzheimer's disease diagnosis have opened new frontiers, and research is moving rapidly. This review summarizes the most recent advancements in deep learning algorithms for automated AD detection. Our major contribution can be summed up as follows in brief [2]:

1. We summarize the latest progress of applying deep learning techniques to AD prediction, especially those appearing only in the past 2 years (2020-2022).
2. We specifically considered MRI for AD prediction.
3. We give a general workflow for AD prediction, based on which the previous studies can be easily classified and summarized. And future studies can refer to the previous work in each workflow step.

This survey's remaining sections are organized as follows: The second section provides an overview of the papers covered; Sect. 3 outlines the key findings in each step of the prediction workflow; Sect. 4 discusses implementation and reproducibility; and Sect. 5 concludes this survey.

2 Methodology

This section gives an overview of the papers reviewed in this study. There are a number of review papers spanning almost two decades, from 2003 to 2021. The least number of papers reviewed was 12, and Borchert et al. [3] covered 252 papers from 2005 to 2021. Almost all the review papers suggest that deep learning gives better results. Thus, in this paper, we focused only on deep learning techniques published between January 2020 and January 2022, covering a span of 24 months, with a total of 244 papers in the initial search. We did systematic elimination and were left with 61 papers[1]. That is roughly 2.5 papers per month.

All the works are searched and collected from Google Scholar using its SERP API, with searching keywords: "Alzheimer", "fMRI", and "deep learning". We have covered the papers published in the last 2 years (January 2020–January 2022).

Initially, 244 articles appeared in the Google Scholar search for 2 years. Out of these, 103 articles are selected based on the title and abstract. Out of 103 articles, 46 articles were rejected for one of two reasons:

1. Unavailability of full article.
2. Book chapters, thesis, and review papers are rejected.

Finally, 61 articles [2] were selected and are included in the survey. Figure 1 shows the detailed steps followed to select 61 papers from 244 papers in search list.

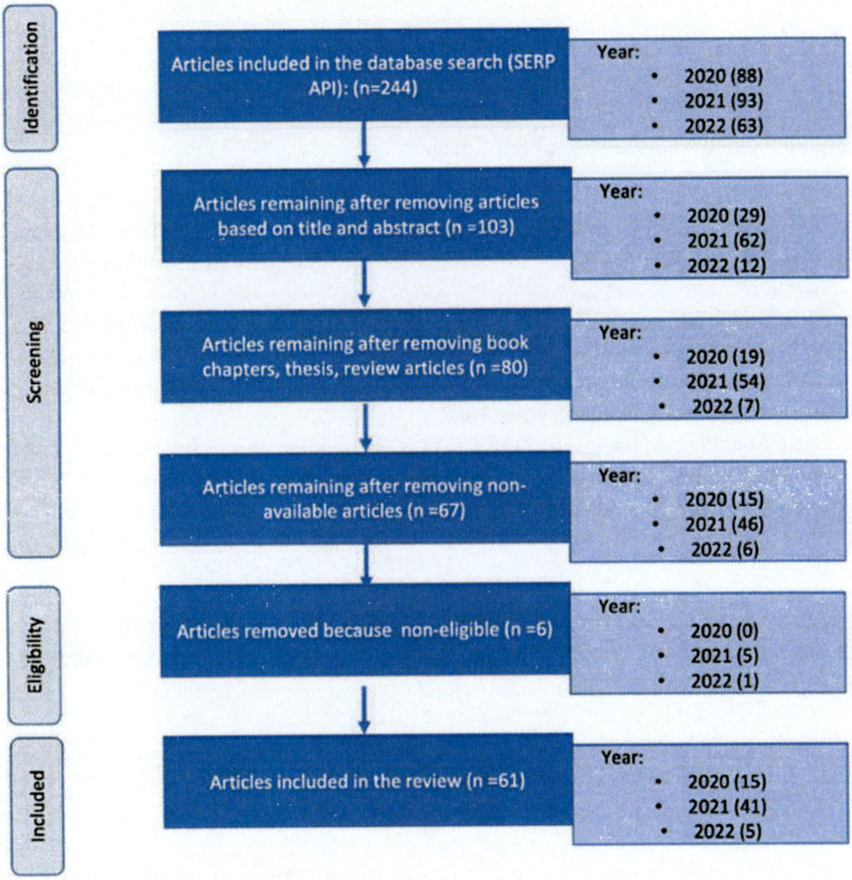

Fig. 1 Article selection criteria for the survey

3 Imaging Techniques, Datasets, and Models

The research papers considered use a varied combination of data sources and models. Therefore, in this section, we summarize the general workflow with four steps that most of the studies follow: Brain Imaging Techniques, Alzheimer Datasets, Prediction Models, and Model Evaluation. In this section, we will discuss each step separately.

3.1 Brain Imaging Techniques

For deep learning, data is a crucial component. Thus, here we list different brain imaging techniques used to get brain images.

1. **Structural Magnetic Resonance Imaging (sMRI)**: MRI is a non-invasive (non-anatomical) imaging (illustration) technique used to evaluate the structural integrity of the affected brain region. Brain scans are performed with MRI scanners.
2. **Functional Magnetic Resonance Imaging (fMRI)**: Functional MRI is also a non-invasive procedure that aids in the diagnosis of AD-related dysfunction. fMRI also permits the observation of oxygen absorption during resting and active states to establish an activity pattern.
3. **Diffusion Tensor Imaging (DTI)**: DTI is an imaging technique based on MRI that depicts minute cross-sectional structural details of brain regions.
4. **Positron Emission Tomography (PET)**: PET scan is a volumetric subatomic illustration technique used to obtain an anatomical and sub-anatomical 3D brain scan.

Figure 2 shows different brain imaging techniques used in the papers considered. We can see that MRI is the most preferred imaging source with only one paper [4] using DTI along with rs-fMRI.

3.2 Datasets

Diverse freely online available datasets have fueled the research in automated AD detection. Below, we describe the datasets used by different research papers.

1. **Alzheimer's disease Neuroimaging Initiative (ADNI):** The ADNI [5] is a database accessible at (adni.loni.ucla.edu) that has been used most frequently in studies pertaining to Alzheimer's detection. ADNI was launched in 2003 by the National Institute on Aging (NIA), the Food and Drug Administration (FDA), the National Institute of Biomedical Imaging and Bioengineering (NIBIB), private non-profit organizations, and private pharmaceutical companies.

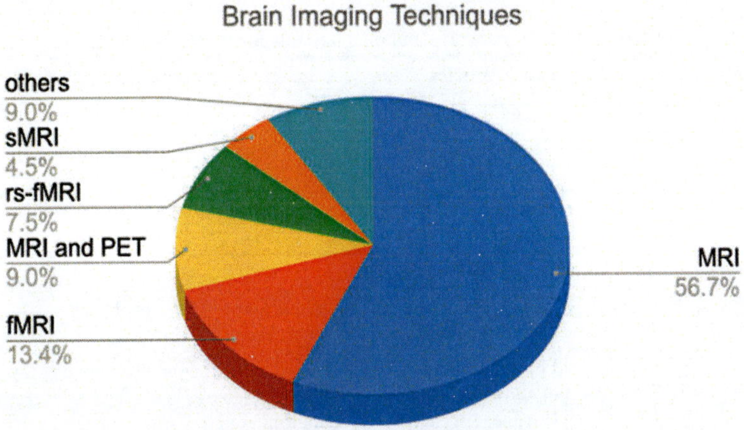

Fig. 2 Brain imaging techniques used across reviewed papers

2. **Open-Access Series of Imaging Studies (OASIS):** OASIS is an open-source repository for Alzheimer's disease classification for the scientific community [6]. The dataset is accessible via the website https://www.oasis-brains.org/. The cross-sectional dataset included MRI images from 416 subjects ranging in age from 18 to 96 years (young, middle-aged, non-dementia, and older adults with dementia).

Besides ADNI and OASIS, one of the considered paper [7] used Multi-Atlas Labelling Challenge (MICCAI 2012) Data [8], and one paper [9] used Minimal Interval Resonance Imaging in Alzheimer's Disease (MIRIAD) [10]. As reported by earlier review papers, ADNI is still the most widely used dataset, however, more and more researchers are now using OASIS, while in 2021 review the Oasis dataset was only between 2 and 10%, in the last 3 years the percentage increased to 26%.

3.3 Prediction Models

Deep learning is one of the most popular techniques in the field of AI today. Deep learning is not one algorithm but a group of algorithms inspired by the workings of biological neural networks. Table 1 briefly lists some deep learning architectures for AD detection. The models are categorized as convolutional neural networks based, using transfer learning, hybrid models, and other deep learning and machine learning techniques. Table 1 lists different types of networks and their variants for classification tasks with the best accuracy.

1. **Convolutional Neural Networks and their variants:** Convolutional neural networks are deep, feedforward artificial neural networks well suited for analyzing visual imagery or image data. CNN is composed of multiple layers:

Table 1 List of different types of networks and their variants for classification task

Prediction model	List of articles	Best accuracy (%)
Convolutional neural networks and its variant		
CNN	Salehi et al. [11], Zubair et al. [12], De Luna et al. [13], Chen et al. [14],	99.3 (100, 100) [12]
3D-CNN	Ramzan et al. [7], Goenka et al. [9], Li et al. [15]	100(−, −100,100,100) [9]
Pre-trained models		
VGG Net	Naz et al. [16], Janghel et al. [17], Mehmood et al. [18], Duc et al. [19]	99.3 [16]
ResNet	Fujibayashi et al. [20], Puente-Castro et al. [21], Odusami et al. 2021 [22], Buvaneswari et al. [23], Ebrahimi et al. [24]	99.8 (100, 99.6) [22]
Hybrid models between deep learning models and traditional models		
PCA based DNN models	Bi et al. [25], Jia et al. [26], Wang et al. [27]	97.01 [25]
SVM based DNN models	Mohammed et al. [28], Ahmadi et al. [29]	98.35 [29]
Other types of models		
Extreme learning machine	Bi et al. [30], Wang et al. [31]	88.7 (79, 83) [31]
Attention networks	Zhang et al. [4]	98.6 [4]
Ensemble networks	Fang et al. [33], Hedayati et al. [32]	99.27 (98.72, 95.89) [33]

*The main metrics is accuracy, in bracket are Specificity, Sensitivity, Precision, Recall, and F1-score in respective order

one or more convolutional layers followed by a non-linear activation function (typically the ReLU function), and ending with an optional pooling layer, typically max pooling. The first section of Table 1 lists the research papers and the prediction CNN model used by them.

2. **Transfer Learning:** Transfer learning is used in training deep neural networks with a small amount of training data. It is a method in which a model trained on one task is repurposed for a second task that is related to the first. It results in reduced training time by tuning the pre-trained model on a larger training dataset, using a method known as "fine-tuning". The second section of Table 1 lists the articles using the pre-trained models and their respective accuracy.
3. **Hybrid Models:** Besides CNN, some other models employed for AD detection include methods like principal component analysis, support vector machine, autoencoder, k-means, and genetic algorithms. The third section of Table 1 lists research papers with hybrid models, that is, they use more than one type of algorithms, mostly a deep learning algorithm along with a machine learning algorithm.

4. **Other Models Used for prediction:** These include other deep learning-based algorithms and machine learning methods employed for AD detection like attention mechanism, capsule network, ensemble learning, fuzzy logic, and many more. The fourth section of Table 1 lists other methods.

The results from the above approaches are interesting, more so because they attempt combine feature selection procedures with classification. Additionally, many of the above papers are attempting three class or more, with an emphasis on improving early MCI (EMCI) detection.

3.4 Model Evaluation

There are several metrics for measuring the performance of various machine learning algorithms. The metrics like accuracy, sensitivity, specificity, precision, F1-score, ROC, Root Mean Square Error (RMSE), Mean Absolute Error (MAE), Mini-Mental State Examination (MMSE), Dice Similarity Coefficient, Positive Predictive Value, Positive predictive value (PPV), Correlation coefficient, and AUC are used in the classification tasks. The detailed list of research papers using different metrics is shown in Fig. 3 and Table 1.

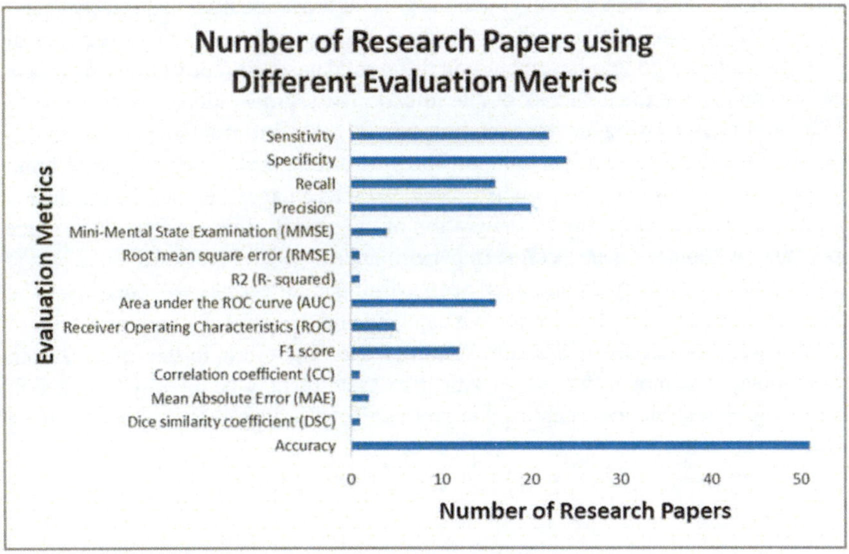

Fig. 3 Number of research papers using different evaluation metrics

4 Implementation and Reproducibility

In this section, we pay special attention to the implementation details of the papers we survey, which is less discussed in previous surveys.

The programming language used to implement machine learning and deep learning models is investigated first. Python has become the dominant choice for model implementation over the past 3 years. It offers a variety of packages and frameworks, such as TensorFlow [34–36], PyTorch [37], and scikit-learn. Other options include R, MATLAB, and Java.

Deep learning models require more computation for training, and GPUs have been used to accelerate the convolutional operations involved. With the need to process multiple types of input data, especially text data, the need for GPU will keep increasing in this research area. GitHub is the mainstream platform for hosting source code in computer science. There are very few articles available with public code repositories, this needs to change.

5 Conclusion and Discussion

Utilizing deep learning to classify AD has made significant progress. Recent research trends in end-to-end deep learning methods present great potential in the early detection of MCI. It was challenging to compare different review papers because of highly variable methodologies, including different datasets, preprocessing methods, selected inputs, and the final task of classification or segmentation.

The future is moving in the direction where more information can be fed to deep neural networks using techniques like spatio-temporal maps, structural brain network, or multimodal models where besides fMRI images, text and voice data is also included. To enable clinical integration of the models, there is a need to reduce model size without a severe decline in performance.

New and increasingly complicated deep learning architectures will continue to be created, and access to higher processing capacity and the precision and resolution of neuroimaging techniques will continue to increase. Therefore, in the future, faster, more precise, and more efficient classification systems may be directly included in neuroimaging techniques, enabling the production of a diagnostic hypothesis from a single brain scan.

References

1. Dementia. https://www.who.int/news-room/fact-sheets/detail/dementia. Accessed 20 May 2022
2. Singh N, Soni N, Kapoor A (2022) Automated detection of Alzheimer disease using MRI images and deep neural networks—a review. Preprint at https://arxiv.org/pdf/2209.11282.pdf
3. Borchert RJ et al (2021) Artificial intelligence for diagnosis and prognosis in neuroimaging for dementia; a systematic review. medRxiv
4. Zhang L, Wang L, Zhu D (2020) Jointly Analyzing Alzheimer's disease related structure-function using deep cross-model attention network. In: 2020 IEEE 17th international symposium on biomedical imaging (ISBI). IEEE, pp 563–567
5. Jack Jr CR, Bernstein MA, Fox NC, Thompson P, Alexander G, Harvey D, et al(2008) The Alzheimer's disease neuroimaging initiative (ADNI): MRI methods. J Magn Reson Imaging: Off J Int Soc Magn Res-Onance Med 27:685–691. https://doi.org/10.1002/jmri.21049
6. Marcus DS, Wang TH, Parker J, Csernansky JG, Morris JC, Buckner RL (2007) Open access series of imaging studies (OASIS): cross-sectional MRI data in young, middle aged, nondemented, and demented older adults. J Cogn Neurosci 19:1498–1507. https://doi.org/10.1162/jocn.2007.19.9.1498
7. Ramzan F, Khan MUG, Iqbal S, Saba T, Rehman A (2020) Volumetric segmentation of brain regions from MRI scans using 3D convolutional neural networks. IEEE Access 8:103697–103709
8. B. Landman S (2012) Warfield, MICCAI 2012 workshop on multi-atlas labeling. In: MICCAI grand challenge and workshop on multi-atlas labeling. CreateSpace Independent Publishing Platform, Nice, France
9. Goenka N, Tiwari S (2021) Volumetric convolutional neural network for Alzheimer detection. In: 5th International conference on trends in electronics and informatics (ICOEI). IEEE, pp 1500–1505
10. Malone IB, Cash D, Ridgway GR, Macmanus DG, Ourselin S, Fox NC, Schott JM (2012) MIRIAD-Public release of a multiple time point Alzheimer's MR imaging dataset. Neuroimage 70C:33–36. https://doi.org/10.1016/j.neuroimage.2012.12.044
11. Salehi AW, Baglat P, Sharma BB, Gupta G, Upadhya AA (2020) CNN model: earlier diagnosis and classification of Alzheimer disease using MRI. In: International conference on smart electronics and communication (ICOSEC). IEEE, pp 156–161
12. Zubair L, Irtaza SA, Nida N, ul Haq N (2021) Alzheimer and mild cognitive disease recognition using automated deep learning techniques. In: International Bhurban conference on applied sciences and technologies (IBCAST). IEEE, pp 310–315
13. Ahmad MF, Akbar S, Hassan SAE, Rehman A, Ayesha N (2021) Deep learning approach to diagnose Alzheimer's disease through magnetic resonance images. In: International conference on innovative computing (ICIC). IEEE, pp 1–6
14. Subaramya S, Kokul T, Nagulan R, Pinidiyaarachchi UAJ, Jeyasuthan M (2021) Detection of Alzheimer's disease using structural brain network and convolutional neural network. In: 10th international conference on information and automation for sustainability (ICIAfS). IEEE, pp 173–178
15. Sadat SU, Shomee HH, Awwal A, Amin SN, Reza MT, Parvez MZ (2021) Alzheimer's disease detection and classification using transfer learning technique and ensemble on convolutional neural networks. In: IEEE international conference on systems, man, and cybernetics (SMC). IEEE, pp 1478–1481
16. Murugan S, Venkatesan C, Sumithra MG, Gao XZ, Elakkiya B, Akila M, Manoharan S (2021) DEMNET: a deep learning model for early diagnosis of Alzheimer diseases and dementia from MR images. IEEE Access 9:90319–90329
17. Naz S, Ashraf A, Zaib A (2021) Transfer learning using freeze features for Alzheimer neurological disorder detection using ADNI dataset. Multimedia Syst 28(1):85–94
18. Janghel RR, Rathore YK (2021) Deep convolution neural network-based system for early diagnosis of Alzheimer's disease. Irbm 42(4):258–267

19. Mehmood A, Yang S, Feng Z, Wang M, Ahmad AS, Khan R, Yaqub M et al (2021) A transfer learning approach for early diagnosis of Alzheimer's disease on MRI images. Neuroscience 460:43–52
20. Rajeswari SS, Nair M (2021) A transfer learning approach for predicting Alzheimer's disease. In: 4th Biennial international conference on nascent technologies in engineering (ICNTE). IEEE, pp 1–5
21. Fujibayashi D, Sakaguchi H, Ardakani I, Okuno A (2021) Nonlinear registration as an effective preprocessing technique for deep learning based classification of disease. In: 43rd annual international conference of the IEEE engineering in medicine and biology society (EMBC). IEEE, pp 3245–3250
22. Puente-Castro A, Fernandez-Blanco E, Pazos A, Munteanu CR (2020) Automatic assessment of Alzheimer's disease diagnosis based on deep learning techniques. Comput Biol Med 120:103764
23. Odusami M, Maskeliūnas R, Damaševičius R, Krilavičius T (2021) Analysis of features of Alzheimer's disease: detection of early stage from functional brain changes in magnetic resonance images using a finetuned ResNet18 network. Diagnostics 11(6):1071
24. Buvaneswari PR, Gayathri R (2021) Deep learning-based segmentation in classification of Alzheimer's disease. Arab J Sci Eng 46(6):5373–5383
25. Fiasam D, Linda R, Yunbo S, Collins A, Osei I, Mawuli CB (2022) Efficient 3D residual network on MRI data for neurodegenerative disease classification. Proc SPIE 12083:120831A-1
26. Bi X, Li S, Xiao B, Li Y, Wang G, Ma X (2020) Computer aided Alzheimer's disease diagnosis by an unsupervised deep learning technology. Neurocomputing 392:296–304
27. Jia H, Wang Y, Duan Y, Xiao H (2021) Alzheimer's disease classification based on image transformation and features fusion. Comput Math Methods Meds
28. Wang Y, Jia H, Duan Y, Xiao H (2021) Applying 3DPCANet and functional magnetic resonance imaging to aided diagnosis of Alzheimer's disease. Res Sq preprint
29. Mohammed BA, Senan EM, Rassem TH, Makbol NM, Alanazi AA, Al-Mekhlafi ZG, Ghaleb FA et al (2021) Multi-method analysis of medical records and MRI images for early diagnosis of dementia and Alzheimer's disease based on deep learning and hybrid methods. Electronics 10(22):2860
30. Bi X, Zhao X, Huang H, Chen D, Ma Y (2020) Functional brain network classification for Alzheimer's disease detection with deep features and extreme learning machine. Cogn Comput 12(3):513–527
31. Wang Z, Xin J, Wang Z, Gu H, Zhao Y, Qian W (2021) Computer-aided dementia diagnosis based on hierarchical extreme learning machine. Cogn Comput 13(1):34–48
32. Saratxaga CL, Moya I, Picón A, Acosta M, Moreno-Fernandez-de-Leceta A, Garrote E, Bereciartua-Perez A (2021) MRI deep learning-based solution for Alzheimer's disease prediction. J Pers Med 11(9):902
33. Ali TWD (2021) Deep convolutional second generation curvelet transform-based MR image for early detection of Alzheimer's disease
34. Basheer S, Bhatia S, Sakri SB (2021) Computational modeling of dementia prediction using deep neural network: analysis on OASIS dataset. IEEE Access 9:42449–42462
35. Fan Z, Li J, Zhang L, Zhu G, Li P, Lu X, Wei W (2021) U-net based analysis of MRI for Alzheimer's disease diagnosis. Neural Comput Appl 33(20):13587–13599
36. Gulli A, Kapoor A, Pal S (2019) Deep learning with TensorFlow 2 and Keras: regression, ConvNets, GANs, RNNs, NLP, and more with TensorFlow 2 and the Keras API. Packt Publishing Ltd
37. Abadi M, Barham P, Chen J, Chen Z, Davis A, Dean JB, Zheng X (2016) {TensorFlow}: a system for {Large-Scale} machine learning. In: 12th USENIX symposium on operating systems design and implementation, vol 16. OSDI, pp 265–283

Skin Disease Classification and Detection by Deep Learning and Machine Learning Approaches

Summi Goindi, Khushal Thakur, and Divneet Singh Kapoor

Abstract One of the most common diseases around the globe is dermatological problem. Despite being frequent, its diagnosis is very challenging because of the complexity of its skin tone, colour and hair existence. This research offers a method to automatically forecast the many types of skin illnesses using a variety of computer vision-based approaches (deep learning). With the many skin photos, the system uses deep learning techniques to train itself. One of the main goals of this approach is improving the accuracy with which skin illness can be predicted and vast majority of the population is healthier. The most common causes of skin disorders include allergies, bacteria, mycosis, viruses, etc. The rapid development of medical and laser technologies that are founded on photonics has made it possible to identify skin issues in a manner that is both timely and precise. Such diagnoses can only be made with expensive, specialized medical equipment. As a result, deep learning algorithms help diagnose skin problems before they become severe. The categorization of skin conditions relies heavily on extracting features. Deep learning algorithms have substantially reduced the requirement for labour-intensive manual operations such as data restoration and extraction of features for classifications.

Keywords Skin disease · Dermatologically · Deep learning

S. Goindi (✉) · K. Thakur · D. S. Kapoor
Chandigarh University, Gharuan, Mohali, Punjab, India
e-mail: summikhurana7@gmail.com

K. Thakur
e-mail: khushal.thakur@cumail.com

D. S. Kapoor
e-mail: divneet.ece@cumail.com

© The Author(s), under exclusive license to Springer Nature Singapore Pte Ltd. 2023

A. Mishra et al. (eds.), *Advances in IoT and Security with Computational Intelligence*, Lecture Notes in Networks and Systems 755,
https://doi.org/10.1007/978-981-99-5085-0_16

1 Introduction

The human skin is by far the most noticeable part of the body. It consists of the dermis, the epidermis and the subcutaneous tissues. The skin can identify its surroundings and shield the body's internal tissues and organs from environmental hazards like bacteria, chemicals and direct sunlight [1]. The skin is susceptible to influences from the inside and out. Experimental skin damage, embryogenic illnesses, environmental pollutants, a compromised immune system and genetic mutations are just some of the causes of skin diseases. An individual's quality of life and health are profoundly impacted by skin issues. People will eventually try home treatments to address their skin conditions. These procedures may have harmful implications if they are not suitable for that skin disease. As skin problems are easily spread from person to person, they must be treated first. Physician often make assumptions about their patients' health based on their experience and their own impressions. It could be detrimental to one's health if the decision is made incorrectly or is delayed [2]. As a consequence, developing efficient strategies for diagnosing and treating skin problems becomes vital and critical. Technological advancement has enabled the design and implementation of a skin monitoring formative days foundational identification of skin issues. There are numerous advancements accessible for pattern-and image-based identification of various skin conditions.

Deep learning is among the disciplines which can help with the practical and exact identification of a various kinds of skin problems. It is possible to use deep learning and image classification to make medical diagnoses [3]. The creation of multiple objective classes and a training model to recognize each type is crucial to the problem of image classification. Deep learning-based technologies could be useful for swiftly recognizing clinical information and providing results. Data therapy is needed due to the obvious complexity of skin infections, the shortage and abuse of qualified physicians, and the urgent need for quick and accurate diagnosis. Recent developments in photonics and laser-based public healthcare technology have made the accurate and expedient diagnosis of skin problems feasible. Even with advances, the price of diagnostic procedures is still prohibitive. In the realm of image and data classification, deep learning systems exceed expectations [4]. In health diagnostics, the reliable acknowledgment of anomalies and categorization of diseases using X-ray, magnetic MRI, CT, PET and signalling data like EEG, EMG and ECG has been recommended. Disease categorization would assist in providing patients with more effective care. DL strategies can confront significant issues and adapt to changes in computational burden by recognizing data input features [5]. Even with just a little bit of computational methods, deep learning models will learn what they need to know to find and study the features in the data patterns they find. This led the researcher to look into the need for a DL classification model wherein a skin disorder is defined based on an image of the affected area. This might assist healthcare practitioners and patients do an effective invasive evaluation of illness with the least amount of cost and work as possible.

1.1 Skin Disease

Skin infections are ailments that affect the skin. These are disorders that can cause skin inflammation, itching, rashes and other symptoms. Certain skin diseases are inherited, whereas others are triggered by environmental causes. Skin problems comprise any diseases that clog, inflame and irritate the skin, resulting in rashes as well as other alterations in the skin's appearance. Skin illnesses are responsible for 1.79% of the global disease burden [6]. The American Academy of Dermatology Association says skin problems affect one out of every four people in the United States. The intensity and symptoms of skin disorders can vary widely, and they can be persistent or transient, severe or painless. Some skin infections are mild, while others are potentially fatal.

1.2 Frequent Forms of Skin Disorders

The following are frequent forms of skin disorders [6]:

- *Eczema or Atopic dermatitis:* Itchy, dry skin that becomes flaky, cracks and swells up.
- *Acne:* Restricted skin follicles cause dead skin, oil and germs to accumulate in the pores.
- *Alopecia Areata:* Hair loss in small spots.
- *Psoriasis:* Skin that is scaly and it may feel warm or bulge.
- *Skin cancer:* Aberrant skin cells that are growing out of control.
- *Raynaud's phenomenon:* This disorder results in a periodic decrease in blood flow to fingers, toes and some other parts of the body, which numbs and changes the appearance of the skin.
- Rosacea: Rosacea causes facial skin that is thick and flushed.
- Vitiligo: Loss of pigment in regions of skin.

1.2.1 Skin Disease Datasets

Infection with skin cancer has nearly doubled in the last 15 years, and skin disorders have grown more widespread in past few decades. Radical treatments for these illnesses are now required because of the severity and broad spread of many different skin conditions [7]. Traditional skin disease diagnosis necessitates a substantial amount of skill in the field as well as extensive information to distinguish between the numerous and differing degrees of skin illnesses. Researchers have recommended using computer-aided methods to identify skin conditions because data science and machine learning applications are becoming more and more important in the medical sector [8]. These systems use diverse machine and deep learning, and data science techniques to uncover objective patterns. Data on skin disease systems can be

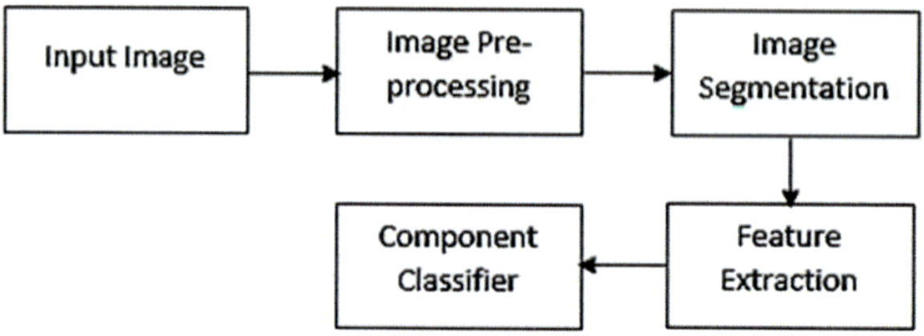

Fig. 1 Basic skin disease classification process

obtained from hospital and healthcare institution datasets or by assembling information from patients independently, that is, both labour-intensive and time-consuming. Figure 1 displays a few representative images taken from the whole dataset [9]. Table 1 contains a compilation of accessible to the public datasets pertaining to skin problems.

1.2.2 Skin Disease Classification

The classical approach to skin disease classification is depicted in Fig. 1.

(a) *Input Image:* Numerous skin disease picture databases are freely accessible. Yet, some are entirely or partly open source while others are offered for purchase. Depending on the employed dataset, the input picture may be dermoscopy or clinical. Image preparation.

(b) *Image pre-processing:* It is a necessary step since an image may comprise several noises, including air bubbles, hairs and dermoscopic gel. Moreover, clinical photos need more pre-processing than dermoscopic images due to the fact that characteristics such as quality, illumination, lighting condition, size of skin region and angle of image capture covered might vary based on the individual recording the image. These collected photos may cause complications in following stages. Various filters, like average or Gaussian filter [10], median [11]; morphological procedures including dilation and erosion; binary thresholding [12]; and software like Dull Razor [11, 13] can be utilized to eliminate skin hairs. Contrast or lesion enhancement methods [14] are beneficial for low-contrast photos. Contrast enhancement with histogram equalization is among the most widely utilized methods described in the literature [11] because it improves image visibility by distributing pixel intensity uniformly all across image.

(c) *Image Segmentation:* Image segmentation could be very useful in identifying skin diseases [11] by separating diseased skin from healthy skin. Image segmentation can indeed be accomplished in three distinct ways: (1) segmentation based on pixels, (2) segmentation utilizing edges and (3) segmentation utilizing regions. Each pixel in an image is categorized into a single object or area in

Table 1 Skin disease datasets

Dataset	Types of skin disease	Samples count	Description
Danderm	Common skin conditions	3000	Niels K. Veien, a dermatologist in medical clinics, captured the images for this clinical dermatological atlas
MED-NODE	Nevi and Melanoma	170	The DDUMCG digital image library contains 100 and 70 naevus and melanoma sample images, respectively
ISIC	Melanoma, benign nevi, seborrheic keratosis	>20,000	The ISIC database was made accessible to the general for the first time in 2016 to be used in dermoscopy image processing benchmark competitions
Dermnet	All sorts of skin illnesses	23,000	Dermnet uses innovative methods to spread knowledge about skin disorders
SD-198	198 classes	6584	The publicly accessible SD-198 dataset6 includes photos of skin disorders
DermIS	Diverse types of skin disorders	1000	DermIS.net seems to be the Internet's most extensive repository of dermatological knowledge
IAD	Benign lesion and melanoma	2000	This IAD database includes 2000 dermoscopy and 800 example images
MoleMap	22 benign and 3 malignant cells, respectively	112,431	MoleMap has over 100,000 images of 25 skin conditions
HAM10000	Pigmented lesions fall into several key diagnostic categories	10,015	This dataset is freely available to the public via the ISIC archive
Hallym dataset	Basal cell carcinoma	152	Patients with basal cell carcinoma were treated in the Hallym dataset

pixel-based segmentation. This is possible with binary thresholding or its variants [12, 15–17]. The edge-based approach identifies and connects edge pixels to create the bounded shape of skin lesions. Examples include the Prewitt, Canny operators, gradient vector flow, Robert, Sobel and adaptive snake [18]. Region-based approaches rely on comparable patterns in the intensity values of nearby pixels and are founded on continuity. The instances are watershed algorithm, region expansion and region mergers [18, 19].

(d) *Feature Extraction:* The most significant characteristics used to visually identify and define skin illnesses are its texture and colour. The colour information serves a crucial function in distinguishing between diseases. Various approaches, including colour correlograms, colour histograms, colour descriptors and GLCM [17, 19], can be employed to derive these colour characteristics. The texture data reveals the intricate visual features of skin infections, including the spatial arrangement of size, colour, shape and brightness. Most of an image's texture comes from how the intensity of each pixel changes. Researchers utilize local binary pattern, SIFT [16] and GLCM to extract texture information from a picture. Depending on the kind of disease and its extent, lesions can vary in shape, colour and size.

(e) *Classification:* Classification is an example of supervised learning in machine learning. To map data into specified classes, a labelled dataset is necessary [20]. Different classification algorithms [21] like k-nearest neighbour (k-NN), feedforward neural network (FFNN), support vector machine (SVM), decision trees (DT), backpropagation neural network (BPN), etc. are employed to classify skin disorder images.

2 Deep Learning Skin Disease Diagnosis

Deep learning techniques have helped with many skin disease diagnosis tasks due to their universal appeal. One can examine at some recent work in skin disease diagnostics that makes use of the DL notion in this section. We first briefly discuss the common data pre-processing and augmentation methods used in DL before conducting a literature review on the numerous applications of deep learning for skin disease diagnosis [22]. Figure 2 displays the taxonomies of the existing literature.

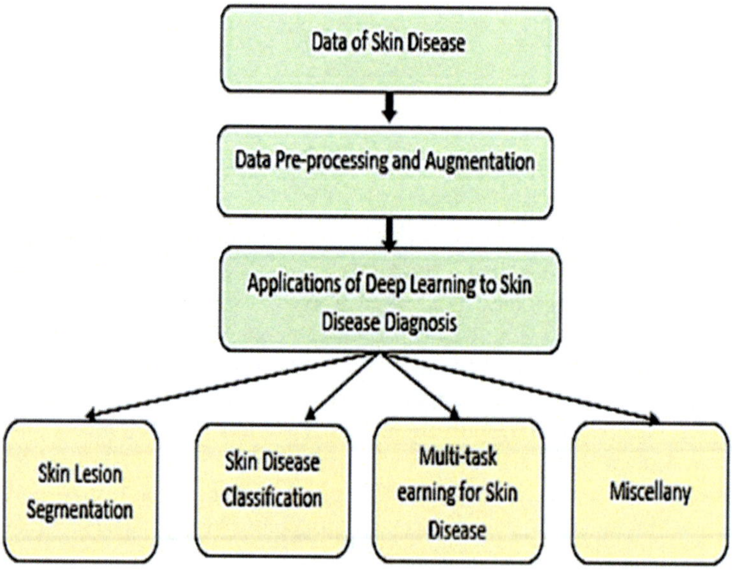

Fig. 2 Deep learning-based accurate classification of skin condition diagnostic literature

2.1 Data Pre-processing and Augmentation

Data preparation plays a significant role in DL for skin disease identification. Images from databases and deep networks frequently have square dimensions because to the vast variety of skin illness photoresolutions; as a result, they should be clipped or shrunk before being used in DL networks. Be warned that directly cropping and shrinking images can cause object distortion or substantial data loss. To solve this issue, it is possible to resize the image's shortest side while maintaining the aspect ratio. Images are commonly standardized by subtracting the mean value and the result is divided by the standard deviation without first being input into a DL network. Medical image processing frequently has constrained data due to the rarity of sickness, patient isolation, requirement for expert medical labelling and high cost of getting medical data. To improve the quality of the original data, a growing number of techniques for data enhancement are being implemented. These include feature expansion, neural style transfer, adversarial training, colour space augmentations, geometric actions (like cropping and flipping), mixing, filtering, meta-learning and random erasing.

2.2 Skin Lesions Segmentation

The objective of segmentation is to divide an image into distinct regions that contain pixels with similar properties. Segmentation is important for skin disease diagnosis since it enables doctors to see the boundaries of lesions. The effectiveness of image analysis depends on how precisely and accurately images are segmented. The scanner's manual boarder identification process takes tumour collisions into consideration when several types of lesions are close to one another and could potentially cause an issue. The typical procedure for segmenting a skin condition is shown in Fig. 3.

First of all, the difference between healthy skin and skin lesions is blurred. Variations in skin tones; old artefacts like hair or ink present on the image; ruler markings or air bubbles; uneven lighting; the actual placement of lesions and variations in lesion colour, sharpness, size, shape and position within image all make segmentation more challenging [23] (Table 2). The skin lesions in Table 3 have been divided using deep learning methods.

2.3 Classification of Skin Conditions

Skin disease classification is the last step in the CAD method for skin disease diagnosis. Based on their objectives, several classification algorithms may produce binary,

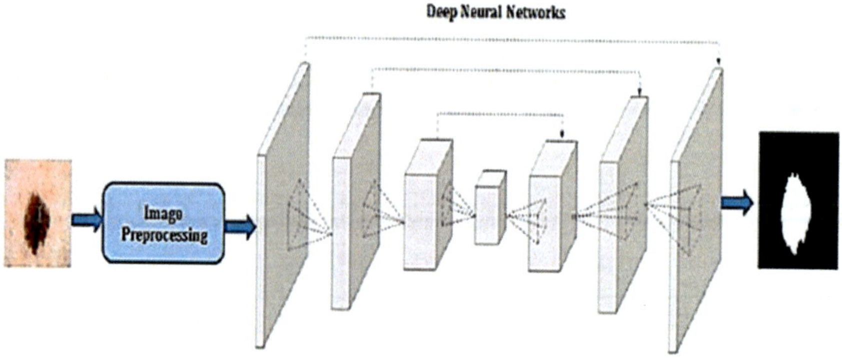

Fig. 3 Process diagram for segmenting skin diseases

trinary or N4 types of skin disease categorization results. Numerous deep learning algorithms have been created in order to summarize skin disease photographs.

Manerkar et al. [19] accomplish automated segmentation of several images of skin cancer using the watershed algorithm and C-means and feature extraction using the Image Quality Assessment (IQA) method and Grey-Level Co-occurrence Matrix (GLCM). Using a multiclass SVM classifier, distinct illness states were classified. Utilizing multiclass SVM, this study classifies several forms of skin illnesses as malignant skin cancer, benign skin cancer and warts. In segmenting images of skin cancer, the C-means algorithm yielded more accurate segmentation results with 98% accuracy in comparison to the watershed method (92% accuracy). Amarathunga et al. [15] have developed a classification system for only three disorders. The system is comprised of two distinct units: image processing and data processing. Images of skin diseases were collected, pre-processed to remove noise, segmented, and features were extracted using the data processing element, while classification and data mining were performed using the data processing unit. AdaBoost, BayesNet, Naive Bayes, J48 and MLP were the five classification methods studied. When compared to the other four classifiers, the MLP classifier performed the best. However, it is not stated where the images and qualities used to classify diseases came from. Chatterjee et al. [24] employed the frequency and spatial domain-based approach to determine if a skin lesion was malignant or benign. The malignant lesions have been further divided into epidermal or melanocytic skin lesions as subcategories. The cross-correlation method is employed to identify local features that are resistant to variations in light intensity and illumination. Additionally, more specific aspects of skin lesions have been retrieved using cross-spectrum-based frequency-domain research.

The SVM classifier was employed for classification with three non-linear kernels [24], with SVM with RBF kernel providing encouraging accuracy in comparison to other kernels. Monisha et al. [25] propose a bio-impedance size methodology for evaluating skin disorders that is used in the early detection of skin illnesses such as cancer, skin tumours and non-cancer. (1) BCC (basal cell carcinoma), (2) malignant cancer skin disease and squamous cellular carcinoma such as zits, scabies, Rubella,

Table 2 Classification methods of skin disorders using conventional and deep learning

References	Type of images	Disease	No. of images	Pre-processing	Segmentation	Feature-based extraction	Classifier	Performance measure
Manerka et al. [19]	Clinical	Benign and Malignant, warts skin cancer	45	Y	Watershed algorithm and C-means clustering	IQA and GLCM	SVM	Accuracy: 96–98%
Amarathunga et al. [15]	Clinical	Impetigo, Eczema, melanoma	–	Y	Thresholding	–	MLP	Accuracy: 90%
Chatterjee et al. [24]	Dermoscopic	Nevus, BCC, Melanoma, SK	6,838	–	–	Cross spectrum, Cross correlation	SVM	Sensitivity: 99.01% Specificity: 95.35% Accuracy: 98.79%
Monisha. et al [25]	Dermoscopic	SA, BCC, Lentigo simplex	–	Y	GMM	DRLBP, GLCMGRLTP	NSGA-II- PNN	–
Chakraborty et al. [16]	Dermoscopic	SA, BCC	–	–	Thresholding	SIFT	NN- NSGA-II	Accuracy:90.56%, F-measure:90.87% Precision:88.26%, Recall:93.64%,
Esteva et al. [26]	Clinical, Dermoscopic	Benign and Malignant skin lesions	Dermoscopic: 3,374 Clinical: 129,450	–	Stanford Hospital [4], ASIC [27], Edinburgh Dermofit Library [14]	Inception V3 with PA (Partition algorithm)	Accuracy: 72.1 ± 0.9% and Mean Sensitivity: 89.4	

(continued)

Table 2 (continued)

References	Type of images	Disease	No. of images	Pre-processing	Segmentation	Feature-based extraction	Classifier	Performance measure
Khan et al. [14]	Dermoscopic	Melanoma Vs other	ISBI 17[27]:2,790 ISBI-16[27]:1,27 HAM10000 [27]: 10,000	Lesion Enhancement	ISBI 2017: 95.60%ResNet50 and ResNet101	ResNet101 and ResNet50	Accuracy: ISBI 2017: 95.60%, ISBI 2016: 90.20%, Ham1000: 89.8%	
Kulhalli et al. [28]	Dermoscopic	Melanoma Nevi, Akiec, BKL SK, BCC, DF	10,015	–	HAM10000 [27]	InceptionV3	Normalized F1 Score: 0.93	
Zhang et al. [14]	Dermoscopic	Melanocyticnevus, BCC, SK psoriasis	DatasetA [5]: 1,067 Dataset b [5]: 522	–	Dataset A [5], Dataset B [5]	Inception v3	Dataset A: Accuracy: 87.25 ± 2.24%, Dataset B: Accuracy:86.63% ± 5.78%	

Disease: SA—Skin angioma, BCC—Basal cell carcinoma, SK—Seborrheic keratoses, Feature extraction: GLCM—Grey-level co-occurrence matrix, GMM—Gradient mixture model, NSGA-II—Nondominated sorting genetic algorithm, PNN—probabilistic neural network, Y-YOLO, IQA—Image quality assessment

Table 3 Some recent skin diseases literature analysis

Author name and yearof publication	Journal/ conference name	Title of paper	Findings	Gaps
Chatterjee et al (2019)	Elsevier	Extraction of features from cross correlation in space and frequency domains for classification of skin lesions	According to the results of the classification performance indices, it is feasible to precisely and sensitively identify all the illness classes under examination using the provided feature extraction method	It would have been more informative to test the method using dermoscopic images of patients suffering from a wider range of skin diseases
Monisha et al.	Journal of Medical Systems	ArtificialIntelligence Based Skin Classification Using GMM	The ANN can assess the disease patterns of a certain ailment and provide a faster prognosis and repute than a human physician	Due to the complexities, identification and diagnosis become more difficult
Liu et al.	Journal of Imaging	Skin Lesion Segmentation Using Deep Learning with Auxiliary Task	The results of the tests indicate that the suggested strategy achieves superior efficiency when compared to methods that are considered state of the art and only use a single integrated model	The employed solutions either demand considerable additional training parameters or additional labelling information that may not be appropriate in practice
Esteva et al. (2021)	HHS Public Access	Dermatologist-level classificationofskin cancer with deep neural networks	In both tasks, CNN matches the ability of all evaluated specialists, demonstrating that AI is able to classify skin cancer at the same level of expertise as that of dermatologists	The CNN scores slightly above the dermatologists' average on the epidermal lesion test and somewhat below the dermatologists' average both on melanocytic lesion tests

(continued)

Table 3 (continued)

Author name and yearof publication	Journal/ conference name	Title of paper	Findings	Gaps
Zafar et al. (2020)	Sensors	Skin lesion segmentation from dermoscopic images using convolutional neural network	The scientists were able to significantly improve the accuracy of their CNN-based skin lesion segmentation system by combining it with an improved hair-removal approach	Hyper-tuning the settings to prevent overfitting and improve training efficiency is essential
NagaSrinivasu et al. (2021)	Sensors	Classification of skin disease using deep learning neural networks with MobileNet V2 and LSTM	In order to reduce difficulties and fatality, the suggested method can aid primary care physicians in making accurate diagnoses of skin disorders	When evaluated on a collection of images taken under different lighting conditions, the model's accuracy drops significantly to slightly below 80%
Thapar, et.al (2022)	Hindwai	A novel hybrid deep learning approach for skin lesion segmentation and classification	The suggested work outperforms the state of the art by 6.12% in classification accuracy and 98.42% in precision, with an MCC of 0.9704. In comparison to the current state of the art, it was discovered that the novel approach improved precision by 5.78%, specificity by 9.21% and F-measure by 8.34%	No conflicts of interest

(continued)

Table 3 (continued)

Author name and yearof publication	Journal/ conference name	Title of paper	Findings	Gaps
Afza et al. (2022)	Sensors	Multiclass skin lesion classification using hybrid deep features selection and extreme learning machine	The results of the experiments demonstrate that the approach that was suggested has an outstanding quality on the data sources that were chosen, particularly with regard to improving the quality of a lesion region, raising the accuracy of classification, and reducing the amount of time needed for computing	Dataset enhancement is a crucial process that can be perfected with modern techniques

sickle-cell anaemia, leprosy and psoriasis. Skin disorders of the foot, mouth and hand. We can deduct from the preceding results that regular skin is more valuable than disease pores and skin. The writers can distinguish between damaged pores and skin and normal skin. Using this dimension, the researchers may readily diagnose and compare afflicted pores and skin to normal skin of any condition. The authors use this strategy to control the framing parameter and avoid the occurrence of many ailments such as early phase pores and also most skin cancers. The first study to describe how the image classifier CNN may perform similarly to the 21 board-certified dermatologists for the diagnosis of malignant lesions was published by Esteva et al. [26]. A skin lesion can be classified as benign, malignant, or non-neoplastic using the three-way disease partition algorithm. In order to place a particular lesion into one of the nine groups specified, a nine-way disease partitioning was also carried out. The most up-to-date InceptionV3 CNN architecture was employed to classify skin lesions [26] and found that if trained on adequate data, the CNN can execute better than human specialists. Brinker et al. [28] provide the first comprehensive evaluation of current studies on categorizing skin lesions with CNNs. Designers restrict the analysis to skin lesion classifiers. In general, approaches that use a CNN solely for segmentation or classification of dermoscopy patterns are excluded from this discussion. Khan et al. [22] suggested an automated approach for skin lesion categorization based on deep neural network (DCNN) features retrieval and kurtosis-controlled principal component analysis (KcPCA)-based optimal feature selection using transfer learning. For extracting features, pre-trained ResNet deep neural networks like RESNET-101 and RESNET-50 are used. The input was then fused and the best features were chosen, which were then fed into supervised learning methods like SVM of radial basis function (RBF) for classification. For experimental data, three databases named ISBI 2017, ISBI 2016 and HAM10000 were used with accuracy of 95.60%, 90.20% and 89.8%, respectively. The total findings indicate that the proposed system's efficiency is reliable when compared to current methodologies. Kulhalli et al. [29] suggested utilizing InceptionV3 CNN architecture, a five-stage, three-stage and two-stage hierarchical technique, to categorize seven illnesses. The authors discussed the class imbalance issue by employing an image augmentation technique to equalize the categorization classes. The five-stage classifier outperformed the two- and three-stage hierarchy classifiers. Zhang et al. [14] also classified four diseases using an InceptionV3 architecture with a customized final layer. The model was trained on two datasets of dermoscopic images that were almost identical. Due to the presence of many disease lesions on a single skin scan, authors [14] determined that misclassification can happen.

3 Conclusion

Important progress has been made in our understanding of skin diseases, and this article provides a concise summary of the diagnostic methods that utilize DL and ML. After a brief review of the background and technical aspects of skin disease, we go into the disease itself for analysis. The methods for collecting skin images and the freely available datasets are outlined in the next section. Third, we have explored the fundamentals of deep learning and its most popular frameworks. Then, we detailed the performance indicators and specific criteria we used to draw conclusions. Deep learning is presently being used to help identify skin conditions. In addition, a thorough evaluation of the field of skin diagnostic testing is provided, with an emphasis on recent deep learning applications. Reading this material in its entirety will empower readers with a firm grounding in the fundamental principles underlying the diagnosis of skin diseases. Deep learning is currently being utilized as a tool to assist in the diagnosis of skin conditions. While several researches used many algorithms, some only used one. Methods of classifying dermatological diseases used multiple algorithms and a hybridization strategy performed better than those that used only one.

References

1. Deepalakshmi P, Lavanya K, Srinivasu PN (2021) Plant leaf disease detection using CNN algorithm. Int J Inf Syst Model Des (IJISMD) 12(1):1–21
2. Anitha J (2018) Identification of melanoma in dermoscopy images using image processing algorithms. In: International conference on control, power, communication and computing technologies (ICCPCCT). IEEE, pp 553–557
3. Roy K, Chaudhuri SS, Ghosh S, Dutta SK, Chakraborty P, Sarkar R (2019) Skin disease detection based on different segmentation techniques. In: International conference on opto-electronics and applied optics (Optronix). IEEE, pp 1–5
4. Kotian AL, Deepa K (2017) Detection and classification of skin diseases by image analysis using MATLAB. Int J Emerg Res Manag Technol 6(5):779–784
5. Sundaramurthy S, Saravanabhavan C, Kshirsagar P (2020) Prediction and Classification of rheumatoid arthritis using ensemble machine learning approaches. In: Proceedings of the 2020 international conference on decision aid sciences and application (DASA). Sakheer, Bahrain, pp 17–21
6. Skin Disease. https://www.medanta.org/patient-education-blog/everything-about-common-skin-disorders/
7. Ronneberger O, Fischer P, Brox T (2015) U-net: convolutional networks for biomedical image segmentation. In: International conference on medical image computing and computer-assisted intervention. Springer, Cham, pp 234–241
8. Yu Z, Jiang X, Zhou F, Qin J, Ni D, Chen S, Wang T (2018) Melanoma recognition in dermoscopy images via aggregated deep convolutional features. IEEE Trans Biomed Eng 66(4):1006–1016
9. Phillips A, Teo I, Lang J (2019) Segmentation of prognostic tissue structures in cutaneous melanoma using whole slide images. In: Proceedings of the IEEE/CVF conference on computer vision and pattern recognition workshops
10. Reddy VJ, Nagalakshmi TJ (2019) Skin disease detection using artificial neural network. Indian J Public Health Res Dev 10(11)
11. Jana E, Subban R, Saraswathi S (2017) Research on skin cancer cell detection using image processing. In: 2017 IEEE international conference on computational intelligence and computing research (ICCIC). IEEE, pp 1–8
12. Zaqout I (2019) Diagnosis of skin lesions based on dermoscopic images using image processing techniques. Pattern Recognit Sel Methods Appl
13. Kumar VB, Kumar SS, Saboo V (2016) Dermatological disease detection using image processing and machine learning. In: 2016 Third international conference on artificial intelligence and pattern recognition (AIPR). IEEE, pp 1–6
14. Khan MA, Javed MY, Sharif M, Saba T, Rehman A (2019) Multi-model deep neural network-based features extraction and optimal selection approach for skin lesion classification. In: 2019 international conference on computer and information sciences (ICCIS). IEEE, pp 1–7
15. Amarathunga AA, Ellawala EP, Abeysekar GN, Amalraj CR (2015) Expert system for diagnosis of skin diseases. Int J Sci Technol Res 4(01):174–178
16. Chakraborty S, Mali K, Chatterjee S, Anand S, Basu A, Banerjee S, Bhattacharya A et al (2017) Image based skin disease detection using hybrid neural network coupled bag-of-features. In: 2017 IEEE 8th annual ubiquitous computing, electronics and mobile communication conference (UEMCON). IEEE, pp 242–246
17. Arifin MS, Kibria MG, Firoze A, Amini MA, Yan H (2012) Dermatological disease diagnosis using color-skin images. In: 2012 international conference on machine learning and cybernetics, vol 5. IEEE, pp 1675–1680
18. Premaladha J, Sujitha S, Priya ML, Ravichandran KS (2014) A survey on melanoma diagnosis using image processing and soft computing techniques. Res J Inf Technol 6(2):65–80
19. Manerkar MS, Snekhalatha U, Harsh S, Saxena J, Sarma SP, Anburajan M (2016) Automated skin disease segmentation and classification using multi-class SVM classifier

20. Barati E, Saraee MH, Mohammadi A, Adibi N, Ahmadzadeh MR (2011) A survey on utilization of data mining approaches for dermatological (skin) diseases prediction. J Sel Areas Health Inform (JSHI) 2(3):1–11
21. Lopez AR, Giro-i-Nieto X, Burdick J, Marques O (2017) Skin lesion classification from dermoscopic images using deep learning techniques. In: 2017 13th IASTED international conference on biomedical engineering (BioMed). IEEE, pp 49–54
22. Zhang X, Wang S, Liu J, Tao C (2018) Towards improving diagnosis of skin diseases by combining deep neural network and human knowledge. BMC Med Inform Decis Mak 18(2):69–76
23. Shi X, Dou Q, Xue C, Qin J, Chen H, Heng PA (2019) An active learning approach for reducing annotation cost in skin lesion analysis. In: International workshop on machine learning in medical imaging. Springer, Cham, pp 628–636
24. Chatterjee S, Dey D, Munshi S, Gorai S (2019) Extraction of features from cross correlation in space and frequency domains for classification of skin lesions. Biomed Signal Process Control 53:101581
25. Monisha M, Suresh A, Rashmi MR (2019) Artificial intelligence-based skin classification using GMM. J Med Syst 43(1):1–8
26. Esteva A, Kuprel B, Novoa RA, Ko J, Swetter SM, Blau HM, Thrun S (2017) Dermatologist-level classification of skin cancer with deep neural networks. Nature 542(7639):115–118
27. Data Science. https://www.medanta.org/patient-education-blog/everything-about-common-skin-disorders/
28. Brinker TJ, Hekler A, Utikal JS, Grabe N, Schadendorf D, Klode J, Von Kalle C (2018) Skin cancer classification using convolutional neural networks: systematic review. J Med Internet Res 20(10):e11936
29. Kulhalli R, Savadikar C, Garware B (2019) A hierarchical approach to skin lesion classification. In: Proceedings of the ACM India joint international conference on data science and management of data. pp 245–250

Symmetric Secret Key-Based Quantum Key and Its Distribution Over the Networks

Avdhesh Gupta, Vishan Kumar Gupta, Dinesh Kumar, and Vimal Kumar

Abstract Quantum key distribution (QKD) is an information-theoretically secure symmetric secret-key negotiation system. QKD stands for "quantum key distribution." Recently, QKD networks have advanced to the point where they have graduated from the domain of theoretical study and are beginning to make their way into a few practical applications. A network is formed by QKD nodes, which can communicate with one another either wirelessly or utilizing light-speed optical connections. Any two QKD nodes have the ability to negotiate the distribution of secret keys to a large number of users located in a variety of different places. This is done to provide both long-term security and forward secrecy. Following the presentation of a foundational overview of QKD and a few developments in QKD, and in this study, we will look into how QKD networks have evolved and where they have found practical use. In conclusion, we make some proposals as to the direction that future research should go, as well as some recommendations for the most effective methods of constructing QKD networks.

Keywords Quantum cryptography · Security · Communication network

---

A. Gupta (✉)
Ajay Kumar Garg Engineering College, Ghaziabad, India
e-mail: avvipersonal@gmail.com

V. K. Gupta
Sir Padampat Singhania University, Udaipur, Rajasthan, India

D. Kumar
KIET Group of Institutions, Ghaziabad, India

V. Kumar
Bennett University, Greater Noida, India

© The Author(s), under exclusive license to Springer Nature Singapore Pte Ltd. 2023

A. Mishra et al. (eds.), *Advances in IoT and Security with Computational Intelligence*,
Lecture Notes in Networks and Systems 755,
https://doi.org/10.1007/978-981-99-5085-0_17

1 Introduction

Almost every facet of modern life involves some type of computer system, and the Internet facilitates the dissemination of a vast amount of information. Researches are increasingly concerned with how to protect sensitive data during online transmission. The proliferation of cyberattacks and the emergence of quantum computing provide new obstacles to the protection of existing networks and systems for the transmission of data and voice. Meanwhile, the rising processing capacity of quantum computers [1] poses a danger to existing cryptosystems. However, Shor's approach (finds an integer's prime factors using quantum computers) can be used to break the security of any system that relies on integer factorization or discrete logarithm problems. In a quantum computer, rendering most cryptosystems insecure once quantum computing reaches maturity. The development of robust information security systems to counteract quantum attacks is, hence, of the utmost importance.

2 Quantum Key Distribution Basics

Quantum key distribution, or QKD, has recently been the focus of extensive research due to the fact that it has the potential to enable the construction of secure networks. Optical fiber networks are utilized by the backbone of all terrestrial, subterranean, metro, and access networks due to the massive amounts of data that these networks are capable of transporting as well as their extensive deployment all over the world.

Quantum key distribution has surfaced as a possible answer to the problem of securing future optical communication networks. Traditional encryption techniques protect data from cyberattacks because they employ public-key cryptography [1, 2]. The degree of safety afforded by such methods is determined by the computing complexity of the mathematical functions employed. Public-key cryptography's security is improving as more powerful processors become affordable. Since traditional encryption methods will be rendered insecure by the advent of quantum computers, QKD is essential for protecting data transmitted over open networks (Fig. 1).

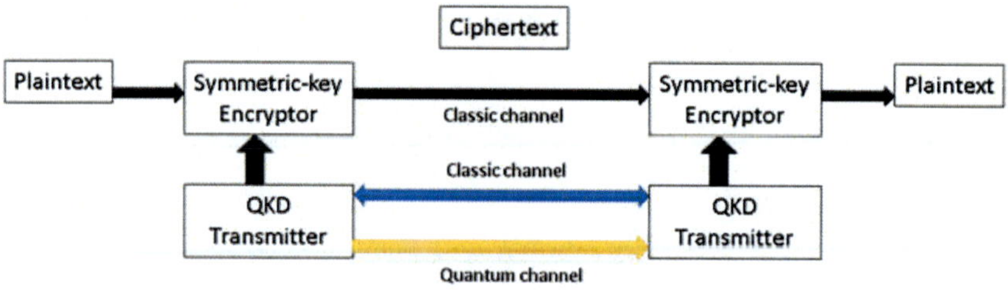

Fig. 1 Quantum key distribution-based cryptography

QKD uses quantum mechanics to securely distribute a secret key. The quantum no-cloning theorem [3] and Heisenberg's uncertainty principle forbid quantum cloning. Heisenberg's uncertainty principle states that two conjugate attributes, such as position and momentum, cannot be measured simultaneously. The "no-cloning" theorem for quantum systems states that particles like photons cannot be accurately cloned [4]. Thanks to the no-cloning theorem and the uncertainty principle, QKD is able to securely generate and transmit private keys to its intended recipients. Data can be encrypted and decrypted using random secret keys, one-time pads, and top-tier encryption (AES). QKD distributes symmetric secret keys using quantum physics. QKD includes a transmitter, receiver, and link. QKD transceiver is a combined transmitter and receiver. The QKD transmitter/receiver isolates QKD-related hardware and software. Quantum and classical channels are used. Alice and Bob must synchronize and distill a key using the classical channel [5, 6].

Since Bob does not receive the quantum states that are collected by an eavesdropper during the transit of single photons across a quantum channel, the eavesdropper is unable to use these quantum states to infer secret keys. This is because Bob does not receive the quantum states. Even if Eve were able to measure quantum states, the very act of measuring or witnessing anything would cause the quantum state to collapse back into the classical world. This would be the case even if Eve had the ability to measure quantum states. As a consequence of this, it is possible to identify any attempts that are made to eavesdrop on QKD discussions. In order for Alice and Bob to encrypt messages, they must first reach a consensus on a secret key. This can be done via QKD or one of the traditional methods depicted in the graphic below. In particular, Alice and Bob are able to make use of the created secret keys with the symmetric-key encryptors and descriptors that are distinctive to each of them.

In order for QKD to function, multiple light particles, also known as photons, are sent between parties using fiber-optic cables. The quantum state of each photon is completely unpredictable, and the combination of all the photons that are transmitted results in a stream of ones and zeros. In a binary system, the ones and zeros that make up this stream are referred to as qubits, which is shorthand for bits. In order for a photon to enter a photon collector, it must first pass through a beam splitter upon arriving at the receiving end of its journey. This device causes the photon to choose at random between two different routes. The original sender then receives a response from the receiver, which contains information about the sequence of the photons that were transmitted. The sender then compares this information with the emitter, which is the component that would have transmitted each photon.

A certain sequence of bits is all that is left after the photons that were gathered by the wrong beam collector have been discarded. Using this particular bit sequence as a key will allow you to decrypt the data that was previously encrypted. During the phase of error correction and other post-processing operations, any errors that occurred as well as any data loss were eradicated. This phase was responsible for the elimination of any errors that occurred. A further post-processing phase known as delayed privacy amplification eliminates any information regarding the final secret key that an eavesdropper may have obtained. This stage is carried out once the generation of the ultimate secret key has been completed.

3 Motivation and Objective of Current Study

The basic motivation behind this study is to find the recent advancements and what we can do in cryptographic algorithm, so make it feasible and easy to float in QKD networks.

The advanced cryptographic algorithm makes data secure and if someone wants to hack the data, they need a high configuration system with sufficient time. If the secret key was made by quantum computers, it is not easy to decode it. The recent decade has proved that CV protocol is much more robust to classical traffic in the same optical fiber because to the filtering effect of the interferometric detection, which can allow for significantly cheaper operational costs.

4 Recent Advancements in QKD

Optical switching, reliable relays, unreliable relays, and quantum repeater-based technologies are frequently used to implement QKD networks. The protocols based on optical switching and trusted relays, as opposed to those based on untrusted relays and quantum repeaters, are further along in terms of development.

An optical switching-based QKD network, which can be swiftly constructed using commercial technologies, connects a pair of QKD nodes by applying several classical optical functions to the quantum signals broadcast across a quantum channel, such as beam splitting and switching. Transmission of quantum signals is possible via short quantum networks without interacting with any nodes that cannot be verified as trustworthy. Since quantum signals have a natural attenuation that can't be overcome by amplification, they are best suited for local and metropolitan networks rather than wide-area ones [7, 8].

In a network that uses trusted relays for QKD connections, local secret keys are generated for each quantum key connection and then saved in the nodes available at both ends of each quantum key links. For QKD with lengthy distances between available nodes, a series of concatenated QKD links and a one-dimensional network of trusted relays are used. End-to-end information-theoretic security is achieved by broadcasting the secret keys hop-by-hop through the QKD path, between the source and the destination nodes.

A QKD network with an untrusted relay allows eavesdroppers to control untrusted relays without compromising QKD. Untrusted relays can potentially increase QKD's secure distance. When TF-QKD protocols are used, the maximum distance that a standalone untrusted relay can travel is restricted to 500 and 600 km. It is not possible to establish a direct connection between two untrusted relays using the

QKD protocol. Therefore, this type of quantum network is better suited for shorter range distance networks; however, in order for its large-scale extension to function correctly, integration with reliable relays is required.

5 QKD Networks Architecture

A QKD network needs verified classical network like an optical network and various other secured applications housed in the said network. QKD networks are being used in communication and safety infrastructures. In light of the range of QKD network architectures offered, we define a simple three-layered architecture from a holistic approach. The architecture is divided into three distinct logical layers [9] (Fig. 2).

Infrastructure Layer: This layer is made up of the physical equipment [10] designed specifically for use in QKD networks, and is part of the QKD network infrastructure. All the devices in the same place are physically safe because they are installed in a trusted node. A QKD node is a specific type of node that satisfies both of these criteria. In the following section, we'll go into depth about how the actual hardware may vary depending on which QKD network implementation choice is chosen. Keys can be generated as symmetric random bit strings by each pair of QKD nodes, and they can communicate with one another through optical fiber links. As a result, it is possible to use the QKD protocols or actual devices created by several vendors in isolation from one another [11]. Since the secret keys that are made up of traditional bit strings, they are easy to store in the QKD nodes [12]. The quantum key node stores its own security-related parameters like secure key, the physical device identifier, and the time stamp when secret keys were made and stored [13]. Each QKD node also keeps track of the error rate of quantum channels and other link metrics like link length and link type.

Control and Management Layer: This layer is made up of the QKD network controller and manager [14], and it is the QKD network controller that is in charge of

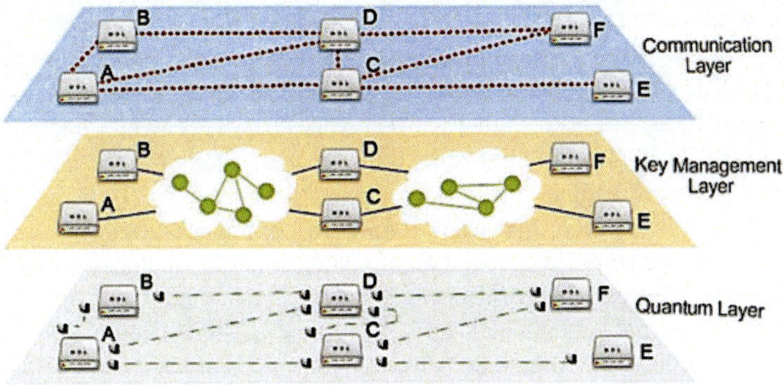

Fig. 2 Architecture of QKD

managing all of the QKD nodes. It is the QKD network controller that is responsible for activating, deactivating, and calibrating the QKD nodes. In contrast, the QKD network manager is in charge of monitoring and directing the QKD network. It monitors the state of the QKD network and all of its connections, as well as the QKD network controller, to ensure that everything is working properly. The data that is gathered through monitoring and administration can be entered into a database and kept there to be updated at predetermined intervals. Original secret keys that are available in quantum key nodes will not be distributed across multiple physical locations, and the QKD network controller or manager will not be able to see these keys.

Application Layer: In this layer, users can access the necessary cryptographic tools. As shown in the figure, a QKD network's service provision pipeline for cryptographic applications is quite straightforward. Before making any security requests, cryptographic applications must notify the QKD network management of details like secret-key size, rate, update period, etc. QKD network management uses these queries to ask QKD nodes for secret keys. If there are real-time secret keys, the quantum key network manager tells the quantum key network controller. The quantum key network controller then tells the right quantum key nodes to give the secret or private keys in the right way. If the supply of secret keys runs out, cryptographic operations must pause until they can continue. After receiving the secret keys, each cryptographic application is solely responsible for their usage; the quantum key nodes and its network management bear no further accountability for the security of these keys. Each QKD network/capacity system is limited by the sum of its users' secret-key needs and the number of secret-key resources accessible in that network.

6 Future Direction for Research

This review lays the groundwork for cross-community discussion across disciplines on how to design the Qinternet and shows how QKD networks can provide compelling applications and new perspectives with robust security into the foreseeable future. Network coding [15] has been studied in traditional networks, however it must be adapted for QKD networks. Network coding can help multicast secret or private keys from different transmitters to numerous receivers [16]. This would enable public multi-user QKD [17]. Quantum network coding is a novel network coding paradigm [18], however most investigations focus solely on its theory [18–20]. A capable area of doing research is conceiving answers for all low confidence-level transmits, such as the trustworthy transmits in figure, and for diverse quantum memory requirements in supporting the Qinternet's evolution.

In terms of network architecture, key generation speed, communication range, and routing algorithm, the existing body of research and testing has shown promising outcomes. But there are a few issues that need to be resolved before we can call it done. This research zeroes in on security concerns (i.e., security assumptions and applications) to highlight the most pressing problems with QKD networks, explains those problems and how to solve them, and gives researchers a solid foundation on which to build effective solutions.

The key distribution service of the existing QKD network only provides point-to-point (P2P) key distribution; however, it does not have the point-to-multipoint (P2M) mechanism. This is due to the absence of point-to-multipoint (P2M) mechanisms in QKD networks.

A multiple-path strategy uses up many of the available resources of a quantum node. The assumption that all quantum nodes must be trusted can be avoided by using a multiple-path strategy; however, the implementation of a multiple-path strategy uses up a large portion of the available resources of a quantum node in order for it to be successful. One example of this is the local keys that are used to assist in the transmission of the session key.

The classical end users and applications cannot communicate securely with the quantum nodes because there is no proper security interface between the two: Working in an environment that utilizes quantum computing poses a substantial problem in terms of providing classical end users and applications with secure access to the key distribution service offered by QKD networks.

Increasing the efficiency of QKD networks is necessary to ensure the safety of future Qinternet users. New protocols and hardware are needed to extend the range and secret-key rate of QKD networks. While TF-QKD [21] and PM-QKD [22] show promising results for addressing the distance limitation of traditional point-to-point QKD protocols, large-scale practical QKD deployment is made possible by quantum key based on chip with integrated devices. Research is needed to construct mathematical models of QKD networks that can precisely define and appraise the presentation of real-world quantum key networks with varying topologies and quantum key protocols.

Battelle [23] has tested custom-made and marketable QKD structures in a measured research laboratory to quantify the presentation realized in real-world urban and long-haul contexts. More study is needed before commercialization of the handheld mobile QKD device family [24]. As an added complication, an attacker might exploit the flaws in the QKD network's implementation in order to render it inoperable, making implementation security a key factor in the industry's reluctance to widely use QKD networks.

As an alternative to QKD networks, post-quantum cryptography uses algorithms that have been confirmed to be secure against existing quantum assaults. Since all post-quantum procedures are realized in package, post-quantum cryptoanalysis can be used in conjunction with any existing security system. The fact is that traditional cryptosystems have features that QKD cannot duplicate just now. Although post-quantum cryptography and QKD solutions are two distinct lines of inquiry, to far they have seen only limited practical use (Fig. 3).

One of the most prominent applications of blockchain technology currently in use is digital currencies such as Bitcoin [25], which have recently risen to popularity. Despite the general notion that this is not the case, blockchain is vulnerable to assaults from quantum computers [26] [Citation needed]. There have been a number of studies [27] that have concentrated on post-quantum blockchain solutions with the

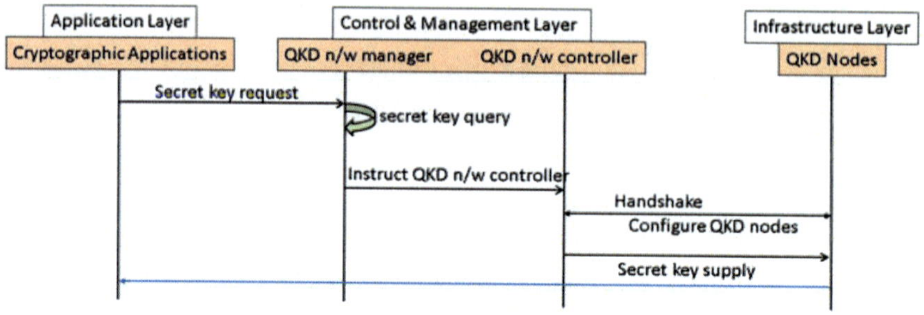

Fig. 3 Simple cryptographic service pipeline diagram

intention of securing the blockchain through the usage of post-quantum cryptography. Nevertheless, QKD is a viable approach for solving the one-of-a-kind challenges that blockchain technology will confront as it enters the quantum age. QKD networks in urban areas have been demonstrated to be a workable solution for the implementation of a quantum-safe blockchain platform for the purposes of authentication [28]. In addition, a framework for a quantum-secured permissioned blockchain is given in [29]. This framework is dependent on the utilization of a digital signature process that is based on the distribution of quantum keys.

The Internet of Things does need a very secure cryptosystem. An important field of IoT research [30–33] is the idea of a post-quantum Internet of Things, which would use post-quantum cryptography to protect IoT systems from the imminent threats posed by quantum computing. In contrast, further study is needed into the quantum IoT, which combines quantum cryptography and the IoT, because it is still in its infancy [34, 35].

Entanglement, one of the most fascinating elements of quantum physics, has a wide range of applications in quantum information science [36], such as the distribution of quantum keys and quantum teleportation [37]. The necessary technology, such as quantum processors and quantum memory, needs to be improved before QKD networks can rely entirely on entanglement.

7 Conclusion

Despite the fact that QKD networks have the ability to provide long-term data protection and security that is future-proof for a wide variety of applications, they also have a number of challenges that have not yet been resolved. This review not only provides an overview of the past achievements in the field of QKD networks but also looks forward to the research that will be conducted in the near future on QKD networks. To begin, we did a high-level overview of the QKD process, including its many possible applications, as well as some of the industry-standard protocols for carrying it out.

Then, after conducting an investigation into the development of QKD network implementations, we classified those that were already in existence in accordance with our findings. Research and development efforts in the domains of physics, computer science, safety, and structures related to communications will all be needed to move forward with the full development of quantum key distribution-based networks.

References

1. Cao Y, Zhao Y, Wang Q, Zhang J, Ng Sx, Hanzo L (2022) The evolution of quantum key distribution networks: on the road to the qinternet'. IEEE Commun Surv & Tutor 24(2):839–94
2. Accessed 13 Dec 2022. https://www.etsi.org/images/files/ETSIWhitePapers/
3. Schreiber LR, Bluhm H (2018) Toward a silicon-based quantum computer. Science (New York, N.Y.)
4. Ladd TD, Jelezko F, Laflamme R, Nakamura Y, Monroe C (2010) JL O/'Brien quantum computers. Nature 464(7285):45–53
5. Debnath S, Linke NM, Figgatt C, Landsman KA, Wright K, Monroe C (2016) Demonstration of a small programmable quantum computer with atomic qubits. Nature 536(7614):63–66
6. Arute F, Arya K, Babbush R, Bacon D, Bardin JC, Barends R, Biswas R et al (2019) Quantum supremacy using a programmable superconducting processor. Nature 574(7779):505–10
7. Gong M, Wang S, Zha C, Chen MC, Huang HL, Wu Y, Zhu Q, et al (2021) Quantum walks on a programmable two-dimensional 62-qubit superconducting processor. Science (New York, N.Y.) 372(6545):948–952
8. Peev M, Pacher C, Alléaume R, Barreiro C, Bouda J, Boxleitner W, Debuisschert T et al (2009) The SECOQC quantum key distribution network in Vienna. New J Phys 11(7): 075001.
9. Tang YL, Yin HL, Zhao Q, Liu H, Sun XX, Huang MQ, Zhang WJ et al (2016) Measurement-device-independent quantum key distribution over untrustful metropolitan network. Phys Rev X 6(1)
10. Briegel H-J, Dür W, Cirac JI, Zoller P (1998) Quantum repeaters: the role of imperfect local operations in quantum communication. Phys Rev Lett 81(26):5932–35
11. Meter R, Touch J (2013) Designing quantum repeater networks. IEEE Commun Mag 51(8):64–71
12. Overview on networks supporting quantum key distribution (2019). Int Telecommun Union
13. Sasaki M, Fujiwara M, Ishizuka H, Klaus W, Wakui K, Takeoka M, Miki S et al (2011) Field test of quantum key distribution in the tokyo QKD network. Optics Express 19(11):10387–10409
14. Tajima A (2017) Quantum key distribution network for multiple applications. Quantum Sci. Technol 2(3)
15. Cao Y, Zhao Y, Lin R, Xiaosong Y, Zhang J, Chen J (2019) Multi-tenant secret-key assignment over quantum key distribution networks. Opt Express 27(3):2544–2561
16. Cao Y, Zhao Y, Wang J, Xiaosong Y, Ma Z, Zhang J (2019) SDQaaS: software defined networking for quantum key distribution as a service. Opt Express 27(5):6892–6909
17. Dynes JF, Wonfor A, Tam WW-S, Sharpe AW, Takahashi R, Lucamarini M, Plews A et al (2019) Cambridge quantum network. NPJ Quantum Inf 5(1)
18. Ahlswede R, Cai, N, Li SYR, Yeung RW (2000) Network information flow. IEEE Trans Inf Theory 46(4):1204–1216
19. Xu FH, Wen H, Han ZF, Guo GC Network coding in trusted relay based quantum network. Accessed 14 Dec 2022
20. Nguyen HV, Trinh PV, Pham AT, Babar Z, Alanis D, Botsinis P, Chandra D, Ng SX, Hanzo L (2017) Network coding aided cooperative quantum key distribution over free-space optical channels. IEEE Access: Pract Innov Open Solut 5:12301–12317

21. Kato G, Owari M, Hayashi M (2021) Single-shot secure quantum network coding for general multiple unicast network with free one-way public communication. IEEE Trans Inf Theory 67(7):4564–4587
22. Shang T, Li J, Liu J-W (2016) Secure quantum network coding for controlled repeater networks. Quantum Inf Process 15(7):2937–2953
23. Satoh T, Ishizaki K, Nagayama S, Van Meter R (2016) Analysis of quantum network coding for realistic repeater networks. Phys Rev A 93(3):032302
24. Lucamarini M, Yuan ZL, Dynes JF, Shields AJ (2018) Overcoming the rate-distance limit of quantum key distribution without quantum repeaters. Nature 557(7705):400–403
25. Ma X, Zeng P, Zhou H (2019) Erratum: phase-matching quantum key distribution [phys. rev. x 8, 031043 (2018)]. Phys Rev X 9(2):029901
26. Li Q, Wang Y, Mao H, Yao J, Han Q (2020) Mathematical model and topology evaluation of quantum key distribution network. Opt Express 28(7):9419–9434
27. Wang Y, Li Q, Mao H, Han Q, Huang F, Xu H (2020) Topological optimization of hybrid quantum key distribution networks. Opt Express 28(18). 26348-26358
28. Yang YH, Li PY, Ma SZ, Qian XC, Zhang KY, Wang LJ, Zhang WL et al (2021) All optical metropolitan quantum key distribution network with post-quantum cryptography authentication. Opt Express 29(16):25859–67
29. Gao Y-L, Chen X-B, Chen Y-L, Sun Y, Niu X-X, Yang Y-X (2018) A secure cryptocurrency scheme based on post-quantum blockchain. IEEE Access 6:27205–27213
30. Li C-Y, Chen X-B, Chen Y-L, Hou Y-Y, Li J (2018) A new lattice-based signature scheme in post-quantum blockchain network. IEEE Access 7:2026–2033
31. Fernandez-Carames TM, Fraga-Lamas P (2020) Towards post-quantum blockchain: a review on blockchain cryptography resistant to quantum computing attacks. IEEE Access 8:21091–21116
32. Kiktenko EO (2018) Quantum-secured blockchain. Quantum Sci Technol 3(3)
33. Sun X, Sopek M, Wang Q, Kulicki P (2019) Towards quantum-secured permissioned blockchain: signature, consensus, and logic. Entropy 21(9):887
34. Fernandez-Carames TM (2020) From pre-quantum to post-quantum IoT security: a survey on quantum-resistant cryptosystems for the internet of things. IEEE Internet Things J 7(7):6457–6480
35. Ebrahimi S, Bayat-Sarmadi S, Mosanaei-Boorani H (2019) Postquantum cryptoprocessors optimized for edge and resource-constrained devices in IoT. IEEE Internet Things J 6(3):5500–5507
36. Liu Z, Choo KKR, Grossschadl J (2018) Securing edge devices in the post-quantum internet of things using lattice-based cryptography'. IEEE Commun Mag 56(2):158–62
37. Ottaviani C, Woolley MJ, Erementchouk M, Federici JF, Mazumder P, Pirandola S, Weedbrook C (2020) Terahertz Quantum Cryptography. IEEE J Sel Areas Commun 38(3):483–495. https://doi.org/10.1109/jsac.2020.2968973

Thermal Management System of Battery Using Nano-coolant

Prakirty Kumari and Manikant Paswan

Abstract It is important to choose an efficient cooling method for thermal control of lithium-ion battery system, so that these strategies should provide cost worthy solutions of energy saving for rise in temperature of the system during the operation of battery. Battery is one of the main parts of electric vehicles and as compared to other batteries like lead-acid, nickel-cadmium, etc., lithium-ion batteries are receiving more attention of automobile industries and other industries due to its high energy density, power density, voltage, life cycle, and low self-discharge rate of energy. The heat generation within the Li-ion battery reduces its performance and life of battery as well. This paper investigates the thermal control of the battery using air, water, and graphene (0.4%)-water nanofluid as coolant having thermal conductivities of 0.0242, 0.60, and 0.6203 W/m·K. In this paper, the NTGK method is used to simulate the thermal analysis of Li-ion batteries under the MSMD model of a battery in ANSYS Fluent. The simulations have been carried out for estimating the maximum temperature of the battery under different velocities of fluid using air, water, and nanofluid coolants. The result shows that the maximum temperatures 308.782, 302.734, and 300.656 K have been obtained after cooling with air, water, and nanofluid with flow velocities of 20, 0.01, and 0.015 m/s. After comparing the maximum temperatures obtained after using different coolants, it is found that nanofluid reduces the maximum temperature more as compared to air and water.

Keywords Electric vehicle · Li-ion battery · Heat generation · ANSYS fluent · MSMD model · Graphene-water-based nanofluid

P. Kumari (✉) · M. Paswan
Department of Energy System Engineering, National Institute of Technology, Jamshedpur, Jharkhand, India
e-mail: prakirtykumari98@gmail.com

M. Paswan
e-mail: mkpaswan.me@nitjsr.ac.in

M. Paswan
Department of Mechanical Engineering, National Institute of Technology, Jamshedpur, Jharkhand, India

© The Author(s), under exclusive license to Springer Nature Singapore Pte Ltd. 2023

A. Mishra et al. (eds.), *Advances in IoT and Security with Computational Intelligence*, Lecture Notes in Networks and Systems 755,
https://doi.org/10.1007/978-981-99-5085-0_18

1 Introduction

In the present scenario, depletion of fossil fuel and emission of greenhouse gases raise the global issue of energy crisis and air pollution. Hence, it is required to move toward the renewable source of energy. The transportation sector is widely acknowledged as a major contributor to air pollution, electric vehicle (EV) and hybrid electric vehicle (HEV) have been designed to minimize some loads on non-renewable source. A power battery is one of the essential parts of EVs and HEVs to store power for smooth working. Several types of batteries like lead-acid, nickel-cadmium, nickel-metal-hydride, and lithium-ion batteries are used in EVs but out of these, lithium-ion batteries are considered as one of the promising batteries because of its advantages such as high energy density, high power density, large power storage, longer life cycle, and low self-discharge rate. In spite of these benefits of LIB, there are some crucial factors that restrict its applicability due to heat generation within the battery and overheating that reduce its performance, efficiency, and life cycle. Also, there have been some recent cases of electric vehicles catching fire and one of the reasons behind which is rise in battery temperature. That's why cooling of battery is an important concern for EVs. Choudharia et al. [1] have described the real concept of generation of heat in battery and its impact on various components of li-ion battery. Also different thermal control techniques such as air, liquid, phase change material (PCM) cooling, and combination of it using different design structures have been studied. Xia et al. [2] have reviewed on the different thermal control systems of the battery for electric vehicle. Various cooling configuration, including air cooling, liquid cooling, and phase change material (PCM) cooling, have been discussed and analyzed. Different configurations for air cooling, such as series, series-parallel mix, and parallel, are discussed and concluded that series-parallel configuration has been proved to be more efficient. Direct and indirect liquid cooling, combination of fins and cold plate, passive cooling method, i.e., PCM-based cooling, other methods of cooling such as cooling using thermoelectric cooler with liquid cooling, cooling with heat pipe are reviewed. Jayamohan et al. [3] have tested thermal behavior of $LiFePO_4$ Li-ion battery 3.2 V/6Ah battery under air, water, and oil cooling. They have also analyzed the charging-discharging characteristics, state of charge (SOC) under different ambient temperatures, and motor torque characteristics and also the validation of experimental results has been done by using MATLAB and Minitab software. Chen et al. [4] have compared the thermal performance of LiB using direct air cooling, direct liquid cooling, indirect liquid (jacket), and fin cooling and have discussed the consumption of parasitic power by different coolants, rise in maximum temperature, temperature difference in a cell, and extra weight used for the cooling systems. Duan et al. [5] have investigated the thermal control of battery modules using phase change materials (PCM) experimentally. Here, they have come up with the two specific thermal control arrangements with PCMs where the first model involves with the heater positioned in a cylindrical box filled with PCM and the second model wraps the heater using a jacket of PCM sheet and found that both designs are impressive. Jaguemont et al. [6, 12] have given the information about the effect of temperature on li-ion batteries in terms of high temperature and safety issues. Different types of thermal management systems such as cooling with air, liquid, refrigerant, PCM, heat pipe, and thermoelectric module have been reviewed. The development made on the future of battery thermal management system and how to overcome the future challenges have been discussed. Sefidan et al. [7] have worked on Al_2O_3-water nanofluid as coolant for battery thermal management system where the cell is immersed into the container of aluminum filled with Al_2O_3-water nanofluid. The generated heat will be extracted by liquid. When the liquid is in direct contact with the battery surface, a lot of heat will enter the intermediate liquid. Then the surrounding thin aluminum will transfer the heat to the air flowing around the module. Using 3D transient CFD simulation, the impact of the existence of the secondary cylinder on the rise in temperature has been analyzed. Tousi et al. [8] have developed a thermal management system for 18,650 /21700 model of li-ion batteries based on AgO nanofluid as a coolant. They have carried out a numerical investigation of the impacts of discharge C-rate, nanofluid volume fraction, and flow speed on the thermal efficiency of the battery percentages at high discharge rates (3C, 5C, and 7C). They have also extensively compared the results of thermal control in 18,650 and 21,700 battery packs. Kumar Bhagat et al. [9] have carried out different types of the thermal management systems for a battery: air cooling, water cooling, cooling with oil, and nanofluid cooling (Al_2O_3) for prismatic pouch li-ion battery and found that the temperature of the battery reduces to 30 °C when cooled with Al_2O_3-water nanofluid. Malik et al. [10] developed the experimental setup of three 20Ah prismatic $liFePO_4$ cells connected in series with cold plates which is in direct contact of each Li-ion cell for liquid cooling at different temperatures of 10, 20, 30, and 40 °C for the thermal analysis of battery under different current discharge rates, i.e., 1C, 2C, 3C, and 4C. It was found that a coolant at 30 °C temperature is appropriate for the battery because at that temperature, the battery's maximum and average temperature is in the range of 20–45 °C.

After going through these papers, it is found that much literature pertaining about the thermal management of Li-ion batteries for electric vehicles is available. Also, numerical investigation and battery design to optimize battery temperature have been done based on air and water as coolants. There are limited works which have been done using nanofluid as a coolant. This paper aims to simulate a Li-ion battery using ANSYS software and investigates the cooling methods of battery using Multi-Scale Multi-Domain (MSMD) model in ANSYS using different coolants based on air and water and graphene-water-based nanofluid.

2 Battery Design

A 3D model of 35Ah pouch cell Li-ion battery with dimension 169 mm × 179 mm × 14 mm [4] is designed using ANSYS software in design modeler for all simulations. The 3D model of single cell Li-ion battery without cooling is shown in Fig. 1a. On both sides of the main cell, an extrusion of 7 mm is performed, representing the fluid

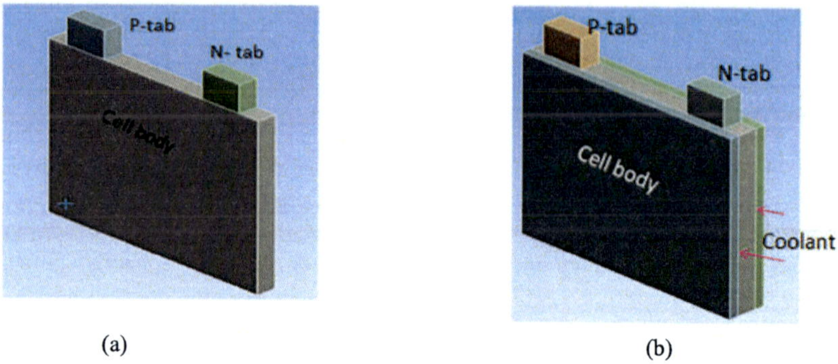

Fig. 1 Li-ion cell design **a** without cooling and **b** with cooling

coolant flowing over the surface of the battery, shown in Fig. 1b to represent the fluid as a flow of coolant on the surface of the body. The model includes three parts: the active volume, positive current tabs, and negative current tabs. The study of the performance of thermal management of the battery is utilized by ANSYS fluent.

The model is solved by some heat transfer equations using MSMD battery module in ANSYS fluent [5]. The electrical and thermal equations solved are given below:

Electrical Equations:

$$\nabla \cdot (\sigma_+ \nabla \phi_+) = -j$$

$$\nabla \cdot (\sigma_- \nabla \phi_-) = -j$$

$$\mathrm{j} = \mathrm{I}/\ \mathrm{Volume} \tag{1}$$

where σ = effective electrical conductivities for electrode, Ø = phase potential for the electrode, + and—signs represent for the positive and negative tabs, and J = volumetric current density.

Thermal Equations:

$$\frac{\partial \rho C_p T}{\partial t} - \nabla \cdot (k \nabla T) = \dot{q}$$

$$\dot{q} = \sigma_+ \nabla^2 \phi_+ + \sigma_- \nabla^2 \phi_- + j\left[U - (\phi_+ - \phi_-) - T\frac{dU}{dT}\right] \tag{2}$$

where T = temperature, q̇ = rate of heat generation during the battery work, k = coefficient of thermal conductivity, C_p = heat capacity, and U = open-circuit voltage of the battery.

3 Cooling Methods

The cooling part of the battery is shown as the extended thickness of 7 mm on both sides of battery as shown in Fig. 1b.The heat is generated during charging/discharging through the cell body that is extracted by coolant inserted in between the battery. The coolant's air and direct liquid cooling flows through the increased thickness direct in contact with both sides of the battery, i.e., the gap between two cells and contact cell. In air, fan is used to blow the air on the surface of the cell body and extract the heat from the surface of the body. Here, water and graphene(0.4%)-water nanofluid are used for direct liquid cooling where cold plate is integrated directly in contact with the cell body for the flow of liquid. Figure 2 illustrates the flowing path of coolant between the cells, including the inlet and outlet points, which facilitate the circulation of the cooling medium through the battery system. Normally for cooling the battery, as shown in Fig. 3, the following arrangement of cooling has been done consisting of a battery pack, a fan or a pump, a heat exchanger, and coolant pipes [4]. In this paper, graphene nanoparticles and water as base fluids are used for making nanofluid. Due to its high electrical and thermal properties, significant energy saving, less consumption of electricity as compared to other compound, and also its antibacterial properties, we have selected graphene to fabricate the nanofluid.

After activating the MSMD model by writing "define module addon-model" in console and selecting the dual potential or now in new version of ANSYS, battery model is already inbuilt in setup in ANSYS fluent. Here, NTGK method is selected for further analysis of battery and model taken in transient state varies with time.

The fluid properties are mentioned in Table 1. The properties of graphene (0.4%)-water-based nanofluid are calculated theoretically by using formulae [11].

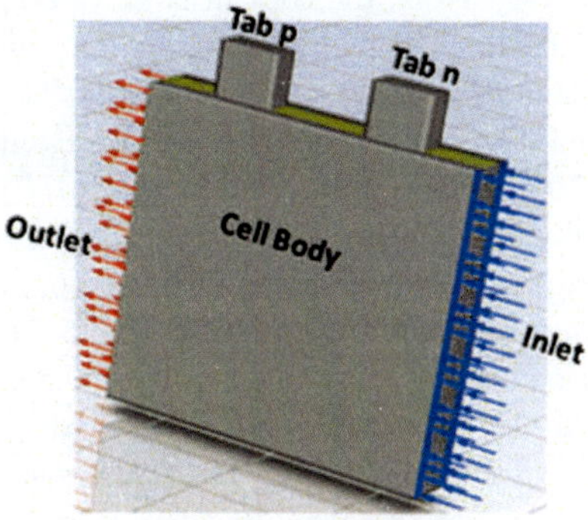

Fig. 2 Flowing path of coolant in battery

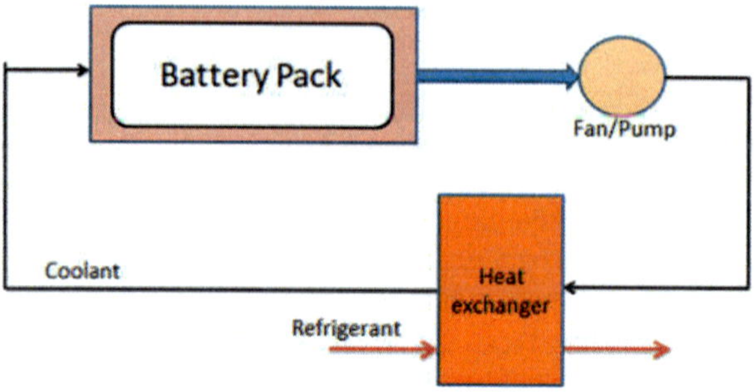

Fig. 3 General schematic diagram of cooling of battery

Table 1 Properties of different fluids used as coolant

Properties	Air	Water	Nanofluid (graphene 0.4%-water)
Density (kg/m^3)	1.225	998.2	1003.2752
Specific heat capacity (j/kg·K)	1006.43	4182	4171.68
Thermal conductivity (W/m·K)	0.0242	0.6	0.6203
Kinematic viscosity (kg/m·s)	1.7894e − 05	0.001003	1.01209e − 3
Electrical conductivity (Siemens/m)	1,000,000	1,000,000	1.01204e + 8

4 Results and Discussion

A various number of simulations were carried out for the estimation of temperature of the battery after cooling with air, water, and graphene-water nanofluid and its effect by changing the flow velocity. Here, the coolant is flowed directly over the surface of cell body from inlet to outlet. The path of fluid flow is considered as laminar.

Different velocity is taken for different coolant used for thermal control of the battery. Air and water cooling simulations were conducted with varying flow velocities, ranging from 0 to 20 m/s for air cooling and from 0 to 0.01 m/s for water cooling [4]. The nano-coolant velocity is taken as 0.015 m/s. The battery discharging rate is taken as 2.71C for all the simulations with the same initial conditions and inlet temperature. The battery maximum temperature without cooling is obtained as 316.229 K at 2.71C discharge rate. When battery was cooled with air, the maximum temperatures of the battery obtained are 310.754, 309.23, 308.782, and 302.889 K, 302.734 K at a flow velocity of 0, 4, 20 m/s for air and 0.005, 0.01 m/s for water and the comparison of results is validated using paper presented by Chen et al. When cooling is carried out with graphene (0.4%)-water nanofluid, the maximum temperature obtained is 300.656 K at a flow velocity of 0.015 m/s. The simulation results for battery cooling with different coolant are shown in Fig. 4.

The contour plot of battery maximum temperature for air, water, and nanofluid with cooling and without cooling is shown in Fig. 5. The variation of temperature at different parts of battery is shown by color spectrum.

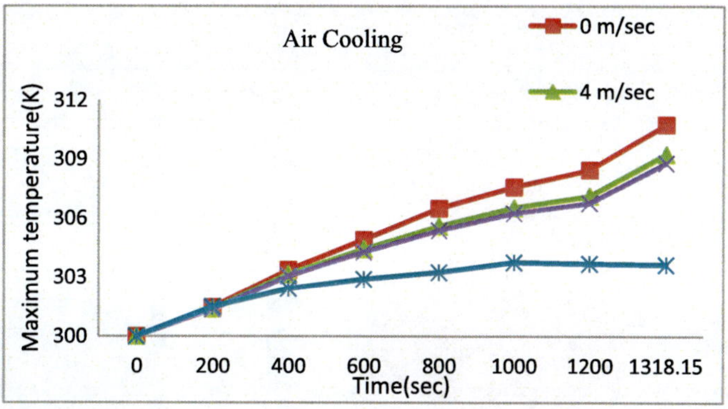

(a)

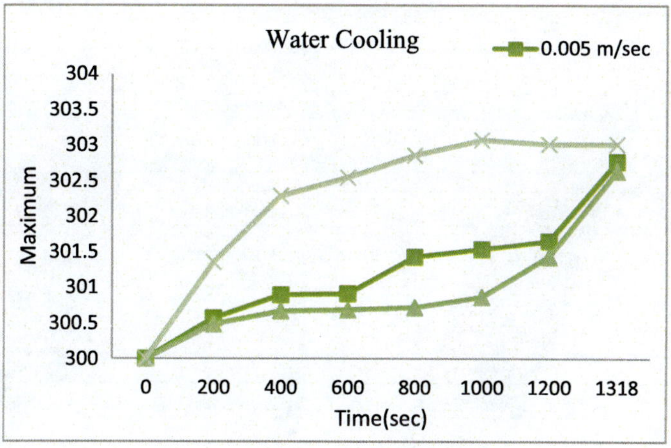

(b)

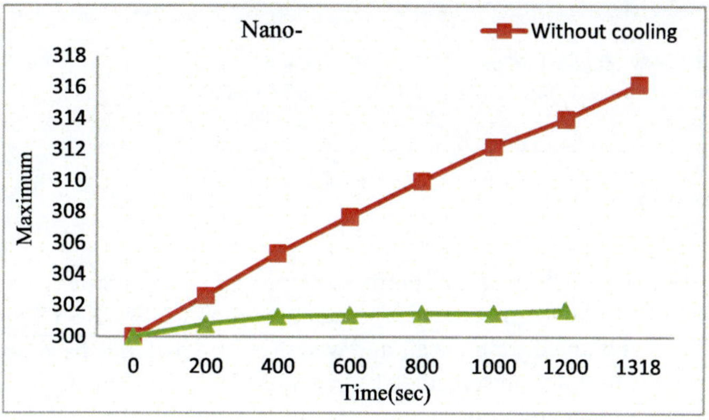

(c)

Fig. 4 Battery maximum temperature at different flow velocity during 2.71C discharge rate and validation of results when cooled with **a** air, **b** water, and **c** nanofluid

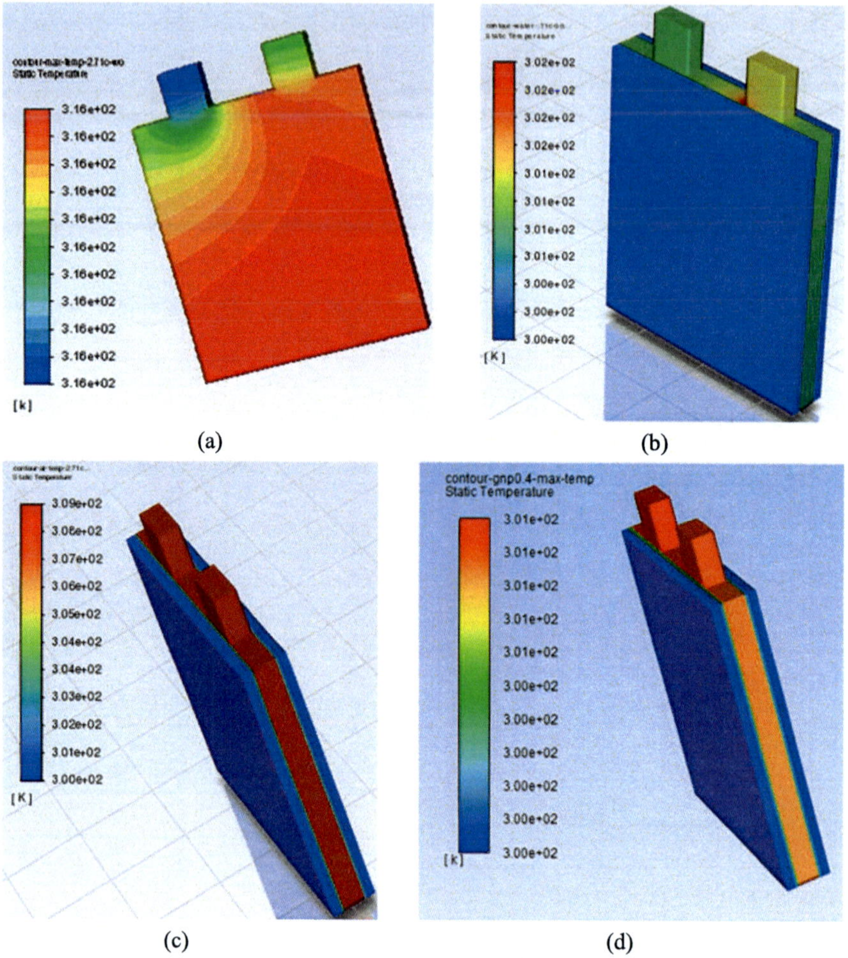

Fig. 5 Battery body temperature profile of **a** without cooling, **b** air cooling at a flow velocity of 20 m/s, **c** water cooling at a flow velocity of 0.01 m/s, and **d** nanofluid cooling at a flow velocity of 0.015 m/s

5 Conclusions

This paper has analyzed the thermal control systems for Li-ion batteries using various coolants: air, water, and graphene (0.4%)-water nanofluid using ANSYS fluent software. The results obtained from the air and water cooling methods have been cross-validated with the findings presented in the paper by Chen et al. Battery temperature is greatly affected by charging/discharging rates.

The maximum temperature of the battery is estimated as 316.229 K during a discharge rate of 2.71C when there is no cooling system constructed with it. But when the battery is cooled by air and water, the maximum temperatures of the battery obtained are 308.754 K and 302.734 K at the flow velocities of 20 m/s for air and 0.01 m/s for water. The maximum temperature of the battery is extensively decreased to 300.656 K when cooled with graphene-water-based nanofluid. By increasing the flow velocity of the fluid, the battery's maximum temperature is reduced more. Also, the extra weight is added when cooled with water and nano-coolant. The leakage of liquid is one of the concerns. Battery cooling with graphene-water-based nanofluid is one of the promising cooling systems for future.

References

1. Choudhari VG, Dhoble AS, Sathe TM (2020) A review on effect of heat generation and various thermal management systems for lithium ion battery used for electric vehicle. J Energy Storage 32:101729
2. Xia G, Cao L, Bi G (2017) A review on battery thermal management in electric vehicle application. J Power Sources 367:90–105
3. Jayamohan D, Venkatasalam R, Thangam C (2022) Experimental analysis of thermal behavior of a lithium-ion battery using constant voltage under different cooling conditions. Int J Electrochem Sci 17(220810):2
4. Chen D, et al (2016) Comparison of different cooling methods for lithium ion battery cells. Appl Therm Eng 94:846–854
5. Duan X, Naterer GF (2010) Heat transfer in phase change materials for thermal management of electric vehicle battery modules. Int J Heat Mass Transf 53(23–24):5176–5182
6. Bandhauer TM, Garimella S, Fuller TF (2011) A critical review of thermal issues in lithium-ion batteries. J Electrochem Soc 158(3):R1
7. Sefidan AM, Sojoudi A, Saha SC (2017) Nanofluid-based cooling of cylindrical lithium-ion battery packs employing forced air flow. Int J Therm Sci 117:44–58
8. Tousi M et al (2021) Numerical study of novel liquid-cooled thermal management system for cylindrical Li-ion battery packs under high discharge rate based on AgO nanofluid and copper sheath. J Energy Storage 41:102910
9. Bhagat VK, Paswan MD (2022) Thermal management analysis of a lithium-ion battery cell using different coolant. J Phys Conf Ser 2178(1) (IOP Publishing)
10. Malik M et al (2018) Thermal and electrical performance evaluations of series connected Li-ion batteries in a pack with liquid cooling. Appl Therm Eng 129:472–481
11. Sheikholeslami M, Rokni HB (2017) Simulation of nanofluid heat transfer in presence of magnetic field: a review. Int J Heat Mass Transf 115:1203–1233
12. Jaguemont J, Van Mierlo J (2020) A comprehensive review of future thermal management systems for battery-electrified vehicles. J Energy Storage 31:101551

Opinion Mining on Ukraine–Russian War Using VADER

Dagani Anudeepthi, Gayathri Vutla, Vallam Reddy Bhargavi Reddy, and T. Santhi Sri

Abstract The purpose of our research paper is to discover the best method to implement opinion mining on Ukraine–Russian war based on the tweets generated by people all over the world on Twitter. We have performed VADER, Naive Bayes, and DistilBERT and have done some relative comparative studies finalizing that VADER is the best approach to perform opinion mining for a vast data like Ukraine–Russian war because VADER gives high accuracy based on emotions just like a human, who understands contexts where machines cannot. We choose DistilBERT and Naive Bayes as they are some of the popular models which give the best output based on sentiment score hence, we figured out that the best model which is optimal to find the sentiment accuracy based on the emotion is VADER when compared with DistilBERT and Naive Bayes. The two main aspects which highly differentiate VADER from all other methods are Polarity and Intensity. VADER being a library has a powerful set of modifiers making it more unique which gives higher accuracy to perform opinion mining on Ukraine–Russian war tweets.

Keywords Opinion mining · Lexicons · Accuracy

1 Introduction

Opinion mining is critical in obtaining customer emotions and opinions from textual data. The process of identifying positive, negative, or neutral attitudes in text is called opinion mining. It is frequently referred to as emotion AI or sentiment analysis. It is used in a variety of applications such as enterprises and service providers to identify user's fulfillment and to understand their demands. Opinion mining looks beyond the polarization of emotions to uncover distinct feelings and emotions such as anger, sadness, happiness, and excitement. It also determines the importance of the text provided as well as the intentions of the individual who composed the material. One

D. Anudeepthi · G. Vutla · V. R. Bhargavi Reddy · T. Santhi Sri (✉)
Koneru Lakshmaiah Education Foundation, Andhra Pradesh, India
e-mail: santhisri@kluniversity.in

© The Author(s), under exclusive license to Springer Nature Singapore Pte Ltd. 2023
A. Mishra et al. (eds.), *Advances in IoT and Security with Computational Intelligence*, Lecture Notes in Networks and Systems 755,
https://doi.org/10.1007/978-981-99-5085-0_19

of the reasons we perform opinion mining is to obtain a sentiment score. It is a score that represents the depth and complexity of the emotions in the textual material. It identifies emotions and assigns a sentiment score, for example, from 1 to 15 where 1 refers to the utmost negative and 15 refers to the utmost positive emotion. The number is determined by the emotional complexity of the feelings depicted in the text by the author. Opinions can be partially positive or partially negative, and these partially positive and partially negative feelings cannot be interpreted as negative or positive. This is made simpler by sentiment score since we have a quantitative measurement for the partial emotions, and if there is a substantial amount of information that needs to be analyzed; sentiment score makes it easier to process. As a result, people's needs are better understood. To comprehend the sentiments behind a text [1–3], we can use NLP approaches to detect patterns and analyze the patterns [4], which will enable us to draw quick conclusions. In general, we use AI and ML to perform opinion mining through feature extraction, training, and prediction. Sentiment analysis cannot be generalized because it is data-specific. There are several methods of sentiment analysis to interpret and analyze based on the information provided [5]. There exist five categories of opinion mining: standard, fine-grained, aspect-based, emotion detection, and intent-based.

The sentiment score is derived by categorizing and computing the negative, positive, and neutral lexicons from the given corpus. The ratio of variations among them and the total word count is then calculated. For computing emotion score, positive, negative, and neutral word counts are used, along with semi-normalization. The emotion score is determined by dividing the positive words by the amount of negative words by the amount of neutral words $+1$. Because we are not using the difference of values, the score will be greater than zero, and because we inserted $+1$ in the denominator, the zero-division error is eliminated. We have the ability to create our customized sentiment score. As a result, you will know what score you will receive since you designed the logic of the sentiment score calculations. It can be adjusted depending upon your logic and demands. When you utilize different logics to create sentiment score, you can have various features for analysis [6]. The challenges of opinion mining originate from the enormous volume and rate of individual-generated public content, paired with the circumstantial sparsity caused by text length and the inclination to utilize abbreviated linguistic conventions to communicate thoughts. A large, top-grade lexicon is frequently required for quick, precise emotion and opinion detection on such big scales [7, 8].

2 Overview of the Data

Nowadays, social media platforms play an essential part in society. Social networking services gives an opportunity for mankind to stay connected and voice their opinions, sentiments, feelings, and emotions. One of the highly used and popular social networking websites is Twitter. Much research has shown that Twitter is the preferred social media. Twitter's accessibility and availability allow us to get structured data

[9]. Twitter has more than 40 million tweets per day. This amount of data will help us examine, analyze, and process data which will give efficient results [10, 11]. Our dataset consists of tweets from Twitter that were taken during Ukraine and Russian conflict. These tweets contained various opinions that ranged from positive to negative and neutral. This data is used in the models for opinion mining.

3 Literature Review

Table 1 lists shows the analysis of literature review

4 Method

4.1 Vader

The evolution, verification, and assessment of VADER (Valence Aware Dictionary for sEntiment Reasoning) are described in this study. We create and scientifically validate an emotion lexicon using a mixture of qualitative and quantitative methodologies. We then combine these lexical elements with five generally applicable guidelines that encapsulate semantical and syntactical patterns that individuals use to convey otherwise emphasize feeling strength. The precision of opinion mining is enhanced by using these heuristics. In the social media platforms sphere, the VADER lexicon functions exceedingly well. When we examine the categorization accuracy further, we observe that VADER exceeds solitary public critics in precisely classifying tweet emotion into positive, neutral, or negative categories. VADER preserves and enhances the advantages of classic emotion lexicons. The VADER emotion lexicon is of the highest caliber and it has been vetted by people. VADER differs from others particularly because of its responses in digital networking contexts while it also generalizes much better. VADER is accessible for free to utilize. This is a text opinion mining model that is sensitive to both sentimental polarity (positive/negative) and depth of the emotion (strong/weak). It is a segment of the NLTK library and can be employed on unlabeled corpus directly [12]. The opinion mining of VADER relies on a dictionary that relates lexical resources to emotional depth defined as sentiment scores. A sentiment score can be calculated by adding the depth of each word contained in the corpus.

Opinion mining is performed in the following way. First, study the text under consideration. A system trained on lengthy paragraphical reviews might be useless. Use a prototype that is befitting the current task. Later, select the type of analysis to perform. Few primary opinion mining algorithms go a stride farther and handle two-word collectives, known as bi-grams [13]. We will begin with complete sentences and use a trained NLTK lexicon known as VADER to perform opinion mining.

Table 1 Various analysis components for literature review

Author	Title	Method/tools	Application/result	Context
Abdullah Al Maruf, Zakaria Masud Ziyad, Md. Mahmudul Haque and Fahima Khanam (2022)	Emotion detection from text and sentiment analysis of Ukraine–Russia war using machine learning technique	Machine learning	Calculating the precision of the models and choosing the best model for identifying racism	Multi-channel social media
Piyush Vyas, Gitika Vyas and Gaurav Dhiman (2023)	RUemo—the classification framework for Russia–Ukraine war-related societal emotions on Twitter through machine learning	Machine learning	It enhances the scant RUW literature by the opinions gained from the people	Twitter
Amrita Shelar and Ching-yu Huang (2018)	Sentiment analysis of Twitter data	Natural language processing	Identified the views of people in the form of polarity	Twitter
Victoria Bobichev, Olga Kanishcheva and Olga Cherednichenko (2017)	Sentiment analysis in the Ukrainian and Russian news	Lexicon-based and machine learning	Developing a collection of information into various divisions by three annotators and experimenting them by using various machine learning methods	News from blogs
Shiv Dhar, Suyog Pednekar, Kishan Borad and Ashwini Save (2018)	Methods for sentiment analysis: a literature study	Lexicon-based, Natural language processing and machine learning	Describes the comprehensive study of various methods and practices used for sentiment analysis	Multi-channel social media
Isaac Geovany López Ramírez and Jorge Arturo Méndez Vargas (2022)	A sentiment analysis of the Ukraine–Russia conflict tweets using recurrent neural networks	Machine learning	Conducted a sentiment analysis of tweets from around the world about the conflict between Ukraine and Russia by using machine learning methods	Twitter

(continued)

Table 1 (continued)

Author	Title	Method/tools	Application/result	Context
Abdullah Alsaeedi and Mohammad Zubair Khan (2019)	A study on sentiment analysis techniques of Twitter data	Lexicon-based, machine learning and natural language processing	Discussion of various sentiment analysis methods on Twitter data	Twitter
Devika M D, Sunitha C and Amal Ganesh (2016)	Sentiment analysis: a comparative study on different approaches	Machine learning and lexicon-based	Discusses numerous sentiment analysis techniques and their levels of analyzing sentiments	Multi-channel social media

We can decipher and quantify the emotions that exist in streaming media, such as audio files, video documents, and text messages according to VADER's material optimal methodology. The speed performance tradeoff for VADER is significantly the same. As per the research, VADER corresponds to authentic facts as well as solitary human critics. When examined more clearly at the F1 ratings (classification accuracy), VADER executes finer at classifying tweet sentiment as positive, neutral, or negative. VADER is responsive to polarities as well as depth of the emotions (whether an emotion is negative or positive). It also holds this into consideration by allocating valence score to phrase under examination [14]. Valence score is a score allotted to a term based on experience and observation instead of just pure logic. The foundation of VADER is a lexicon that maps data terms and other classes that are frequently found in sentiment representation in short texts. These elements include a wide range of Western emoticons, emotional acronyms, and frequently used nomenclature with emotional value.

Applying the Wisdom of the Crowd method, VADER proved the broad application of lexicons that are involved for different mindsets. It structures its business template on the idea that, in place of skillful expertise, the accumulated knowledge of a clique of entities may be dependent upon. This allowed them to estimate a meticulous point for a text's sentiment valence score irrespective of context. The impact of each hidden meaning of words on the overall effectiveness of emotion in paragraph is considered by VADER using a group of guidelines. These guidelines are called heuristics. These heuristics go above and beyond what is traditionally captured by a pouch paradigm. They involve relationships between terms that are word-order sensitive. The addition of valence scores of all the words in the corpus, fine-tuning them according to the guidelines, validating them among negative end (−1) and positive end (+1) [15]. Figure 1 shows the process of opinion mining.

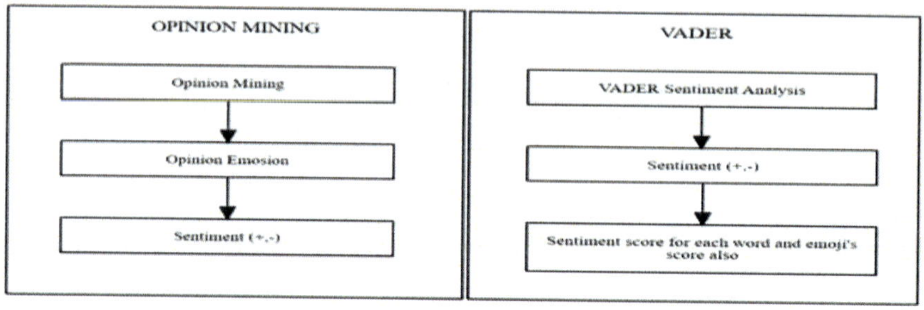

Fig. 1 The figure above shows the process of opinion mining and how it is done when VADER model is taken

4.2 DistilBERT, A Distilled Version of BERT

DistilBERT is a compressed, quick, affordable, and insubstantial Transformer prototype which was created by distilling BERT. It contains 40% less variables than uncased BERT-base, runs 60% quicker, and preserves more than 95% of BERT's execution on the GLUE language understanding standard. The standard BERT model is taken, and knowledge distillation is performed. Knowledge distillation is a model dimensionality reduction strategy where a compact prototype (the learner) is instructed to mimic the characteristics of a bigger prototype (the scholar) or even a collection of prototypes. A classification model is often trained in supervised learning to forecast a specific sample class by maximizing the expected likelihood of winning tags [16]. Reducing the log loss amidst the framework's projected distribution and the one hot empirical distribution of training tags is thus a basic training target. A model that performs effectively on the training dataset will anticipate an outcome distribution with high-level confidence in the relevant class and practically nil confidence in the remaining classes.

Training loss, the learner is conditioned along a distilling loss amid the scholar's soft goal likelihoods, which are predicted via the scholar (respective to the learner). By utilizing the entire scholar distribution, this goal yields a rich training signal. It employs a Soft Max-temperature as proposed by Hinton et al. (2015). The last training goal is a linear amalgamation of distilling loss LCE and the supervised training loss, which in this case is the masked language modeling loss LMLM [17]. Cosine embedding loss (Lcos) is useful for aligning the objectives of the learner and scholar's hidden states [18].

4.3 Naïve Bayes Algorithm/Classifier

According to Bayes' theorem, the Naive Bayes algorithm is a supervised method for machine learning. It is the most basic and fast categorization algorithm for vast amounts of data. It engages Bayes' probability theorem. When previous knowledge is

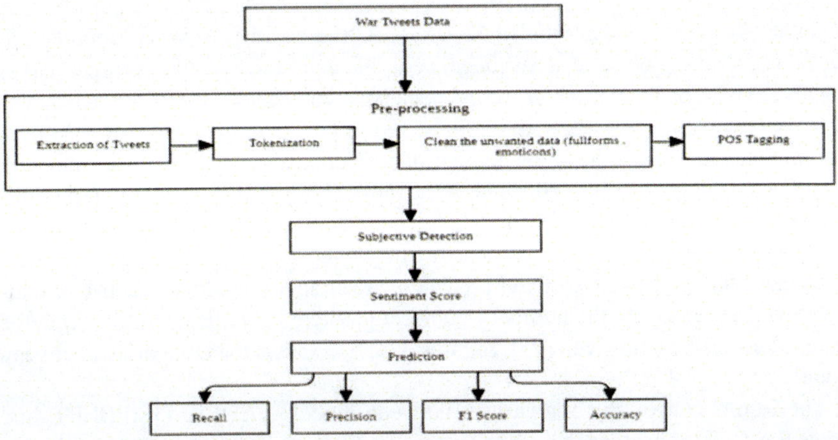

Fig. 2 The flowchart figure above depicts how our models perform opinion mining and the performance measures used

obtainable, Bayes' theorem is utilized to determine the likelihood of an interpretation. It is resolute by conditional probabilities. Naive Bayes is a probabilistic model that uses Bayes' theorem with a significant prediction of autonomy among the attributes [19]. It generates optimal results if aimed at linguistic analysis of corpus. The Bayes' theorem is used by Naive Bayes classifier to anticipate membership probability for every class, like the probability of a particular data element corresponding to that specific class. The most probable class is the one with the most likelihood. It is also known as Maximum A Posteriori (MAP) [8].

The Naive Bayes classifier presumes that all of the features are unrelated. The existence or lack of a feature has zero influence on the existence or lack of other features. Given various evidence regarding features in real-life data, we evaluate an assumption. As an outcome, the calculations get largely competitive. To simplify the process, the attribute independence methodology is used to detach different elements of proof and deal with them as distinct objects. There are three kinds of Nave Bayes algorithms. They are Gaussian Naïve Bayes, Multinomial Naïve Bayes, and Bernoulli Naïve Bayes [20]. Figure 2 showcases methodology of opinion mining.

5 Results

F1-score, precision, recall, and accuracy are called performance quantifiers in machine learning. They are employed to achieve an understanding of the robustness and limitations of models for devising predictions. Model's performance is crucial, and these performance quantifiers help accomplish it. The basic of precision, recall, and F1-score and accuracy is the concept known as True Positive, False

Positive, True Negative, and False Negative. A true positive is when ground positive is forecasted as positive, a true negative is when ground negative is forecasted as negative, a false positive is when ground negative is forecasted as positive, and a false negative is when ground negative is forecasted as positive. Precision is the value or measure or calibrated value of the number of correct positive predictions formulated. Recall is the value or calibrated value of the number of accurate positive predictions when compared to all the positively labeled data. F1-score is computed by using the reciprocal of the average of reciprocals of both precision and recall, simply put it is the harmonic mean of recall and precision. Accuracy is a value or metric or calibrated value to measure the number of correct predictions in the overall formulated predictions for the data given [21]. Figures 3, 4, 5, 6 shows the comparison between models.

The results convey the comparison between models VADER, DistilBERT, and Naïve Bayes. It is observed that when comparing F1-scores of the three models Naïve Bayes has a much better F1-score followed by VADER. When looking at precision, VADER has the upper hand followed by Naïve Bayes. The graph of recall depicts that Naive Bayes is better at formulating accurate predictions. But a change is observed when accuracy score is taken, it is seen that VADER still gives high results, but Naïve Bayes doesn't produce results at the required accuracy. The results above depict that

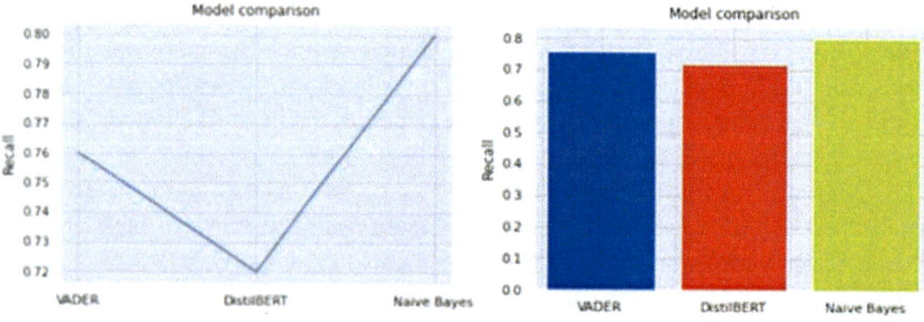

Fig. 3 The above line graph and bar graph show the difference between models when recall value is considered over the whole data

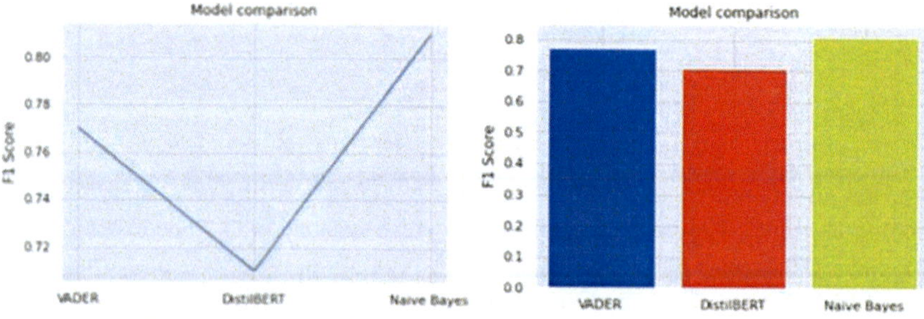

Fig. 4 The above line graph and bar graph show the difference between models considered and F1-score for the dataset taken

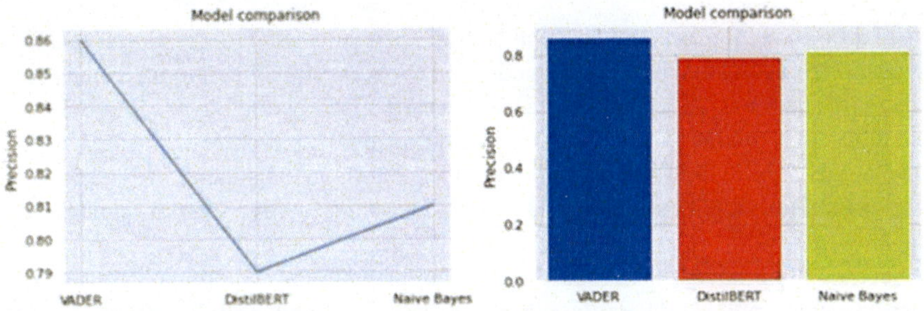

Fig. 5 The above line graph and bar graph show the difference between models when rrecision value is considered for data given

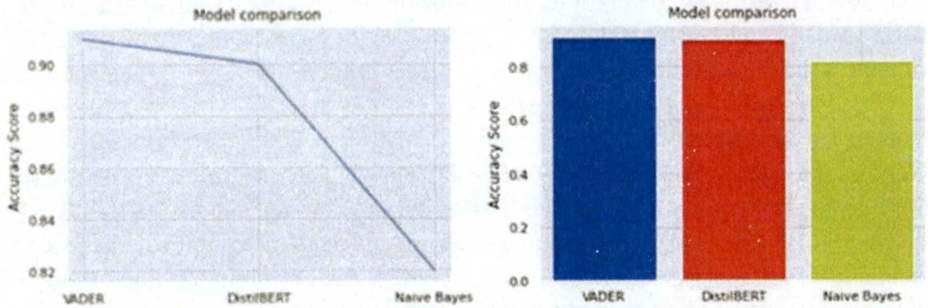

Fig. 6 The above line graph and bar graph depict the difference among models considered and accuracy score for overall data taken

both F1-score and recall of Naïve Bayes are greater than VADER, while VADER dominates in precision and accuracy. While DistilBERT is a good contender when compared to the other models, the decision is between Naïve Bayes and VADER due to the results shown by performance quantifiers. Even though Naïve Bayes is scalable and useful for real-time prediction analysis when considering things like context, tone, and other linguistic characteristics that make language complicated. Naïve Bayes uses N-grams and dimension reduction to achieve results. The ability of VADER to utilize high-quality lexicons along with language rules with proper accuracy makes VADER a much more accurate model. It is observed that VADER does significantly well than other models when performing opinion mining.

6 Conclusion

This research paper provides information on opinion mining on a social media platform. The research conducted referred to information from various published research papers, journals, articles, and studies. This paper makes the following contributions. Firstly, this paper shows you three different methods that can be used to perform opinion mining on any data taken from different social media platforms. The process of choosing an appropriate model and the accuracy of the model depends on the data user collects. The above models are DistilBERT, Naïve Bayes, and VADER. The things that should be taken into consideration are the structure of the data taken, and the size of the data. When choosing a model contingent on dataset, prefer lexicon-based models for smaller dataset and machine learning or deep learning models for bigger dataset [22].

Secondly, we have showcased three different methods which can be used for opinion mining and produce effective results. We have gathered data from Twitter. Twitter allows people all around the world to voice their opinions. This leads to data that is verified and helps to comprehend data and users efficiently [23]. The three models were taken into consideration to depict how the VADER model is more efficient in delivering the accuracy and precision needed. The data shown in the results can be used to conclude why VADER is preferable. In Naïve Bayes, the wide ranges of features are reduced by taking term frequency which results in dimensionality reduction and N-grams are used to keep the context of the data. This leads to results that aren't accurate enough [24].

The next model we have used is DistilBERT which shows us promising outcomes but the main problem this method faces is the internal error that happens in the model. This could lead to various problems and when data is huge or complex it can lead to a decrease in accuracy. VADER is a dictionary that assigns predetermined score to individual features. The superiority of VADER is it can assign sentiment score to any word, acronym, or emoji. The valance-based lexicon is able to detect both intensity and polarity which helps in the effectiveness of the model. The predetermined scores are evaluated and changed based on the modifiers in the model. All of these features available in VADER contribute to the success of VADER. The accuracy and precision are always estimated to be higher and more efficient. Our results depict why the VADER model should be preferred when implementing opinion mining [25].

References

1. Vinodhini G, Chandrasekaran RM (2012) Sentiment analysis and opinion mining: a survey. Int J Adv Res Comput Sci Softw Eng 2(6). ISSN: 2277 128X
2. Sarlan A, Nadam C, Basri S (2014) Twitter sentiment analysis. In: International conference on information technology and multimedia (ICIMU)
3. Prakruthi V, Sindhu D, Kumar SA (2018) Real time sentiment analysis of twitter posts. In: 3rd IEEE international conference on computational systems and information technology for sustainable solutions
4. Liu B (2009) Sentiment analysis and opinion mining. 5th text analytics summit. Boston
5. Agarwal A, Bhattacharyya P (2005) Sentiment analysis: a new approach for effective use of linguistic knowledge and exploiting similarities in a set of documents to be classified. International conference on natural language processing (ICON 05). IIT Kanpur, India
6. Rahman R et al (2017) Detecting emotion from text and emoticon. Lond J Res Comput Sci Technol
7. Pang B, Li L (2008) Opinion mining and sentiment analysis. Found Trends Inf Retr 2(1–2):1–135

8. Amolik A et al (2016) Twitter sentiment analysis of movie reviews using machine learning techniques. Int J Eng Technol 7.6
9. Mittal A, Patidar S (2019) Sentiment analysis on twitter data: a survey. Delhi Technological University, New Delhi, India. ACM. New York, ACM
10. Rosenthal S, Noura F, Nakov P (2017) SemEval-2017 task 4: sentiment analysis in Twitter. In: Proceedings of the 11th 2017 international workshop on semantic evaluation (SemEval-2017)
11. El Rahman SA, AlOtaibi FA, AlShehri WA (2019) Sentiment analysis of twitter data. In: The 2019 international conference on computer and information sciences (ICCIS)
12. Agarwal A, Xie B, Vovsha I, Rambow O, Passonneau R (2011) Sentiment analysis of twitter data. In: Proceedings of WLSM-11s
13. Sadeghi SS, Khotanlou H, Rasekh Mahand M (2021) Automatic persian text emotion detection using cognitive linguistic and deep learning. J AI Data Min 9(2):169–179
14. Ahmed K, El Tazi N, Hossny AH (2015) Sentiment analysis over social networks: an overview, systems, man, and cybernetics (SMC). In: IEEE international conference on, IEEE
15. Healy M, Donovan R, Walsh P, Zheng H (2018) A machine learning emotion detection platform to support affective well being. In: 2018 IEEE international conference on bioinformatics and biomedicine (BIBM). IEEE, pp 2694–2700
16. Hasan MR, Maliha M, Arifuzzaman M (2019) Sentiment analysis with NLP on twitter data. In: Proceedings of the 2019 international conference on computer, communication, chemical, materials and electronic engineering (IC4ME2). Rajshahi, Bangladesh, pp 1–4
17. Hasan M, Agu E, Rundensteiner E (2014) Using hashtags as labels for supervised learning of emotions in twitter messages. In: ACM SIGKDD workshop on health informatics, vol 34, no 74. New York, USA, p 100
18. Jabreel M, Moreno A (2019) A deep learning-based approach for multi label emotion classification in tweets. Appl Sci 9(6):1123
19. Mahesh B (2020) Machine learning algorithms-a review. Int J Sci Res (IJSR), [Internet], 9:381–386
20. Wongkar M, Angdresey A (2019) Sentiment analysis using Naive Bayes algorithm of the data crawler: twitter. In: Fourth international conference on informatics and computing (ICIC)
21. Kim S-M, Hovy E (2018) Determining the sentiment of opinions. In Proceedings of the 20th international conference on computational linguistics, pp 1367, 2004; Classification, Comput Mater Continua 55(2):243–254
22. Sarlan A, Nadam C, Basri S (2014) Twitter sentiment analysis. In: Proceedings of the 6th international conference on information technology and multimedia. IEEE, pp 212–216
23. D'souza SR, Sonawane K (2019) Sentiment analysis based on multiple reviews by using machine learning approaches. In: 2019 3rd International conference on computing methodologies and communication (ICCMC). IEEE, pp 188–193
24. Le B, Nguyen H (2015) Twitter sentiment analysis using machine learning techniques. In: Advanced computational methods for knowledge engineering. Springer, pp 279–289
25. Mishra N, Jha CK (2012) Classification of opinion mining techniques. Int J Comput Appl 56(13)

Gait Recognition Using Activities of Daily Livings and Ensemble Learning Models

Sakshi Singh, Nurul Amin Choudhury, and Badal Soni

Abstract Gait, a biological trait, is extremely valuable for identifying personnel. Identifying gait is a complex process as it needs a combination of sensor data and efficient learning algorithms. In this paper, we incorporated activities of daily living data for recognizing gait using multiple shallow and ensemble learning models. The proposed model includes various state-of-the-art human activity recognition datasets in order to create a gait recognition database based on daily activities like—walking and running. The dataset is processed using different pre-processing techniques like data normalization, solving class imbalance problems, and removing noise, corrupted data, and outliers, as well. Different shallow and ensemble models are incorporated for training and testing, and the performance is measured using the evaluation metrics like Precision, Recall, F1-Score, and Accuracy. The included model achieved the highest accuracy of 99.60% with CatBoost Classifier. Also, other ensemble models managed to perform optimally for both walking and running activity instances to recognize gait.

Keywords Gait recognition · Machine learning · Ensemble learning · Activities of daily livings (ADLs) · Human activity recognition (HAR)

1 Introduction

One of the most emerging research areas in biometric identification is gait recognition. A person's gait has a distinctive pattern that may be recognized and verified. Researchers [3, 14] around the globe practice gait recognition based on image clas-

S. Singh · N. A. Choudhury (✉) · B. Soni
Department of Computer Science and Engineering, National Institute of Technology Silchar, Silchar 788010, Assam, India
e-mail: nurul21_rs@cse.nits.ac.in

S. Singh
e-mail: sakshi_pg_21@cse.nits.ac.in

B. Soni
e-mail: badal@cse.nits.ac.in

© The Author(s), under exclusive license to Springer Nature Singapore Pte Ltd. 2023

A. Mishra et al. (eds.), *Advances in IoT and Security with Computational Intelligence*, Lecture Notes in Networks and Systems 755,
https://doi.org/10.1007/978-981-99-5085-0_20

sification, sensor data analysis, and video analysis. Most of these studies are based on machine vision but are hindered by factors like illumination, shadows, moving objects, etc., which lowers their effectiveness. Nowadays, recognizing individuals is also done by analyzing the time and frequency domain of gait acceleration signals. Initially, a microcontroller-based gait recognition system was designed that recognizes gait using acceleration data of walking ADL [12]. As the use of smartphones is growing at a rapid pace, the ability to use a smartphone's built-in accelerometer for gait recognition is an ideal data collection, and processing module [6].

Machine and deep learning algorithms are very popular for various application domains like Gait recognition, HAR, Gait Analysis, and other biometric and medical fields [8, 11]. Deep learning models are essentially incorporated for getting efficient classification results using massive data instances. The need for feature set analysis is not required as the deep learning model extracts and selects the features on its own using its hidden layer functions and synaptic weights. But, it requires a huge dataset and rigorous hyperparameter tuning for optimal results. On the contrary, shallow and ensemble learning models need manual feature engineering and efficient data pre-processing for optimal results. Also, they generally tend to use fewer data instances for model training and testing.

The field of Human Activity Recognition (HAR) [1, 4, 13] has seen significant advancement in the past few years. The data captured using smartphones for HAR is much less complex than the data captured for gait recognition as it exploits the daily living activities in an uncontrolled environment. The use of HAR datasets can make gait recognition tasks uncomplicated and is yet to explore its full potential. On this ideology, this paper incorporated HAR datasets for generating gait data of different user sets using the subjects' labels and ADLs like walking and running. The main contribution and novelties of this paper are as follows:

1. Different shallow and ensemble learning models are incorporated for successfully recognizing the gait, and an analytical framework is proposed with a detailed analysis of results for comprehensive understanding.
2. Detailed data pre-processing and exploration have been done to make the dataset robust for classification.
3. Gait Recognition dataset is constructed by incorporating HAR datasets and ADLs like walking and running. Three publicly available datasets are used for detailed analysis and comparisons.

This paper is divided into multiple sections. Section 1 introduces the field of study; Sect. 2 outlines the recent state-of-the-art gait recognition works; Sect. 3 introduces the learning models with the proposed methodology; and Sect. 4 reports the experimental findings. Finally, Sect. 5 concludes the paper in a concise way with future directions.

2 Literature Survey

Researchers emphasize the study of sensor-based gait and activity recognition using optimal learning algorithms. Multiple approaches for sensor-based activity and gait recognition have been proposed over the years, and some of the most recent methods are discussed in this section.

Rong et al. [12] designed a portable microprocessor-based data collection device to measure three-dimensional gait acceleration. The device consists of a tri-axial accelerometer, a microcontroller unit, 32 bytes of RAM, and a data transfer module. They proposed the gait analysis on the time and frequency domain and used the 1-nearest neighbor for the recognition. Preliminary results indicate that it is possible to recognize users based on their gait acceleration signals. The method was reasonably robust to changes in speed when dynamic time wrapping (DTW) was used. Makhdoomi et al. [9] proposed a model that aims to recognize humans based on their body shape and walking pattern from different viewing angles. They used the CASIA gait database: Dataset and incorporated similarity comparison and optimum threshold experimentation. For viewing angle, gallery step information cum data, ranging from 54 to 144°, was incorporated, and the best result was at 54, 72, and 108°.

Yulita et al. [14] proposed a method that focuses on training Support Vector Machine (SVM) as a base learner on the AdaBoost ensemble classifier. The proposed model used 200 iterations of poly kernel SVM, and various algorithms were incorporated to compare the result of the proposed framework. Upon comparing various incorporated algorithms, Random Forest achieved efficient accuracy with a performance difference of 3% compared to the proposed model in this study. The HugaDB database, which collects data on human gait for analysis and activity recognition, has been the subject of research by Kececi et al. [7] using machine-learning algorithms for user authentication. Six wearable inertial sensors on the right and left feet, shins, and thighs were used to form a body sensor network to gather data. Running, standing, walking, and sitting activities were considered from the HugaDB for the research work in this paper. Three algorithms, IB1, Random Forest, and Bayesian Net, achieved an accuracy rate higher than 99% from the selected machine learning algorithms.

The authors in [15] experimented deep learning-based gait recognition system using inbuilt smartphone sensors. They analyzed the hybrid models are more accurate towards recognizing gait and managed to achieve 93.75% with the hybrid CNN-LSTM model. Chao et al. [3] experimented that normal walking activity contributes more towards gait recognition than other human activities. Giorgi et al. [5] used filtering and buffer mechanism with CNN for recognizing gait. They incorporated 175 user data and used CNN for training and testing. On testing, they managed to achieve an accuracy of 95.8%.

3 Proposed Model

In order to analyze the effect of different ADLs on gait recognition, we build a framework from data collection to gait classification. The proposed method consists of the following steps—data collection, data pre-processing, training, and testing the model with specific human activities and also with a combination of multiple ADLs. The architecture of our proposed model is shown in Fig. 1.

3.1 Data Collection

In order to build the gait dataset, we used three publicly available HAR datasets. All the datasets have wearable sensor data at different sampling rates and mounting locations. mHealth [2] dataset comprises body motion and vital signs recordings for ten volunteers of the diverse profile while performing several physical activities. Sensors were positioned on the subject's right wrist, chest, and left ankle to gather data for 12 actions performed by subjects. These sensors track various body parts' acceleration, rate of rotation, and magnetic field orientation. The sensor positioned on the chest also took readings of the 2-lead ECG.

HARSense [4] dataset is a collection of subject-specific daily life activity data from the cellphones' built-in accelerometer and gyroscope sensors. Smartphones were put in the user's waist and front pockets. Twelve subjects carried out six activ-

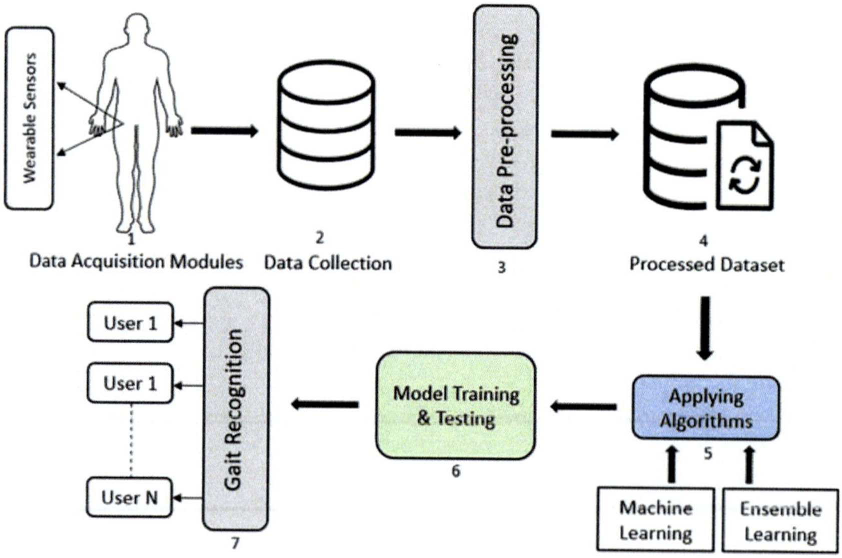

Fig. 1 Architecture (gait recognition chain) of our proposed framework

ities of daily living. Finally, the MotionSense [10] dataset includes time series data generated by accelerometer and gyroscope sensors. An iPhone 6s was used for data collection from different mounting locations. A total of 24 participants performed 15 trials of 6 activities in the same environment and conditions.

We used data on subjects' running and walking activities from mHealth, HARSense, and MotionSense datasets. After gathering the necessary data from the above datasets, seven separate datasets are created. From mHealth, three datasets are produced, one for running activity, one for walking activity, and the final one that combines both running and walking. In the same way, three datasets are formed from the HARSense dataset. But for the MotionSense dataset, only one dataset is created for walking activity as it does not have running activity instances.

3.2 *Data Pre-processing*

All three datasets were pre-processed to make them robust towards classification, as shown in Fig. 2. First, we checked for noise and corrupted values and concluded that all three datasets were free from noise and corrupted instances. Then we checked for missing value instances. Few were present in the datasets, and we removed them by dropping them from the datasets as augmenting it statistically destroys the feature dependencies. Later, we checked for we checked outliers and managed to find very minimal outliers in all three datasets. We did not remove them as it would make the classification challenging, like a real-life scenario.

$$S_s = \frac{x_i - \text{Mean}\ (\mu)}{\text{Standard Deviation}\ (\sigma)} \tag{1}$$

Exploratory data analysis with data visualization was also done to check the dataset's quality and for detailed understanding. We found out that the 3D features are highly dependent on one another, and the removal of one feature will impact the

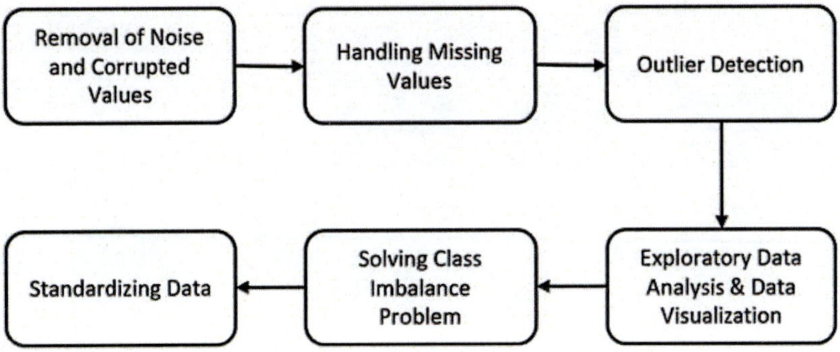

Fig. 2 Data pre-processing pipeline

overall model performance. Also, the intensity of walking and running features are different from one another, and the sampling rate of running is always higher than walking activity. Then we checked for class imbalance problems in all three datasets and found out that the HARSense dataset has the problem of class imbalance with huge data instance mismatching. We addressed this problem using random over-sampling and fixed the issue for optimal model training. Finally, data was standardized using *standard scalar* (S_s) method as mentioned in Eq. 1 for all three datasets to bring all the features to a common scale. Also, we did not perform feature extraction and selection as these datasets are standardized, and all the features incorporated by the researcher [2, 8, 10] are already showing the best results for HAR. With this intuition, the features were kept the same for our experiment as well.

3.3 *Model Training and Testing*

In order to analyze the effect of different learning models, we used multiple shallow and ensemble learning models based on their learning paradigms which are as follows:

1. **Decision Tree (DT)**: DT classifier is a tree-based classifier that uses impurity-based splitting information for classifying different tasks. DT generates a human-readable form of rules and regulations in the model training phase and tests the data on those rules (Fig. 3).
2. **Random Forest (RF)**: RF is one of the most used bagging-based ensemble classifiers that works on the principle of the DTs. It uses multiple DTs with various data

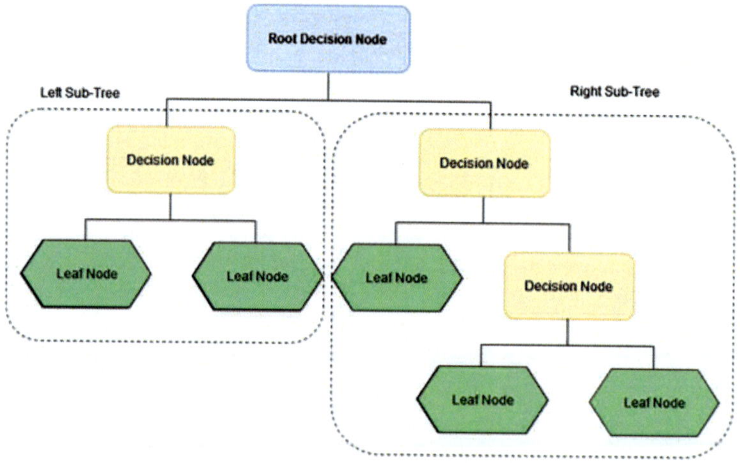

Fig. 3 Basic architecture of decision tree

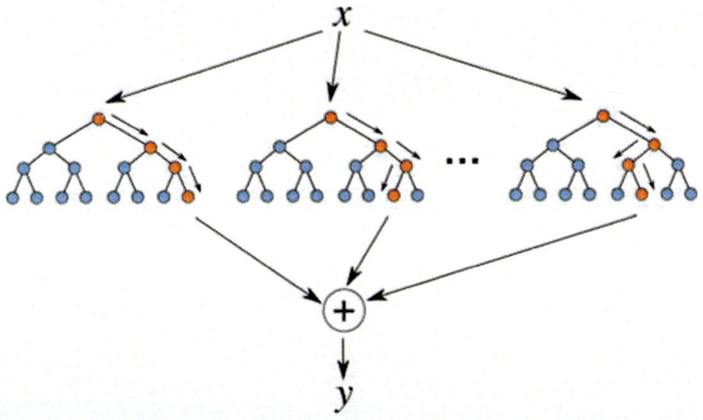

Fig. 4 Basic architecture of random forest

sampling to build base learners and compute its final result using the aggregation method (Fig. 4).

3. **Support Vector Machine (SVM)**: SVM classifier is a decision boundary-based classifier that segregates different class labels using best-fit lines. It is computationally expensive but generally yields good performance results because of its optimal hyperplane generation between different classes.
4. **Extreme Gradient Boosting (XGB)**: XGB in a boosting-based ensemble approach that uses tree-based structure for classification. Internal modes are used for attribute storage, and each leaf node store or classify different labels.
5. **Extra Tree Classifier (ETC)**: ETC is incorporated for its continuous optimal performance in multiple-application domains. ETC uses multiple DTs that are de-correlated in nature and calculates the results using the aggregation method (Fig. 5).
6. **CatBoost (CTB)**: CTB generates boosting method on decision trees. It uses multiple DTs for model training using boosting of base learners and dataset subsampling of model training.

4 Experimental Results

All the models are trained and tested on *Google Collab*, and the models were built in Python 3.8 using various libraries like *sklearn, NumPy, Pandas*, etc. The dataset splits into two parts for training and testing with the ratio of 75:25, i.e., 75% training data and the remaining 25% as test data. The holdout method was used for training and testing the model performance with various performance parameters that are defined as follows:

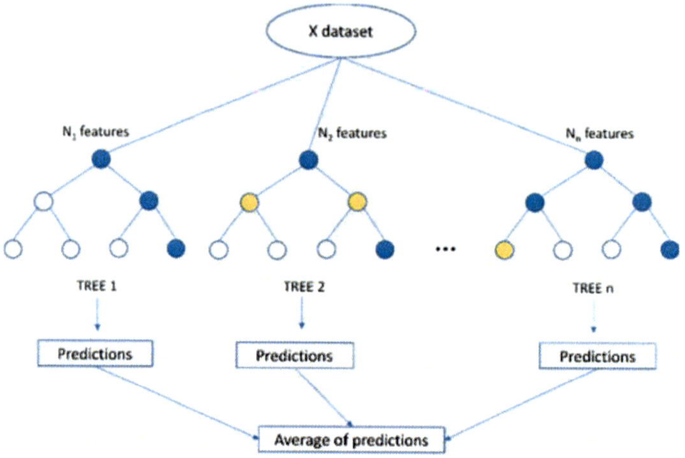

Fig. 5 Basic architecture of ETC

1. **Accuracy** is used to find out the proportion of correct predictions out of all the predictions made by the model. Mathematically, it is given by Eq. 2:

$$A = \frac{1}{n}\sum_{i=1}^{n}\left(\frac{T_{p_i} + T_{n_i}}{T_{p_i} + T_{n_i} + F_{p_i} + F_{n_i}}\right) \tag{2}$$

2. **Precision** is the ratio of true positives over the sum of false positives and true negatives. It is used to show how accurate the prediction was out of those predicted as positive. Mathematically, it is given by Eq. 3:

$$P = \frac{1}{n}\sum_{i=1}^{n}\left(\frac{T_{p_i}}{T_{p_i} + F_{p_i}}\right) \tag{3}$$

3. **Recall** is the ratio of correctly predicted outcomes to all predictions. Mathematically, it is given by Eq. 4:

$$R = \frac{1}{n}\sum_{i=1}^{n}\left(\frac{T_{p_i}}{T_{p_i} + F_{n_i}}\right) \tag{4}$$

4. **F1-score** is the harmonic mean of precision and recall. Mathematically, it is given by Eq. 5:

$$F1 = \frac{1}{n}\sum_{i=1}^{n}\left(2 * \frac{P_i * R_i}{P_i + R_i}\right) \tag{5}$$

4.1 Analysis of Shallow Learning Algorithms

Once the dataset is pre-processed, and all classification matrices are defined, we classify gait using shallow learning algorithms. Both DT and SVM show very less accuracy on testing data after training, and other performance metrics were also not satisfactory. The lack of base learners and optimal dataset sub-sampling hampers the overall performance in a negative way. The achieved accuracy of DT and SVM are shown in Fig. 6. Other performance matrices are described in Table 1.

4.2 Analysis of Shallow Learning Algorithms

In the next phase of gait recognition, we trained all the mentioned four ensemble classifiers and tested them with the test dataset. On evaluation, we analyzed that all the classifiers are performing significantly better than shallow learning models. The other performance matrices are also optimal and higher compared to shallow models. The accuracies achieved by all four ensemble approaches are mentioned in Fig. 7. CTB performs better than other approaches as it segregates the datasets by using the

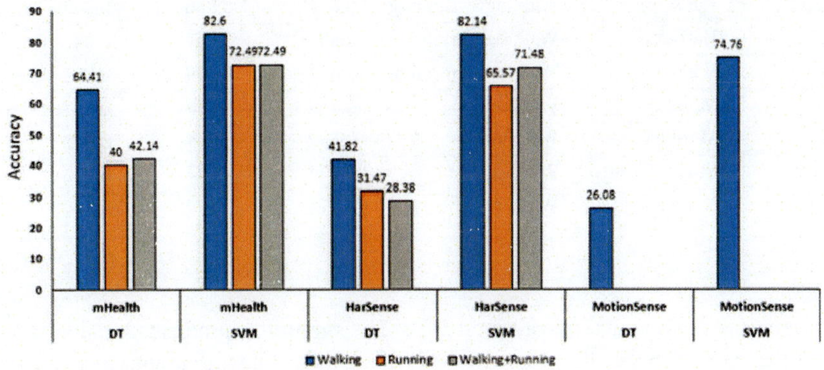

Fig. 6 Accuracy comparison of shallow learning algorithms

Table 1 Performance comparison of shallow learning algorithms

Classifier	Dataset	ADL	Precision	Recall	F1-score
DT	mHealth	Walking + running	0.295	0.421	0.327
SVM	mHealth	Walking + running	0.738	0.726	0.722
DT	HarSense	Walking + running	0.284	0.283	0.228
SVM	HarSense	Walking + running	0.727	0.715	0.706
DT	MotionSense	Walking	0.141	0.413	0.157
SVM	MotionSense	Walking	0.803	0.747	0.742

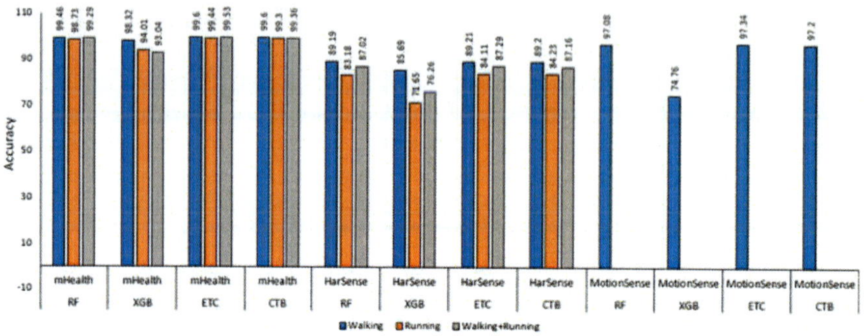

Fig. 7 Accuracy comparison of ensemble learning algorithms

Table 2 Performance comparison of ensemble learning algorithms

Classifier	Dataset	ADL	Precision	Recall	F1-score
RF	mHealth	Walking + running	0.995	0.995	0.996
XGB	mHealth	Walking + running	0.983	0.984	0.983
ETC	mHealth	Walking + running	0.996	0.996	0.996
CTB	mHealth	Walking + running	0.996	0.996	0.996
RF	HarSense	Walking + running	0.88	0.88	0.8791
XGB	HarSense	Walking + running	0.847	0.846	0.842
ETC	HarSense	Walking + running	0.88	0.881	0.88
CTB	HarSense	Walking + running	0.881	0.879	0.88
RF	MotionSense	Walking	0.859	0.857	0.857
XGB	MotionSense	Walking	0.749	0.747	0.742
ETC	MotionSense	Walking	0.974	0.974	0.973
CTB	MotionSense	Walking	0.974	0.974	0.4

sub-sampling method and boosting on DTs. The support of gradient decision trees learns the patterns in the data more optimally using its feature importance mechanism and yields efficient results. P, R, and F1-Score values of CTB are also optimal as compared to all other classifiers.

Other ensemble approaches are learning the pattern in an efficient way, and they managed to outperform all the shallow learning models. This model performs result aggregation using maximal voting or averaging, where the miss classification rate becomes minimal because of the support of multiple model building and optimal hyperparameter tuning. On the contrary, these classifiers take a bit more extra computational time for model training, but the performance makes the trade-off almost negligible. The detailed comparison of all the ensemble approaches is described in Table 2.

5 Conclusion and Future Works

In this paper, we analyzed the effect of different shallow and ensemble learning models for gait recognition using wearable sensors like accelerometers and gyroscopes. Three publicly available HAR datasets are considered for creating a gait database, and different pre-processing methods were applied to make the dataset robust and efficient towards classification. On studying the effect of both shallow and ensemble approach, we conclude that ensemble algorithms are far superior to shallow learning algorithms and yields better performance results. CatBoost managed to achieve the highest accuracy of 99.6% with the mHealth dataset, while the shallow model managed to get the highest accuracy of 82.6% only.

In future, we will try to create our own gait recognition dataset with hundreds of users to make the classification task challenging. Also, lightweight deep learning models will be incorporated for automatic feature engineering and efficient predictions.

References

1. Amin Choudhury N, Moulik S, Choudhury S (2020) Cloud-based real-time and remote human activity recognition system using wearable sensors. In: 2020 IEEE international conference on consumer electronics-Taiwan (ICCE-Taiwan), pp 1–2
2. Banos O, García R, Holgado-Terriza J, Damas M, Pomares H, Rojas I, Saez A, Villalonga C (2014) mhealthdroid: a novel framework for agile development of mobile health applications, vol 8868, pp 91–98
3. Chao H, Wang K, He Y, Zhang J, Feng J (2022) Gaitset: cross-view gait recognition through utilizing gait as a deep set. IEEE Trans Pattern Anal Mach Intell 44(7):3467–3478
4. Choudhury NA, Moulik S, Roy DS (2021) Physique-based human activity recognition using ensemble learning and smartphone sensors. IEEE Sens J 21(15):16852–16860
5. Giorgi G, Martinelli F, Saracino A, Sheikhalishahi M (2017) Try walking in my shoes, if you can: accurate gait recognition through deep learning. In: Tonetta S, Schoitsch E, Bitsch F (eds) Computer safety, reliability, and security. Springer International Publishing, Cham, pp 384–395
6. Huan Z, Chen X, Lv S, Geng H (2019) Gait recognition of acceleration sensor for smart phone based on multiple classifier fusion. Math Probl Eng 1–17(06):2019
7. Kececi A, Yildirak A, Ozyazici K, Ayluctarhan G, Agbulut O, Zincir I (2020) Implementation of machine learning algorithms for gait recognition. Eng Sci Technol Int J 23(4):931–937
8. Lyngdoh AC, Choudhury NA, Moulik S (2021) Diabetes disease prediction using machine learning algorithms. In: 2020 IEEE-EMBS conference on biomedical engineering and sciences (IECBES), pp 517–521
9. Makhdoomi NA, Gunawan TS, Habaebi MH, Rahim RA (2014) Human gait recognition: viewing angle effect on normal walking pattern. In: 2014 international conference on computer and communication engineering, pp 104–106
10. Malekzadeh M, Clegg R, Cavallaro A, Haddadi H (2019) Mobile sensor data anonymization, pp 49–58
11. Rajesh S, Choudhury NA, Moulik S (2020) Hepatocellular carcinoma (HCC) liver cancer prediction using machine learning algorithms. In: 2020 IEEE 17th India council international conference (INDICON), pp 1–5

12. Rong L, Jianzhong Z, Ming L, Xiangfeng H (2007) A wearable acceleration sensor system for gait recognition. In: 2007 2nd IEEE conference on industrial electronics and applications, pp 2654–2659
13. Vaka A, Soni B, Raman M (2022) Effective breast cancer classification using SDNN based e-health care services framework, pp 1–6
14. Yulita IN, Paulus E, Sholahuddin A, Novita D (2021) Adaboost support vector machine method for human activity recognition. In: 2021 international conference on artificial intelligence and big data analytics, pp 1–4
15. Zou Q, Wang Y, Wang Q, Zhao Y, Li Q (2020) Deep learning-based gait recognition using smartphones in the wild. IEEE Trans Inf Forensics Secur 15:3197–3212

A Conceptual Prototype for Transparent and Traceable Supply Chain Using Blockchain

Pancham Singh, Mrignainy Kansal, Shatakshi Singh, Shivam Gupta, Shreya Maheshwari, and Sonam Gupta

Abstract Over the past several years, blockchain technology has gained prominence and acceptance. However, there hasn't been much empirical study on the scope of blockchain technology in the supply chain. As a result, data traceability and transparency are also restricted. To illustrate the development of supply chain management using blockchain in the real world, various real-time use cases are shown. To compare our findings to earlier findings and determine what is the impact of blockchain on different sectors, a research direction has been considered. Creating a conceptual framework for supply chain management utilizing blockchain technology is the goal of this research. The framework suggests an idea to use cryptographic hash function to generate a unique ID which will help us to uniquely identify any product throughout the supply chain by enhancing supply chain traceability with the application of machine learning and internet of things. This study takes into account ongoing blockchain developments and also suggests that in comparison with permissionless blockchain, permissioned blockchain is a better match.

Keywords Supply chain management · Blockchain · Traceability · Transparency · Consensus algorithm · Smart contract · Machine learning · Internet of things and message-digest 5

1 Introduction

A distributed, decentralized public ledger system is what blockchain is. Blockchain is used in corporate networks to record transactions and monitor assets. Blockchain is a database that uses a consensus mechanism and contains blocks of digitally

P. Singh (✉) · M. Kansal · S. Singh · S. Gupta · S. Maheshwari · S. Gupta
Department of Information Technology, Ajay Kumar Garg Engineering College, Uttar Pradesh, Ghaziabad, India
e-mail: singhpancham@akgec.ac.in

M. Kansal
e-mail: kansalmrignainy@akgec.ac.in

© The Author(s), under exclusive license to Springer Nature Singapore Pte Ltd. 2023

A. Mishra et al. (eds.), *Advances in IoT and Security with Computational Intelligence*, Lecture Notes in Networks and Systems 755,
https://doi.org/10.1007/978-981-99-5085-0_21

signed transactions that are connected sequentially and cryptographically. Any node in the blockchain network can read, create and verify transactions that are stored in a distributed ledger but can't update them, due to the features provided by blockchain like integrity and availability [1]. To protect the blockchain system, there are many different cryptographic mechanisms, such as digital signatures, hash functions, etc. A consensus protocol is a collection of guidelines that enables various parties to concur on a single record, which is necessary to validate the transactions. In recent years, blockchain has offered consumers desirable qualities such as decentralization, autonomy, fault-tolerance, integrity, immutability, verification as well as anonymity, auditability, transparency, and traceability [1].

Supply chain management is responsible for the circulation of goods and services from the producers to the customers including some intermediaries [2]. Supply chain management is useful for the company's supply chain process of products that helps customers to find it uncomplicated to track their product throughout the whole process. To produce and provide goods and services to customers, several supply chain partners must cooperate [3]. The idea of supply chain management profoundly alters the character of a corporation since control doesn't depend on the direct observation of internal company processes but rather on integration across member businesses.

Current supply chain management lacks various facilities that can be revamped by implementing blockchain. According to a resource-based perspective (RBV) in this research, blockchain may be seen as a technology to support the firm's capacity, such as the ability of integration and collaboration [2]. Blockchain technology must resolve supply chain current problems or result in process improvements in order to be useful economically. The use of Smart Contracts and the transparent blockchain database are the two blockchain technology features that have been identified as having a high scope for supply chain management. The immutable database provides a cost-effective auditing process assuring the integrity and trust to the business [4]. Blockchain can improve supply chain by creating and sharing data records among parties, resulting in enhancing traceability and transparency and quicker and more affordable product delivery [2]. Blockchain enables parties to comprehend how products are transferred through each intermediary, reducing gray market trade and eliminating the impact of counterfeit products. Blockchain also promotes trust by enabling information exchange between supply chain parties via a distributed network [5].

2 Application of Blockchain Technology in Supply Chain Management

Till now, we have discussed the characteristics of blockchain separately. The leading companies are using blockchain with supply chains to enhance their efficiency. This has brought great change in the market and has created many customers. This helps us in understanding the need of an improved framework for blockchain that can keep up with modern tasks.

2.1 Present Uses

1. Amazon.com, Inc.: The biggest corporations in the world are now using blockchain technology. A scalable blockchain network can be set up and maintained by users and organizations using the thoroughly managed service known as Amazon Managed Blockchain.
2. Microsoft Corporation: The necessary infrastructure is provided by Microsoft Corporation's Azure Blockchain Workbench for the creation of a consortium network. Blockchain is being used by businesses to digitize workflows for procedures like distributing physical goods across the supply chain.
3. Huawei Technologies: The Blockchain Service (BCS) from Huawei is a low-cost, user-friendly service with applications in supply chain tracing and banking. In order to assure fairness and transparency and to reduce the cost of risk management, it uses decentralized, shared, and tamper-proof ledgers.
4. IBM: IBM: In blockchain supply chain solutions, smart contracts are automatically executed when established business logic is encountered, providing real-time visibility into the processes and enhancing the capacity to act promptly.
5. Accenture, Plc.: Blockchain solutions from Accenture provide an unchangeable audit trail that demonstrates the legitimacy and compliance of documentation, software and hardware throughout supply chains.
6. Oracle Corporation: The Oracle Blockchain Cloud, Oracle Transportation Management, Oracle Supply Chain Management Cloud, Oracle IoT Cloud, and third-party enterprise systems make up Oracle's Digital Supply Chain with Blockchain solution.
7. SAP SE: SAP's supply chain management solutions have been specifically designed to be compatible with blockchains [6]. With the use of SAP's blockchain, supply chain management, objects can be tracked and processed along the supply chain, and product origins may be determined.

This has inspired new businesses, and this industry is seeing new advancements. Numerous answers are being offered by blockchain, which may very likely determine how technology develops in the future [7].

3 Methodology

An interpretative method was used as part of the qualitative study design to comprehend the key characteristics of the blockchain for supply chain collaboration and integration [2]. We discussed the effects of blockchain in many industries and gathered the main conclusions in light of that. Next, we created a conceptual framework for Supply Chain Management. It is shown in Table 1.

3.1 Research Direction

To better understand the connection, application, and bloom of blockchain technology in the area of supply chain management and to address the research objectives, we did a Systematic Literature Review (SLR). Because blockchain and the related technology, smart contracts, are in demand in the market, we found that a lot of work has been done in terms of research, technological standards, and cutting-edge applications. We selected the most recent 4 years, 2018–2021 (extended to early 2022), as

Table 1 Impact of blockchain in different sectors

Integration	Impact	References
Blockchain and IOT	Internet-connected devices with the help of IOT can transmit data to private blockchain networks so as to produce immutable records of shared transactions	[8]
Blockchain and RFID	Blockchain and RFID technology working together can offer businesses incredible benefits. RFID transmits and collects product data, which the blockchain then secures	[9]
Blockchain and ML	ML uses a reasonable amount of data to generate accurate predictions	[10]
Blockchain and security	It has been used commercially, shaped global currency markets, and aided the growth of nefarious dark web marketplaces	[11]
Blockchain and SCM	To manage the supply chain more efficiently, and keep track of price, date, location, quality, certification, etc., using a blockchain enabled supply chain	[12]
Blockchain and Information Sharing	It is a type of distributed ledger technology that can help with reliable information exchange by giving each network member a lasting digital footprint	[13]

the time window for our literature review in order to cover the most ongoing use of these technologies in supply chain management. To examine and evaluate as much existing research regarding blockchain and supply chain as we could, we looked into the IEEE Xplore, ACM, Google Scholar, and other key digital libraries from different, academic publishers.

Working implementation of Blockchain. While we were doing our research on supply chain management, we went through several areas where blockchain has shown its impact. Some of them are listed below to make our research direction clearer.

Blockchain Built-in analytics. Blockchain technology is now used as a transaction repository, and technologies for data-parallel processing like map-reduce, dry-ad Spark, or Flink are not accessible. The utilized blockchain must be thoroughly examined before any analytics are performed. The execution engine and input reader's integration might contribute to the Blockchain's in-built analytics. It's also possible to incorporate edge analytics functionalities [14].

Smart contract template development. Smart contracts are built for use on the blockchain after much reading and research and are handmade. This method takes far too long, and it occasionally leads to errors because it uses powerful natural language processing algorithms. A uniform semantic framework for smart contracts must be created in order to fully automate the process. [14].

Analysis of Blockchain Technology in Supply Chain Management in different sectors. In the last few years, tremendous work has been done in different sectors. Some industrial supply chain management include business process management, security and privacy, business model and operation, quality management, ownership management, manufacturing, physical distribution and logistics, transaction processing, internet of things, etc. [15].

Public Blockchain versus Private Blockchain. Anyone can join a public blockchain (permissionless) as a normal user or miner. Any participant may conduct transactions or operate under rules that have been predetermined by smart contracts [16].

Only authorized nodes that have been previously registered by authorities are permitted to be the part of the blockchain network in a private (permissioned) blockchain (Tables 2 and 3).

3.2 Conceptual Framework

The conceptual framework will include a supply chain system that uses blockchain, and all necessary supply chain participants (i.e., the producer, supplier, distributor, consumer, and retailer) will join the blockchain network. The supply chain system with blockchain will do the whole procedure of data management where data input, data sharing, and data monitoring will be done in the whole supply chain network to

Table 2 Important features that make private blockchain better than public blockchain [17]

Ability to prevent unauthorized access	Sensitive data stored on private blockchains can be effectively protected by limiting access for the general public by allowing only authorized individuals to utilize the network with distinct responsibilities allocated to each participant
Enhanced scalability	Private blockchains limit the number of validators, improving the network's effectiveness. As a result, high scalability and transaction throughput can be attained at significantly faster rates
Regulatory compliance	A private blockchain enables the enterprise to establish and enforce regulations in accordance with the laws, frameworks, and policies of authorities
Cost-efficiency	As the number of nodes is significantly limited it substantially reduces the costs of maintaining the network
Energy consumption	Private blockchains consume a lot less energy and power

maintain the integrity of the system. It will first verify the present participants and then give read-only access to them whenever necessary and write access so that they can perform a transaction between each other [18]. It is illustrated in Fig. 2.

Product Identification. When we talk about identifying a product on a dedicated blockchain, it is fairly easy. But when a group of fundamentally different commodities are involved the level of complexity rises exponentially. We are establishing a direct link between producers and consumers so that customers may follow a product's journey beginning with its manufacture [19]. Thus, the system's traceability may be improved, Not only for customers but also for other intermediaries. This keeps the product genuine for everyone involved in the supply chain, regardless of whether the actual product is delivered to the customer or not and regardless of whether the customer returns the real product or not.

Considering the big count, to recognize these products over the blockchain we need an identification system [20]. Here, we can use various details of a product and make them go through a hash function which generates a fixed unique code. The various attributes of the product are:

- Product Code (derived from product name);
- Manufacturing date (DD/MM/YYYY);
- Manufacturing time in hundredths of second (HH:MM: SS: NN);
- Industry Code (unique for every industry).

There are several hash functions but the two most commonly used are message-digest 5 (MD5) and Secure Hash Algorithm (SHA256). Both algorithms certainly have their advantages and disadvantages. The complexity of the Algorithms MD5 and SHA256 is the same, i.e., (N), yet it is shown that MD5 is faster than SHA256 [21]. So, the hash function that we are utilizing here is message-digest MD5, which will provide us with a unique hash code for each input string after merging all of this data to generate a final input string. This hash code will serve as the product's unique ID and be consistent across the board (Fig. 1).

Table 3 Test results after applying hash function message-digest MD5

S. No	Product name	Product code	Manufacturing date	Manufacturing time	Industry code	Final input string	Hash code
1	Shirt	SH	12/05/2022	11:55:24:94	027	SH12052211552494027	4c19ae79be66438cee07ff382b0f9bd4
2	Bag	BG	02/08/2021	14:22:11:57	143	BG02082114221157143	611a8f92192ace21df8a4b24027b892c
3	Cookie	CO	25/02/2020	08:32:45:76	089	CO25022008324576089	62bc491c4158fd5b33d668d0cd481c17
4	Shoes	SO	10/12/2022	03:27:56:12	231	SO10122203275612231	f427b36adb64692cafe1641c7f9db463

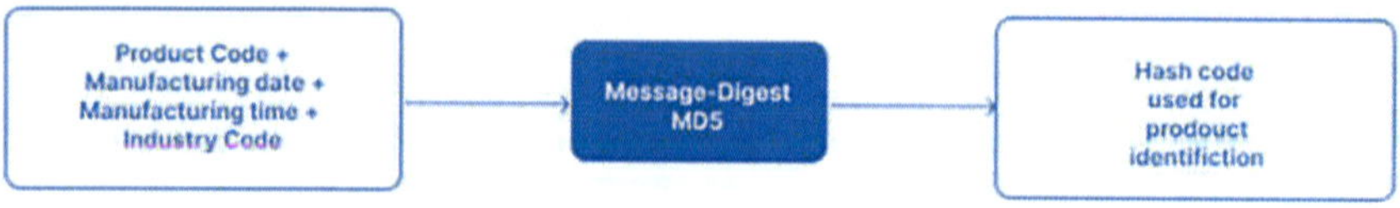

Fig. 1 Hash code generation using message-digest MD5

Transaction Flow. The requirement is that the node will confirm the sender's identity who submits the transaction request. A transaction request submitted by a user will be added to the blockchain database and then prior to submitting the transaction data to the consensus mechanism, it will authenticate. The database will append the transaction data once it has been signed using the consensus algorithm, and supply chain management will alert the user and display the results. When necessary, the processors connected to the same blockchain network will enquiry the transaction data. Prior to obtaining the transaction data, the processor will still need to be verified [18].

Customers can get the information of a product using a variety of methods, such as QR code, barcode, RFID tag, etc. Consumers can also request transaction information, which will then generate a query through a supply chain management application. The information about a product is disclosed to the consumer is up to the other parties who sold the goods to the customer. As a result, the consumer can access product information without having to prove their identification. Identity authentication is not required during the transaction process, and the customer will be shown the information immediately (Fig. 2).

Transparency and Traceability. In a traditional supply chain system, i.e., centralized, the third-party service providers receive all information from supply chain participants. It is possible that irrelevant parties could change the data in such a

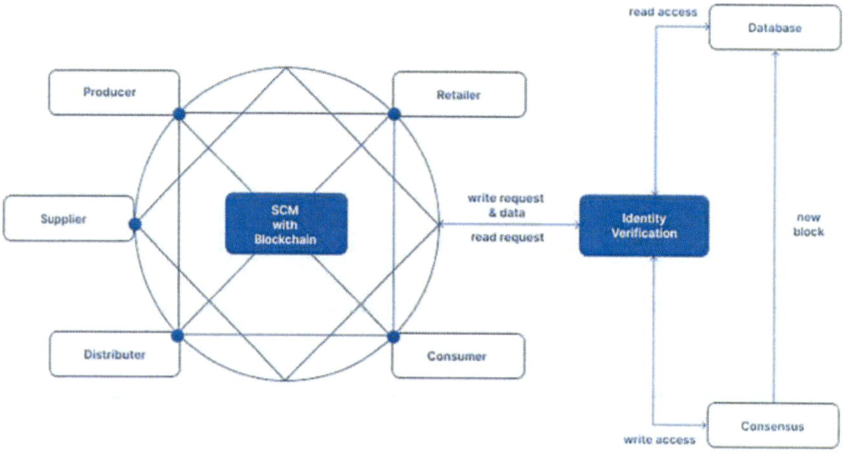

Fig. 2 Illustration of conceptual framework of supply chain using blockchain

system for questionable reasons. Additionally, the supply chain party is intended to own all transactional data in the system, yet a third party like itself is in charge of that data. As a result, the transparency and traceability of the data have been impacted by the reliability of the data in the system [18].

With the use of blockchain in supply chain management, no transaction has gone via a non-relevant party. Before being added to the blockchain database, the information in the blockchain is automatically signed and encrypted to assure its veracity. Each transaction flow is open and visible to all supply chain participants. The access to the blockchain data's information is based on the permission provided by the supply chain management manager. As a result, it also preserved the data's confidentiality. The system's traceability will operate as it should if it is transparent [22].

Blockchain with ML and IoT. We can make our supply chain network an astute one by introducing various technologies, e.g., artificial intelligence, machine learning, 3-D printing, IoT devices, 5G, Big Data, etc. [2]. There are various AI technologies available like robotics, machine learning, vision, optimization and planning, scheduling, etc., that can help to make the business deemed. ML helps in last-mile tracking by considering various data points about how people provide their geographical locations and the overall time taken to bring the goods to specific locations and preventing fraud by automating inspection and auditing processes to detect irregularity or deflections from normal trajectories [23]. Inventory would be marked using GS1-compliant electronic product codes or RFID tags as part of supply chain operations (globally accepted instructions for managing supply chain data) and then to create a complete record of transactions, link the ERP systems of a company with those of its suppliers [2].

These technologies will help to eliminate the errors and enhance the traceability. Thus, customers can easily track their product during the whole timeline and there will be no fake delivery of the product at customer's doorstep.

4 Future Scope

In contrast to traditional supply chain management, this study paper explains how blockchain technology is better appropriate for supply chain management. The use of blockchain technology may perform verification, fault-tolerance testing, maintaining anonymity, auditability, transparency, immutability, and costs at each level of the supply chain while avoiding delays and human error. According to the statistics, this technology will transform the game in the near future [24]. According to a Gartner report, it will be utilized by several innovative firms. Records indicate that over 30% of the world's customers may embrace blockchain technology as their primary technology in the next few years. In a very short period of time, blockchain has gone from being a specialized technology known only to a small number of ardent cryptographers and experienced computer scientists to being a popular issue, and the narrative does not end here. This technology is being used into a number of further fresh technologies to increase security and transparency [25].

5 Conclusion

In this research paper, we have concluded how blockchain has developed concerning its application in the supply chain industry. We have gone through several parameters and done research about in which sectors blockchain has shown its interest and how many organizations tried to innovate the field of blockchain and become successful in their attempts. Appending a method to identify products over a blockchain technology—supply chain management uniquely using cryptographic hash functions hence aggrandizing the traceability and our main motivation to do this research is to show how blockchain with supply chain management will help in decreasing the frauds related to product replacement and how the realness of the product is maintained throughout the supply chain management cycle. But the implementation of product identification is underrepresented and it will take a few years to come into practical use for supply chain management.

This paper also explains why permissioned blockchains are preferable to permissionless ones since they increase the system's overall security, which is what we are primarily trying to achieve. But permissionless blockchain defeats the real purpose of blockchain which is to provide anonymous participation for every node. The adoption of technologies like ML and IOT can be a game changer outside of the areas that are already in place because it has several advantages. But because it may increase the cost of the finished project, this asset has to undergo a thorough and in-depth study process.

References

1. Guo H, Yu X (2022) A survey on blockchain technology and its security. Blockchain: Res Appl 3(2) ScienceDirect
2. Wang M, Wu Y, Chen B, Evans M (2020) Blockchain and supply chain management: a new paradigm for supply chain integration and collaboration. Oper Supply Chain Manag: Int J 14(1):111–122
3. Basuki M (2021) Supply chain management: a review. J Ind Eng Halal Ind 2:9–12
4. Song JM, Sung J, Park T et al (2019) Applications of blockchain to improve supply chain traceability. Procedia Comput Sci 162(1877–0509):119–122. ScienceDirect
5. Berneis M, Bartsch D, Winkler H (2021) Applications of blockchain technology in logistics and supply chain management—insights from a systematic literature review. Logistics
6. Yadia K, Damle M (2020) Adoption of blockchain technology in indian public distribution system challenges and solutions. PalArch's J Archaeol Egypt/Egyptol 17(6):4753–4766
7. Top Companies in the Blockchain Supply Chain Industry I Blockchain Supply Chain Market. Emergen Research (2022). https://www.emergenresearch.com/blog/top-7-leading-companies-in-the-blockchain-supply-chain-industry. Accessed 14 Dec 2022
8. Alam T (2019) Blockchain and its role in the internet of things (IoT)
9. Anand A, Seetharaman A, Maddulety K (2022). Implementing blockchain technology in supply chain management. 147–160
10. Tanwar S, Bhatia Q, Patel P, Kumari A, Singh P Hong WC (2020) Machine learning adoption in blockchain-based smart applications: the challenges, and a way forward. IEEE Access 474
11. Taylor PJ, Dargahi T, Dehghantanha A, Parizi RM, Choo KKR (2020) A systematic literature review of blockchain cyber security. Digit Commun Netw 6(2):147–156. ISSN 2352–8648

12. Using Blockchain to Drive Supply Chain Transparency and Innovation. Deloitte (2022). https://www2.deloitte.com/us/en/pages/operations/articles/blockchain-supply-chain-innovation.html. Accessed 14 Dec 2022
13. Wan PK, Huang L, Holtskog H (2020) Blockchain-enabled information sharing within a supply chain: a systematic literature review. IEEE Access 8:49645–49656
14. Jabbar S, Lloyd H, Hammoudeh M et al (2021) Blockchain-enabled supply chain: analysis, challenges, and future directions. Multimedia Syst 27:787–806
15. Blockchain in Supply Chain Management | Real World Blockchain Use Cases. ConsenSys (2022). https://consensys.net/blockchain-use-cases/supply-chain-management/. Accessed 14 Dec 2022
16. Vitaris B, Vitáris B (2021) Public vs. Private Blockchain: What's the Difference? Permission.io. https://permission.io/blog/public-vs-private-blockchain/. Accessed 14 Dec 2022
17. Difference between Public and Private blockchain. GeeksforGeeks (2022). https://www.geeksforgeeks.org/difference-between-public-and-private-blockchain/. Accessed 14 Dec 2022
18. Chan KY, Abdullah J, Khan AS (2019) A framework for traceable and transparent supply chain management for agri-food sector in malaysia using blockchain technology. Int J Adv Comput Sci Appl (IJACSA) 10(11)
19. Siby LT et al (2022) Blockchain based Pharmaceutical Supply Chain Management System. Int J Eng Res Technol (IJERT) 10(4):6
20. CHART: Count of Products in Amazon.com for Top 10 categories (2016). ScrapeHero. https://www.scrapehero.com/chart-count-of-products-in-amazon-com-for-top-10-categories-may-2016/ . Accessed 14 Dec 2022
21. Dian R (2018) et al A comparative study of Message Digest 5(MD5) and SHA256 algorithm. J Phys: Conf Ser 978
22. Aegi MAN, Jha AK Blockchaintechnology in the supply chain: An integrated theoretical perspective of organizational adoption. Int J Prod Econ 247
23. Ariwala P 9 Ways machine learning can transform supply chain management. Maruti Techlabs. https://marutitech.com/machine-learning-in-supply-chain/. Accessed 30 Jan 2023
24. Taylor P (2022) Global use cases for blockchain technology 2021. Statista. https://www.statista.com/statistics/878732/worldwide-use-cases-blockchain-technology/. Accessed 14 Dec 2022
25. Makridakis S, Christodoulou K (2019) Blockchain: current challenges and future prospects/applications. Futur Internet 11:258

Fresh and Rotten Fruit Detection Using Deep CNN and MobileNetV2

Abhilasha Singh, Ritu Gupta, and Arun Kumar

Abstract Accurately recognizing rotten fruits has been important, particularly in the agricultural sector. Generally, manual efforts are utilized to classify fresh and rotting fruits which can be tedious sometimes. Unlike humans, machines do not get tired from performing the same task repeatedly. This study suggested a technique for detecting flaws in fruit images, which might reduce human effort, slash manufacturing costs and save time. The rotten fruits may contaminate the good fruits if the flaws are not found within time. Therefore, to prevent the spread of rottenness, a model has been put forward. Based on the fruit images provided as input, the recommended system can identify rotten and fresh fruits. Images of oranges, bananas, and apples have been considered in this paper. Using a convolutional neural network along with max pooling and MobileNetV2 architecture, the features from the input images are collected, based on which images are subsequently classified. On a Kaggle dataset, the suggested model's performance is evaluated, and by using MobileNetV2, it gets the greatest accuracy in training data (99.56%) and validation set (99.69%). The max pooling had a validation accuracy of 95.01% and a training accuracy of 94.97%. According to the results, the suggested CNN model can differentiate between fresh and rotten apples in an efficient manner.

Keywords Rotten fruits · Convolutional neural networks · Deep learning · Max pooling · MobileNetV2

A. Singh (✉) · A. Kumar
Department of Computer Science and Engineering, SRM Institute of Science and Technology, Delhi-NCR Campus, Ghaziabad, Uttar Pradesh, India
e-mail: abhilass1@srmist.edu.in

R. Gupta
Department of Information Technology, Bhagwan Parshuram Institute of Technology, IP University, New Delhi, India

A. Singh
Department of Electronics and Communication Engineering, SRM Institute of Science and Technology, Delhi-NCR Campus, Ghaziabad, Uttar Pradesh, India

© The Author(s), under exclusive license to Springer Nature Singapore Pte Ltd. 2023
A. Mishra et al. (eds.), *Advances in IoT and Security with Computational Intelligence*, Lecture Notes in Networks and Systems 755,
https://doi.org/10.1007/978-981-99-5085-0_22

1 Introduction

In today's world, health is an important issue. It doesn't matter what age group an individual belongs to, may be a small child, a young hard worker, or an aged person. A healthy life is what everyone cares about. In India, every year, a very large number of people get infected due to bacterial and rotten fruit consumption. Healthy and fresh fruit is in demand for everyone and more than that every fruit organization is working to provide good quality products to the consumers [1, 2]. Also, fruit and food manufacturing industries are becoming more dynamic in the twenty-first century.

Freshness of the fruits is greatly affected by climate and temperature changes. Hence, detection of fresh and rotten fruits is very necessary. Humans can easily identify rotten fruits, but it's not so easy for the technology. The model proposed in this paper intends to address this issue. A division of machine learning, deep learning, can predict and address a wide range of issues. As they developed over time in the area of image processing, Convolutional Neural Networks (CNNs) have made significant progress. Deep Neural Networks (DNNs) are superior to all other methods and algorithms in terms of accurately identifying, categorizing, and separating diverse fruit categories. Fruit classification can be tricky because of similarity in shape, colur, and texture of fruits that belong to the same class.

This is the point that the proposed work is concerned about. The proposed system, on the basis of input images of fruit, detects whether the fruit is fresh or decayed with the help of Artificial Neural Network (ANN) [3]. ANN works with neural net which consists of neuron layers. One should have to train the network to get the optimistic results. In the proposed system, fruit images are provided as input, and are checked for their state, being fresh or rotten. Two main parameters considered are a number of regions along with the intensity. Numerous morphological and segmentation functions have been utilized to compute the uneven shade areas on fruit. Multiple images of fruits from different angles have been considered in order to get the precise result [4].

The suggested system's primary goal is to automate the manual identification and characterization of rotten or fresh fruit. It can be of great use to fruit industry for delivering quality products to consumers. It can be helpful in increasing the speed of fruit segregation and also lowering down the cost due to reduced manual labor requirement. The proposed system may provide a new pathway and part of this automating process.

2 Literature Review

Mandeep Kaur et al. [5] presented the categorization and division of fruits on the basis of observation and experience. The system uses the image processing techniques to reveal the quality of the fruits. 2D images of fruit are classified on color- and shape-based analysis. Ashwani Awate and Gayatri Amrutkar [6] presented system for fruit disease detection which shows various diseases of fruits. The proposed system includes K-means for image segmentation, diseases are categorized on the basis of four features—color, structure, morphology, and texture. ANN is used for the identification of disease and pattern matching.

Using a dataset called Fruits-360 that can be obtained from the internet, Horea et al. [7] performed fruit classification using neural networks (NN). The authors explained their decision of choosing the classifier to identify the fruits' freshness. To identify fruit quality, A. Bhargava et al. [8] employed four algorithms: KNN, SVM, SRC, and ANN. SVM outperforms the others with a fruit identification accuracy of 98.48% and average accuracy of 95.72% over a database of four different fruits.

Santi et al. [9] employed machine learning and artificial intelligence to describe, categorize, and grade fruits using the RF method together with additional classifiers including KNN and SVM. The RF method with SIFT features outperforms other classifiers with a 96.97% accuracy rate. For the categorization of multiple fruits, Hasan et al. [10] employed the deep learning technique. Using the quicker R-CNN and MobileNet techniques of TensorFlow, the author achieved 99% accuracy. A system was created by Shamim et al. [11] using two models: 16-layered deep learning model and a light model with 6 convolutional neural network layers.

A single feature descriptor with two distinct kinds of characteristics was proposed by Susovan et al. [12]. Using features, SVM creates a classification model that combines color and texture information to identify the fruit in a dataset. The accuracy came out to be 83.33% overall. The key contributions of their study are (1) an enhanced system for identification and classification of fruits and vegetables and (2) a segmentation method that performs better for multicolored fruits. Roy et al. [13] determined the fruit's freshness or decay based on the fruit's skin flaws. Using enhanced UNet (En-UNet), promising results were achieved with segmentation. The suggested model worked well on real-time segmentation, recognition, and labeling of fresh and rotten apples are the areas where it excels. In order to distinguish between fresh and rotting samples, Diclehan et al. [14] explored a dataset including sample images of three different varieties of fruits.

A control system-based convolutional neural network was developed by Zaw et al. [15] for identifying objects. Their test accuracy for fruit recognition and classification for the 30 classes of 971 photos was close to 94% after parameter adjustment. CNNs were utilized by Siyuan et al. [16] to classify fruits. A CNN containing six layers were built using totally connected layers, pooling layers, and convolution layers. The outcomes demonstrated that the method outperformed three cutting-edge methodologies with a 91.44% accuracy rate. Senthilarasi et al. [17] suggested a system to categorize banana fruit ripening into three stages: ripe, overripe, and unripe. In their studies, the fuzzy model surpassed cutting-edge methods, averaging a classification rate of 93.11%. A convolutional neural network with 13 layers was made by Yu-Dong et al. [18]. Their solution outperformed five cutting-edge techniques, with a 94.94% total accuracy. Mohammed et al. [19] presented a system that distinguishes between pink and white grapefruit using deep learning. They employed a Kaggle dataset of 1,312 photos. All of the photographs belonged to two different species of grapefruit. They achieved 100% accuracy rate in identifying the type of grapefruit. In order to categorize four different types of potatoes, Abeer et al. [20] worked with a model which exhibited 97.45% training accuracy, 100% validation accuracy, and 99.5% test set precision. A powerful machine vision framework for robots that collect dates was put out by Hamdi et al. [21].

3 Proposed Methodology

In this study, convolutional neural networks were employed to distinguish between fresh and rotten apples. Classical neural networks are highly challenging for image recognition. CNN is employed to overcome this issue and challenge. The CNN algorithm is effective for computer vision and can identify the input image based on certain parameters. The primary benefit of CNN is that it simply needs a few parameters which reduces the amount of data needed for the training of desired model and learning time becomes short [15]. CNN has multiple layers. The convolution layer is the top layer of a CNN. It operates in 32 dimensions using weighted input images that might be two- or three-dimensional. When essential information is maintained, the effective layout is enhanced. Every research investigation needs data preparation. Images were rescaled first. The dataset comprises of training and testing set, the ratio of which is 80/20 in this experimental setup. In addition, the images were resized to $256 \times 256 \times 3$ [22, 23]. Figure 1 shows a few of the fresh and rotten fruit photographs we have used.

Fig. 1 Some images from dataset

3.1 Preprocessing and Augmentation of Data

For this study, we have used data from kaggle.com [24]. This dataset contains images of fresh fruits and decaying fruits. 13,599 photos from the dataset were used for training and validation. Figure 2 displays the types of images in the dataset utilized.

Data preparation is the procedure through which unprocessed information is altered, transformed into a suitable format for further processing. The resizing of images to 256 × 256 pixels is done using the Image Data Generator tool in Keras. After resizing, images were normalized. Images are transformed into NumPy arrays for speedier computation. Few operations like rotation, zooming, shearing, and flipping horizontally may increase the amount of data. The images are then resized so that the second convolution layer will use 128 × 128 pixels and the third convolution layer will use 64 × 64 pixels.

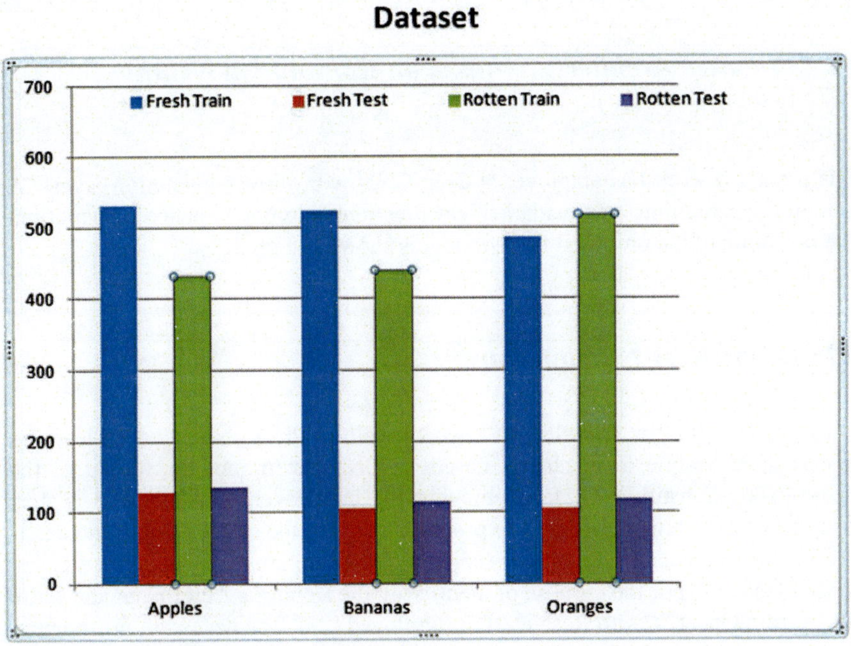

Fig. 2 Types of images in dataset

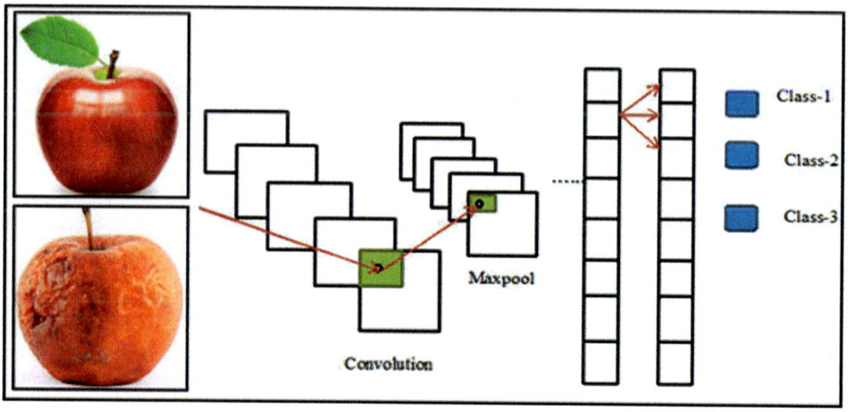

Fig. 3 Proposed CNN architecture

3.2 Proposed Architecture Based on Convolution Neural Network (CNN)

This proposed work has employed a deep CNN with three convolution layers. A method for combining two mathematical functions into one is called convolution. The operation of the proposed CNN model is shown in Fig. 3.

3.3 MobileNetV2 Architecture

For classifying images, MobileNetV2 works really well. A compact deep learning model called MobileNetV2 that is based on CNN determines the weight of the picture using TensorFlow. After removing the base layer, MobileNetV2 is then trained on a new trainable layer. The program analyzes the collected data and determines which aspects of images are most linked. There are 19 levels of bottleneck in MobileNetV2. For the purpose of identifying the front of a fruit image, the Caffe model from OpenCV with ResNet-10 is employed [23]. The necessary information is then extracted and sent to the layer of the fruit classifier. To handle the situation of overfitting, dropout layer has been used. With MobileNetV2 and base layer (include top = False) has been eliminated. The images have been altered. The proposed model has been constructed with a pool size max pooling technique and has 256 hidden layers. For greater precision, a learning rate of 0.001 is established. The stochastic gradient descent approach developed by Adam aids the model in comprehending picture characteristics. The working layer for MobileNetV2 is seen in Fig. 4.

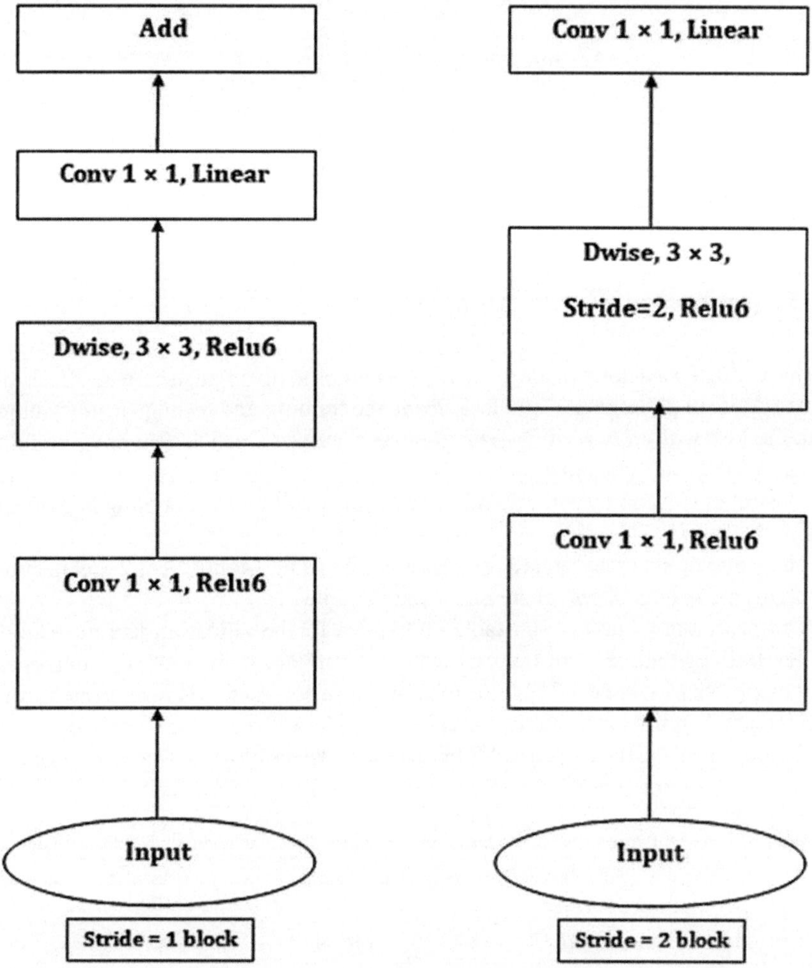

Fig. 4 Architecture of MobileNetV2

3.4 Performance Evaluation Through Matrices

We have evaluated the effectiveness of the models using various evaluation parameters accuracy, recall, precision, and F1-score for training as well as testing phases. Mathematical formulas for the above-mentioned parameters are as follows:

$$\text{Precision} = \frac{TP}{TP + FP} \tag{1}$$

$$\text{Recall} = \frac{TP}{TP + FN} \tag{2}$$

$$\text{Accuracy} = \frac{TP + TN}{TP + FP + TN + FN} \quad (3)$$

$$\text{F1-score} = \frac{\text{Recall} \times \text{Precision}}{\text{Recall} + \text{Precision}} \quad (4)$$

4 Experimental Result Analysis

We have utilized a dataset made up of 13,599 photos to distinguish between fresh and rotting fruits in photographs. Table 1 shows the training and testing accuracy along with the loss with max pooling. The greatest accuracy was 94.79% in the training set and 95.01% in the validation set.

The training accuracy and validation accuracy graph for max pooling is shown in Fig. 5.

The value of accuracy greatly increased while using MobileNetV2 architecture, accuracy, the results of which are displayed in Table 2.

The maximum accuracy from Table 2 is 99.56% for the validation data and 99.69% for the training data. Figure 6 presents the training accuracy and validation accuracy graph for MobilenetV2 architecture. A confusion matrix is also created using MobilenetV2 architecture. Table 3 accurately explains the confusion matrix.

Figure 7 displays the results of MobileNetV2's detection of fresh and rotten fruits.

Table 1 Max pooling training loss and accuracy, validation loss and accuracy for different epochs

Epoch	Training loss (%)	Training accuracy (%)	Validation loss (%)	Validation accuracy (%)
1	44.31	87.87	14.34	90.19
2	12.23	88.47	10.02	90.47
3	10.45	88.69	9.16	91.26
4	9.81	90.78	6.56	91.68
5	8.34	90.95	6.34	91.84
6	7.80	91.14	6.06	92.43
7	7.20	91.32	5.88	92.73
8	6.61	91.98	5.76	93.86
9	6.56	93.16	5.04	93.92
10	6.33	93.67	4.94	94.28
11	5.87	94.24	4.74	94.41
12	5.69	94.68	4.47	94.69
13	5.44	94.79	4.02	94.89
14	5.34	94.97	3.89	95.01

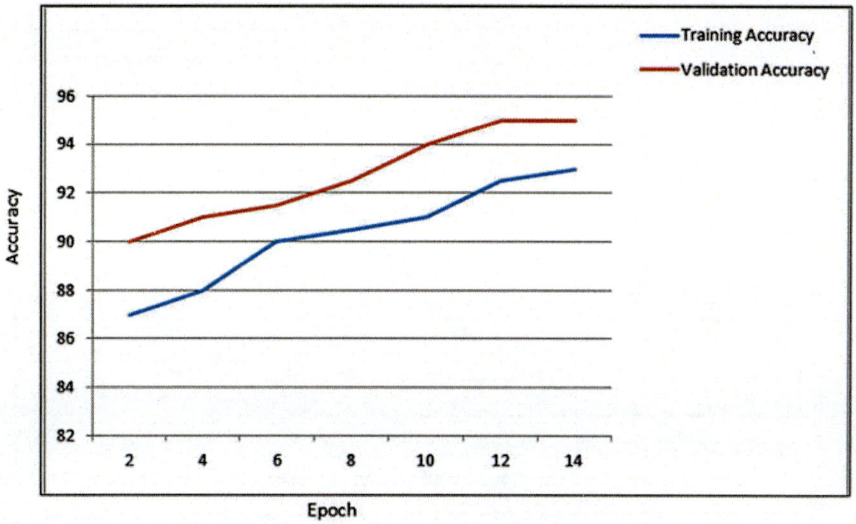

Fig. 5 Test accuracy and training accuracy with max pooling layer for CNN

Table 2 MobileNetV2 training loss and accuracy, validation loss and accuracy with different epochs

Epoch	Training loss (%)	Training accuracy (%)	Validation loss (%)	Validation accuracy (%)
1	4.48	98.06	4.31	98.65
2	4.39	98.04	4.22	98.79
3	4.28	98.16	4.12	98.86
4	4.21	98.28	3.96	99.02
5	3.38	98.44	3.82	99.11
6	3.84	98.53	3.71	99.20
7	3.56	98.68	3.66	99.26
8	3.53	98.87	3.51	99.32
9	3.42	98.98	3.32	99.41
10	3.36	99.02	3.26	99.44
11	3.34	99.43	3.21	99.48
12	3.32	99.48	3.16	99.51
13	3.25	99.51	3.16	99.58
14	3.06	99.56	3.12	99.69

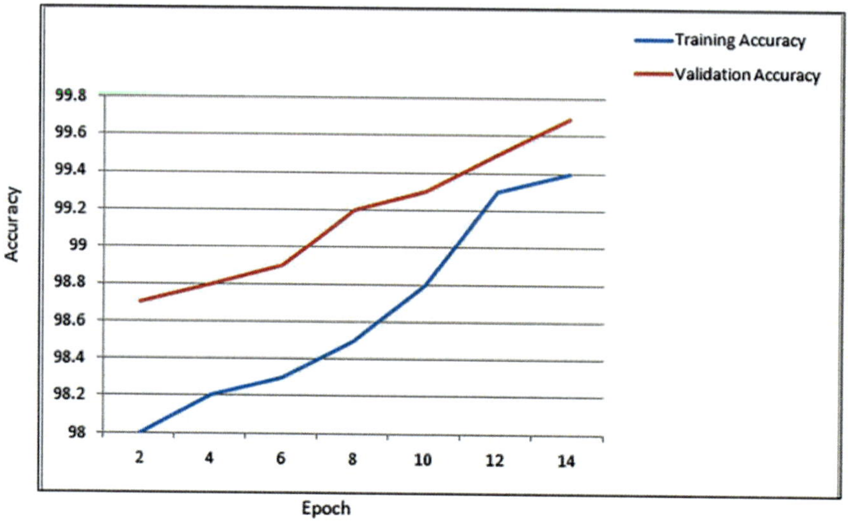

Fig. 6 MobileNetV2 with average pooling layer test accuracy and training accuracy

Table 3 Confusion matrix with MobilenetV2

Class	F1-score (%)	Recall (%)	Precision (%)
0 (Fresh Apples)	98	99	97
1 (Fresh Bananas)	99	97	98
2 (Fresh Oranges)	98	97	99
3 (Rotten Apples)	99	98	97
4 (Rotten oranges)	98	97	99
5 (Rotten Bananas)	97	99	98

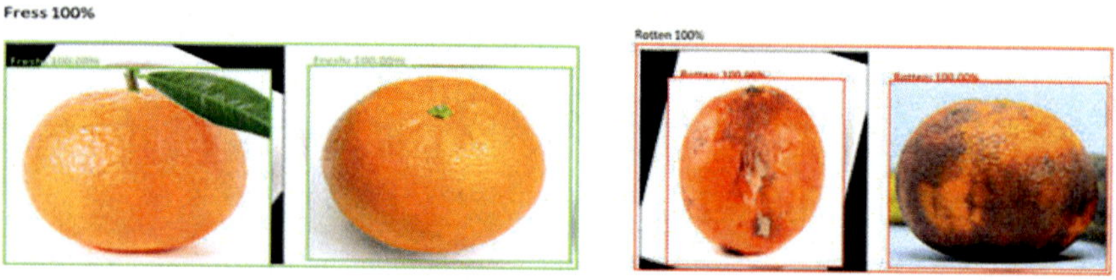

Fig. 7 Outcome of the model for detection of fresh fruits

The max pooling has a validation accuracy of 95.01% and a training accuracy of 94.97%. Additionally, MobileNetV2 architecture achieved the greatest accuracy of 99.69% for validation and 99.56% for training. In Table 4, a brief explanation is included.

Table 4 Comparison between MobileNetV2 and deep CNN model results

	Max pooling (%)	MobileNetV2 (%)
Epochs	14	14
Training loss	5.34	3.06
Training accuracy	94.97	99.56
Validation loss	3.89	3.12
Validation accuracy	95.01	99.69

Table 5 State-of-the-art comparison

Reference	Purpose	Techniques/models used	Highest accuracy achieved (%)	Dataset size
Matboli et al. [25]	Fruit disease's identification and classification	CNN, VGG16, Inception-V3, MoblieNet-v2, MobileNet	99.16	150
Gawas et al. [26]	Fruit freshness detection	Inception-V3	99.17	13,599
Foong et al. [27]	Rotten fruit detection	CNN, ResNet50	98.89	2100
Miah et al. [28]	Fresh and rotten fruit detection	CNN, InceptionV3, Xception, VGG16, MobileNet, NASNetMobile	97.34	5658
Proposed	Fresh and rotten fruit detection	CNN, MoblieNet-v2	99.69	13,599

Following Table 5 shows the comparison between the existing models in literature. The table clearly shows that the proposed model outperforms the other models in terms of accuracy.

5 Conclusion

Numerous uses for computer vision exist in the fruit processing industry, allowing for the automation of processes. Fruit quality classification and subsequent grading are essential if the industry production unit is to supply the best raw fruits that can be sold in the market. Two models based on CNN architectures (deep CNN and MobilenetV2) were employed in this investigation. The main aim was to provide an appropriate model with high precision so that fruit detection in the agriculture industry could be made simpler. As a future extension of the proposed work, more models can be added to evaluate performance with a larger dataset.

References

1. Awate A, Deshmankar D, Amrutkar G, Bagul U, Sonavane S (2015) Fruit disease detection using color, texture analysis and ANN. In: 2015 international conference on green computing and internet of things (ICGCIoT), 2015, pp 970–975. https://doi.org/10.1109/ICGCIoT.2015.7380603
2. Tripathi MK, Maktedar DD (2016) Recent machine learning based approaches for disease detection and classification of agricultural products. In: International conference on computing communication control and automation (ICCUBEA), pp 1–6. https://doi.org/10.1109/ICCUBEA.2016.7860043
3. Gavhale KR, Gawande U, Hajari KO (2014) Unhealthy region of citrus leaf detection using image processing techniques. In: International conference for convergence for technology 2014, pp 1–6. https://doi.org/10.1109/I2CT.2014.7092035
4. Iqbal SM, Gopal A, Sankaranarayanan PE, Nair AB (2015) Estimation of size and shape of citrus fruits using image processing for automatic grading. In: 2015 3rd international conference on signal processing, communication and networking (ICSCN), pp 1–8. https://doi.org/10.1109/ICSCN.2015.7219859
5. Kaur M, Sharma R (2015) ANN based technique for vegetable quality detection. IOSR J Electron Commun Eng 10(5):62–70. https://doi.org/10.9790/2834-10516270
6. Abdelsalam AM, Sayed MS (2016) Real-time defects detection system for orange citrus fruits using multi-spectral imaging. In: 2016 IEEE 59th international midwest symposium on circuits and systems (MWSCAS), pp 1–4. https://doi.org/10.1109/MWSCAS.2016.7869956
7. Mureşan H, Oltean M (2018) Fruit recognition from images using deep learning. Acta Univ Sapientiae, Inf 10(1):26–42
8. Bhargava A, Bansal A (2019) Automatic detection and grading of multiple fruits by machine learning. Food Anal Methods 13(3):751–761
9. Behera SK, Rath AK, Mahapatra A, Sethy (2020) Identification, classification and grading of fruits using machine learning and computer intelligence: a review. J Ambient Intell Humaniz Comput
10. Basri H, Syarif I, Sukaridhoto S (2018) Faster R-CNN implementation method for multi-fruit detection using tensorflow platform. In: International electronics symposium on knowledge creation and intelligent computing (IES-KCIC)
11. Hossain MS, Al-Hammadi M, Muhammad G (2019) Automatic fruit classification using deep learning for industrial applications. IEEE Trans Industr Inf 15(2):1027–1034
12. Jana S, Basak S, Parekh R (2017) Automatic fruit recognition from natural images using color and texture features. In: Devices for integrated circuit (DevIC)
13. Roy K, Chaudhuri SS, Pramanik S (2020) Deep learning based realtime Industrial framework for rotten and fresh fruit detection using semantic segmentation. In: Microsystem technologies
14. Karakaya D, Ulucan O, Turkan M (2019) A comparative analysis on fruit freshness classification. In: 2019 innovations in intelligent systems and applications conference (ASYU)
15. Khaing ZM, Naung Y, Htut PH (2018) Development of control system for fruit classification based on convolutional neural network. In: 2018 IEEE conference of Russian young researchers in electrical and electronic engineering (EIConRus)
16. Lu S, Lu Z, Aok S, Graham L (2018) Fruit classification based on six layer convolutional neural network. In: IEEE 23rd international conference on digital signal processing (DSP)
17. Marimuthu S, Roomi SMM (2017) Particle swarm optimized fuzzy model for the classification of banana ripeness. IEEE Sens J 17(15):4903–4915
18. Zhang Y-D, Dong Z, Chen X, Jia W, Du S, Muhammad K, Wang S-H (2017) Image based fruit category classification by 13-layer deep convolutional neural network and data augmentation. Multimed Tools Appl 78(3):3613–3632
19. Abu-Saqer MM, Abu-Naser SS, Al-Shawwa MO (2020) Type of grapefruit classification using deep learning. PhilPapers, Int J Acad Inf Syst Res (IJAISR). ISSN: 2643-9026
20. Elsharif AA, Dheir IM, Mettleq ASA, Abu-Naser SS (2020) Potato classification using deep learning. Int J Acad Pedagog Res (IJAPR). ISSN: 2643-9603

21. Altaheri H, Alsulaiman M, Muhammad G (2019) Date fruit classification for robotic harvesting in a natural environment using deep learning. IEEE Access 7:117115–117133
22. Palakodati SSS, Chirra VR, Dasari Y, Bulla S (2020) Fresh and rotten fruits classification using CNN and transfer learning. Rev d'Intell Artif 34(5):617–622
23. Rahman MM, Manik MMH, Islam MM, Mahmud S, Kim JH (2020) An Automated System to Limit COVID-19 Using Facial Mask Detection in Smart City Network. In: IEEE International IOT, Electronics and Mechatronics Conference (IEMTRONICS), pp 1-5. https://doi.org/10.1109/IEMTRONICS51293.2020.9216386.
24. https://www.kaggle.com/sriramr/fruits-fresh-and-rotten-for-classification
25. Matboli MA, Atia A (2022) Fruit disease's identification and classification using deep learning model. In: 2nd international mobile, intelligent, and ubiquitous computing conference (MIUCC), pp 432–437. https://doi.org/10.1109/MIUCC55081.2022.9781688
26. Gawas A, Tolia K, Gaikwad S, Deshpande K, Upadhyaya K, Jain S (2021) E-fresh: computer vision and IOT framework for fruit freshness detection. In: 2021 international conference on advances in computing, communication, and control (ICAC3), pp 1–6. https://doi.org/10.1109/ICAC353642.2021.9697306
27. Foong CC, Meng GK, Tze LL (2021) Convolutional neural network based rotten fruit detection using ResNet50. In: 2021 IEEE 12th control and system graduate research colloquium (ICSGRC), Shah Alam, Malaysia, pp 75–80. https://doi.org/10.1109/ICSGRC53186.2021.9515280
28. Miah MS, Tasnuva T, Islam M, Keya M, Rahman MR, Hossain SA (2021) An advanced method of identification fresh and rotten fruits using different convolutional neural networks. In: 2021 12th international conference on computing communication and networking technologies (ICCCNT), Kharagpur, India, pp 1–7. https://doi.org/10.1109/ICCCNT51525.2021.9580117

IoT-Based Smart Drainage System

Akshansh Jha, Pitamber Gaire, Ravneet Kaur, and Monika Bhattacharya

Abstract In order to establish a smart city, there are several aspects which are required to be taken care of, including solid waste management and sanitation, energy management and power distribution, health care, traffic management, and urban development. In addition to all these application areas, a smart drainage system is of immense importance in today's time, especially in metropolitan cities, where people face many issues related to the drainage system on an everyday basis. Currently, monitoring of drainage systems is done manually, which is not so reliable and accurate. It has also been shown in various studies that an inaccurate and inefficient drainage system results in the loss of many precious human lives every year. The present article deals with designing an IOT-based smart drainage system which can help in overcoming various issues associated with a normal manual drainage system. A smart drainage system can be used to closely monitor and solve various problems which include detecting water leakage through drainage pipes, reducing exposure to harmful gases, and monitoring the water level of the drainage system. This IOT-based intelligent device collects device location details along with all the relevant real-time data which is required to analyze and troubleshoot various common drainage-related problems. This information is then directly transmitted by the device to the nearest municipality office which ensures quick and timely resolution of the problems. The proposed smart drainage monitoring system can greatly help in the prevention of flooding and clogging of drainage system. Further, it can prove to be very useful in removing toxic materials and disease-causing organisms.

A. Jha
Bhaba University, Bhopal, Madhya Pradesh 462026, India
e-mail: akshansh.jha31@gmail.com

P. Gaire
Dr. A.P.J. Abdul Kalam University, Indore, Madhya Pradesh 452016, India
e-mail: gairepitamber96@gmail.com

R. Kaur · M. Bhattacharya (✉)
Acharya Narendra Dev College, University of Delhi, Delhi 110019, India
e-mail: monikabhattacharya@andc.du.ac.in

R. Kaur
e-mail: ravneetkaur@andc.du.ac.in

© The Author(s), under exclusive license to Springer Nature Singapore Pte Ltd. 2023
A. Mishra et al. (eds.), *Advances in IoT and Security with Computational Intelligence*,
Lecture Notes in Networks and Systems 755,
https://doi.org/10.1007/978-981-99-5085-0_23

Keywords Smart drainage system · IOT · NodeMCU board · Sensors · LED · ThingSpeak

1 Introduction

In today's time, the primary focus across the world is on enhancing quality and performance of various urban services by moving towards automated systems which are intelligent and smarter. One such vital urban service which can be considered as one of the major lifelines of a city is the drainage system. There are many common problems which are generally encountered in drainage system, including clogging and blockage which prevents the normal flow of wastewater, the leakage from drainage pipes, and the presence of harmful gases and toxic elements. The conventional drainage system is not automated. It has to be managed and operated manually which makes overall maintenance very inconvenient and tedious. There are many problems which are very difficult to solve in manual drainage system. One major issue is the safety of the sanitation workers who are employed for the maintenance of these drainage systems. There are many safety precautions which are required to be followed, whenever a worker has to go inside a manhole for cleaning it or for any other work related to maintenance and repair. Inside a manhole, different types of toxic gases such as methane (CH_4), carbon monoxide (CO), sulfur dioxide (SO_2), nitrogen dioxide (NO_2), etc., are present which could be extremely dangerous and hazardous for human health. Exposure to these toxic gases can lead to serious health problems like typhoid, fever, and even death [1]. Manual cleaning of drainage systems can prove to be a major health hazard and should be completely avoided.

According to the Guardian report, since 2017 at least one Indian worker die [2] while cleaning sewer or septic tanks every 5 days [3].

In a manual drainage system, the overall process of detecting and troubleshooting a problem is very inaccurate and time-taking. On the other hand, through an IOT-based smart drainage system, all the work related to maintenance and troubleshooting drainage-related problems can be done in a more accurate and timely manner without risking human lives.

2 Research Methodology

Schematic Design and Working

The schematic diagram of the proposed IOT-based smart drainage system is shown in Fig. 1. It comprises of a gas sensor which can detect the presence of harmful gases (CH_4, CO, etc.), a water level-monitoring sensor which can closely monitor the water level, and a flow sensor which can detect the presence of any blockage or leakage in the drainage pipes. Another vital component is the NodeMCU board

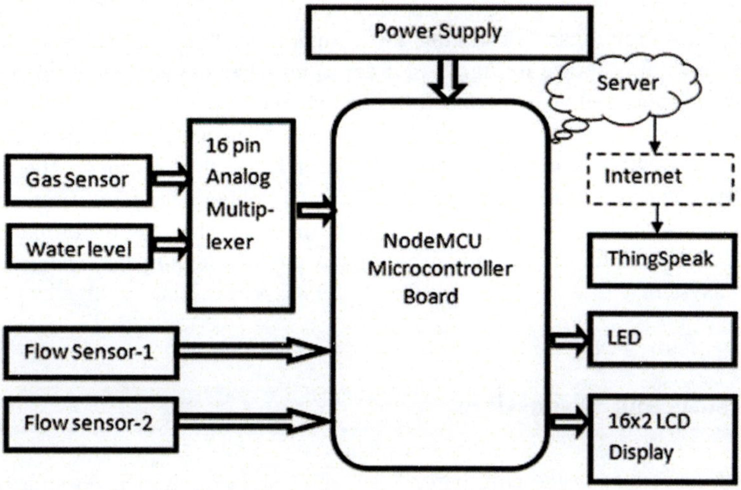

Fig. 1 Block diagram illustrating different modules of smart drainage system [5]

(microcontroller). All the real-time data collected by the sensors is processed by the NodeMCU board and displayed on a 16 × 2 LCD display. The data is then also uploaded to the Things Speak server through WiFi (ESP8266WiFi Module [4]). By analyzing the real-time data uploaded on the Things Speak server, the municipality office can keep close track of problems occurring in the drainage system.

The primary purpose of the proposed IOT-based smart drainage system is to enable easy maintenance of drains and ensure a clean and healthy environment. Poor maintenance of drains is a serious issue in cities today. One major problem is leakage from drainage pipes. The wastewater gets mixed with the pure water that leads to many health problems and life-threatening diseases.

Subsequent climate changes in different seasons foster in undesirable rains which further results in several problems like water blockage and harmful gases in the drainage system. Another major issue is the manual cleaning of drainage pipes and manholes by sanitation workers which can be very hazardous and should be completely avoided. In order to solve all these issues, an IoT-based smart drainage system has been created.

Many sensors have been used in the proposed smart drainage system which are capable of detecting water leakage, checking water flow, monitoring water level, and checking for harmful gases like CH_4, SO_2, CO, etc. Water level sensor and gas sensor are connected to the 16-pin analog multiplexer [6] as both sensors are analog and in NodeMCU board. The system is connected to the Things Speak Server [7], where the real-time value of sensor can be checked very easily and anywhere and the system will generate an alert message using LED blink which helps the department to take correct actions on it. The real-time data is analyzed through various sensors and is updated automatically on server with the exact system location.

A detailed flow chart illustrating the complete working of IOT-based smart drainage system is shown in Fig. 2. In order to initialize the hardware module [5], the system is powered up by connecting it to the power supply. On start-up, "Smart drainage System" is displayed on the LCD display. Four sensors have been used in this system: water sensor, gas sensor, flow sensor-1, and flow sensor-2. The micro-controller controls all the sensors and continuously obtains the input from sensor. LED is used as an indicator to produce an alert message whenever the output of the sensor exceeds the threshold value. ThingSpeak aids in tracking the location of the drainage system.

3 Results and Discussion

The four sensors form the core part of the smart drainage system through which various drainage-related problems are detected. The water level-monitoring sensor senses the level of water and gas sensor senses the harmful gases. Both sensors send all the data to the NodeMCU board [8]. When the water level sensor and gas sensor cross the threshold value then a message with a sensor value displays on the LCD display, and LED makes a glow alerting indication and the same process is carried out when the flow sensors sense the blockage and water leakage in manhole.

When water flows, the flow sensor system is able to readily determine the presence of any water leakage in manhole by comparing the water pressure from flow sensors. All system modules and outputs of sensors are closely monitored through the ThingSpeak website. Exact location of the hardware can also be specified accurately which further helps in detecting and troubleshooting the problems easily.

Water flow sensor is a kind of sensor which measures the rate of flow of water and converts it into a corresponding analog signal. This sensor has hydraulic input and hydraulic output. When water passes through the sensor [9], it causes the rotation of the turbine [10].

The output of Flow Sensor-1 and Flow Sensor-2 obtained through ThingSpeak is shown in Figs. 3 and 4. The variation in the output of flow sensors (1 and 2) with time gives an indication presence of any blockage or leakage in the drainage pipe. When the output of both the sensors is equal, if the flow rate value (from flow sensors 1 and 2) at a particular instant of time is same [11], this means that there is no blockage or leakage in the system [2].

The working formula to evaluate the water flow rate from pulse frequency measured by the flow rate sensor is given in Table 1. Water flow rate calculated for different pulse frequencies has also been calculated.

Measurements from Flow Sensor 1 and Flow Sensor 2 taken at different instances of time are tabulated in Table 2. As shown in the table, a large difference between the output of Flow Sensor 2 and Flow Sensor 1 is an indication of leakage in the pipe.

Figure 5 indicates the rate of water leakage in drainage pipes [2]. When the water flow rate measured by Flow Sensor-2 is less than the water flow rate measured by Flow Sensor-1 [11]. Then this means that there are some types of water leakage, and

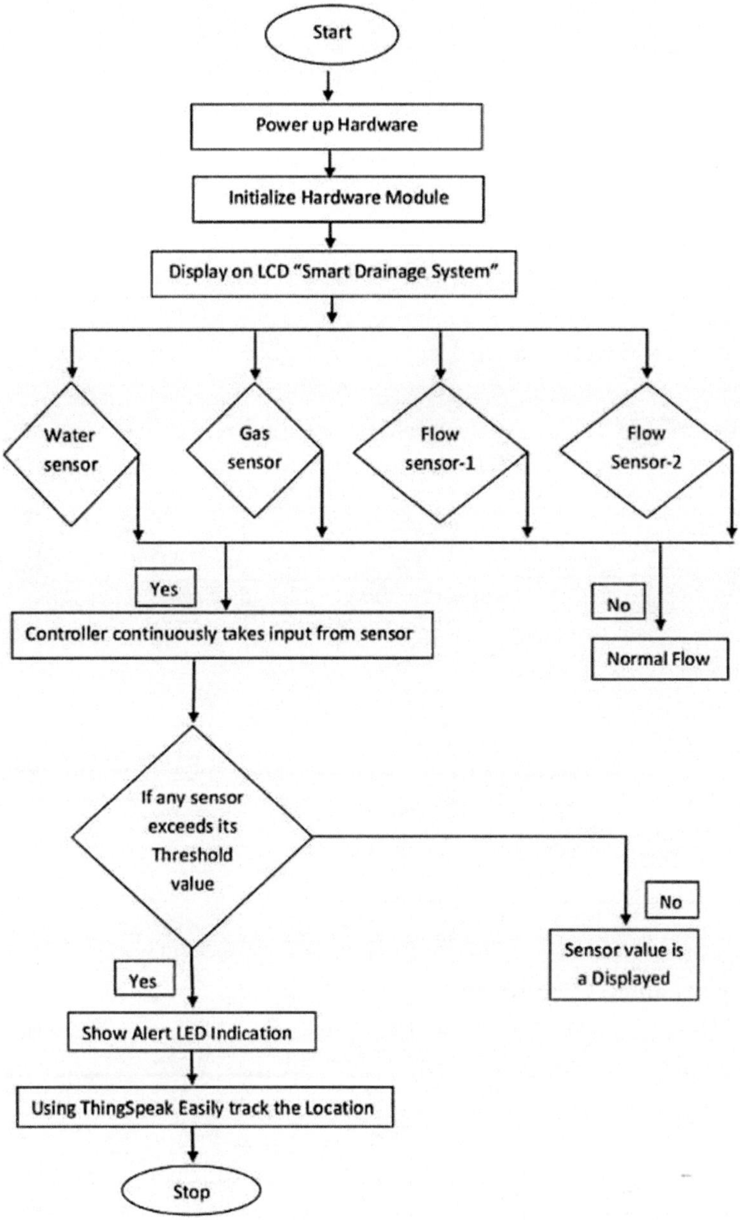

Fig. 2 Flow chart illustrating complete working of IOT-based smart drainage system

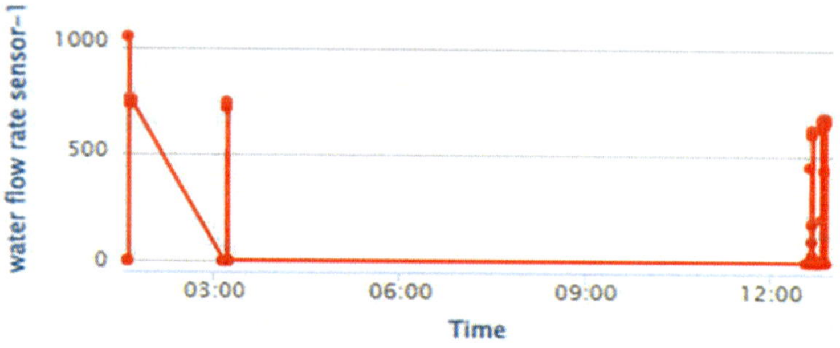

Fig. 3 Water flow rate (flow sensor-1)

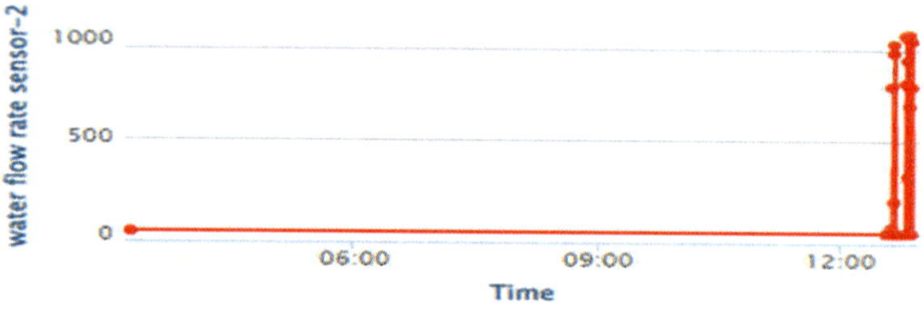

Fig. 4 Water flow rate (flow sensor-2)

Table 1 Flow rate

S. no	Frequency (Hz)	Flow (Lit./Hour) $= \frac{\text{Pulse Frequency (Hz)} \times 60}{7.5}$
1	15.6	125
2	31.25	250
3	46.8	375
4	62.5	500

Table 2 Characteristics of flow sensor

S. No	Input voltage (V)	Flow sensor-1 (F1)	Flow sensor-2 (F2)	Difference F1–F2	Output
1	5	490	485	5	No leakage
2	5	505	350	155	Leakage
3	5	520	352	136	Leakage
4	5	488	476	12	No leakage
5	5	386	322	54	Leakage
6	5	460	320	140	Leakage
7	5	510	505	5	No leakage
8	5	480	470	10	No leakage

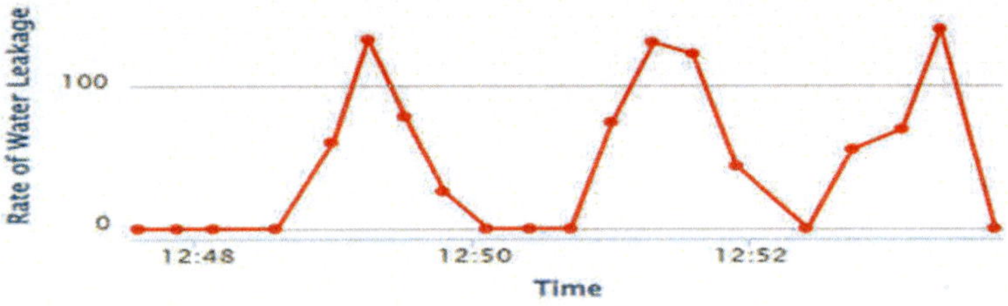

Fig. 5 Rate of water leakage

if the output of both the flow sensor is the same, then this means that there is no leakage in the drainage pipe.

Another major module which has been employed in the system is the gas sensor which is capable of detecting and measuring the concentration level of different types of toxic gases present in the atmosphere [6]. The gas sensor used in the system can detect a very broad range of gases [9]. Some important characteristics of the gas sensor used in the system are given in Tables 3 and 4.

The output of gas sensor which indicates the presence of harmful gases in the drainage pipe is shown in Fig. 6. When the gas sensor detects high concentration of harmful gases, then we can block air vents and dry pumping.

A water level sensor has also been used to continuously monitor the water level. Characteristics of water level sensor are given in Table 5. The water level-monitoring sensor which indicates the variation of water level of the drainage system is shown in Fig. 7. Water Level > 500 is considered to be "Very High". Whenever the water depth level in the drainage pipe gets greater than 500, an alert message is generated. It is an indication that there may be a blockage in the pipe causing obstruction in the normal flow of water.

Table 3 Gas sensor values [12]

1	Heater voltage (V_Z)	5.0 V ± 0.2 V AC or DC
2	Load resistance (R_L)	Adjustable
3	Heater resistance (R_H)	31 Ω ± 3 Ω
4	Heater consumption (P_H)	≤900 mW
5	Sensing resistance (R_S)	2–20 KΩ (in 2000 ppm C3H8)
6	Loop voltage (V_c)	≤24 V DC
7	Preheat time	Over 48 h

Table 4 Indication of toxic gases from gas sensor output

Input voltage (V)	Gas level (PPM)	Output (high/low)
5	0–50	Low
5	50–100	Medium
5	100–150	High

Fig. 6 Monitoring the presence of toxic gases

Table 5 Characteristics of water level sensor

Input voltage (V)	Water level	Output
5	0–100	Low
5	100–250	Medium
5	250–500	High
5	500–	Very high

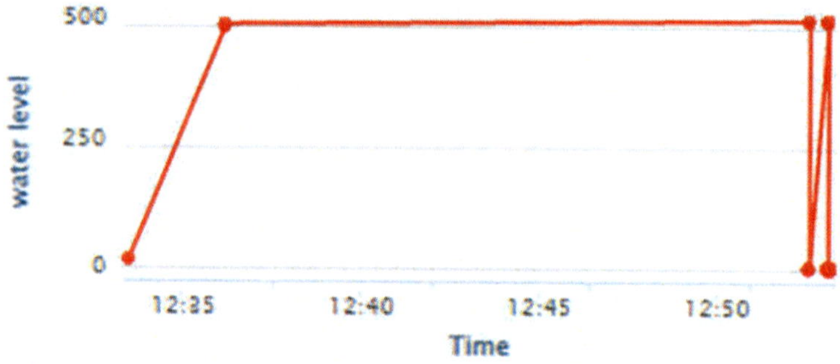

Fig. 7 Continuous monitoring of Water level

In the ThingSpeak server, there are location fields for tracking a device's location. These are used for statics devices [7]. By specifying the latitude and longitude [11], we can easily track the system location (Fig. 8).

Location Details from where measurements were made: **Latitude**: 19.1139729, **Longitude**: 19.1139729.

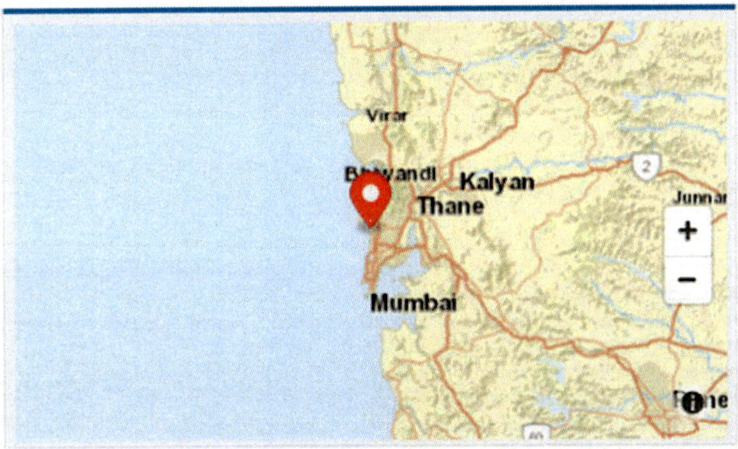

Fig. 8 Location details

4 Conclusion

This proposed IoT-based smart drainage system has been designed with a vision to make detection and troubleshooting of common drainage-related problems, less tedious and less hazardous for human workers. A smart drainage system can prove to be extremely vital and useful in realizing the concept of a smart city. The proposed system can help in continuous monitoring of drainage system in a more accurate and efficient manner. The primary objective of this system is to make the overall process of maintenance of drainage system more efficient and safer. The system also helps in curtailing plenty of accidents and disasters which may happen when a sanitation worker goes inside a manhole to manually detect and repair drainage problems.

In this system, various types of sensors have been used to keep track of various drainage issues, including gas sensor, water level-monitoring sensor, and flow sensor. Data collected by these sensors are processed by NodeMCU microcontroller board and uploaded on the ThingSpeak server. Data uploaded on the server can then be accessed and monitored by the city municipality offices for quick and immediate resolution of the problems. This system assists in abating death of sewage workers. We can use this method in various places such as smart water management, rain and stormwater management, etc.

Acknowledgements The authors would like to thank the Department of Biotechnology, Ministry of Science and Technology, (Govt. of India) for providing the necessary assistance to carry out this research work.

References

1. Sonawane G, Mahajan C, Nikale A, Dalvi Y (2018) Smart real-time drainage monitoring system using internet of things. IRE J 1(11)
2. Patel A, Dave P, Patel A (2020) Drainage monitoring system using IoT. Int J Res Eng Appl Manag 6(8)
3. Arulananth T, Ramya Laxmi G, Renuka K, Kartik K (2019) Smart sensor and arm based drainage monitoring system. Int J Innov Technol Explor Eng 8(11)
4. ESP8266EX Datasheet (2015) https://components101.com/sites/default/files/component_datasheet/ESP8266-NodeMCU-Datasheet.pdf
5. Rewatkar HP, Pathak T, Deshmukh S, Reddy P (2021) Smart drainage monitoring and controlling system using IoT. Int J Res Eng Sci 9
6. Waluyo BD, Hutahaean Desmon H, Junaidi A (2020) Multiplexer performance testing for IoT based air quality monitoring system. J Mantik 4(1)
7. ThingSpeak. Retrieved from Wikipedia. https://en.wikipedia.org/wiki/ThingSpeak. Last accessed 21 April 2022
8. Arduino (2022) Arduino. Retrieved from Wikipedia. https://en.wikipedia.org/wiki/Arduino
9. Tina SJ, Kateule BB, Luwemba GW (2022) Water leakage detection system using Arduino. Eur J Inf Technol Comput Sci 2(3)
10. Karale A, Dhurjad S, Lahamage S, Chaudhari M, Gend A (2020) Smart underground drainage blockage monitoring and detection system using IoT. Int Res J Eng Technol 7(2)
11. Vijayan A, Narwade R, Nagrajan K (2019) Real time water leakage monitoring system using IoT based architecture. Int J Res Eng Appl Manag 5(8)
12. Ajiboye A, Opadiji JF, Yusuf AO, Poppla JO (2021) Analytical determination of load resistance value for MQ-series gas sensor. TELKOMNIKA Telecommun Comput Electron Control 19

Evaluation of an Enhanced Secure Quantum Communication Approach

Neha Sharma and Vikas Saxena

Abstract The gathering and sharing of information is the foundation of our society. A message must be transmitted safely in a number of situations. QKD is a modern security innovation that helps to solve security problems and defend against hackers. QKD is already inseparable from the BB84 procedure; hence, academics are continually working on it. The purpose of this research article is to enhance existing security using polynomial key generation. As a result, this study will analyse the BB84 protocol's key generation process with the polynomial-based key. Further, researchers will get insight into the various challenges involved in secure communication.

Keywords Quantum computing · Quantum communication · Quantum key distribution · Quantum cryptography · Quantum information

1 Introduction

There are multiple instances in which encrypted communication has been substantial, ranging from confidential business, institutional or healthcare data to essential infrastructure information [1]. Whereas different cryptographic algorithms and methods for cracking them have been deployed throughout history, most cybersecurity rules depend upon the computation time of one-way operations. Furthermore, advances in quantum computing may severely compromise safety strategies, including the Rivest–Shamir–Adleman (RSA) method which is focused on factoring large numbers of substantial populations, a problem that future quantum computer could solve efficiently using Shor's algorithm [2]. Furthermore, while traditional cryptographic

N. Sharma (✉) · V. Saxena
Department of Computer Science and Information Technology, Jaypee Institute of Information Technology, Noida, India
e-mail: nehasharmablp461@gmail.com

V. Saxena
e-mail: Vikas.saxena@jiit.ac.in

© The Author(s), under exclusive license to Springer Nature Singapore Pte Ltd. 2023

A. Mishra et al. (eds.), *Advances in IoT and Security with Computational Intelligence*, Lecture Notes in Networks and Systems 755,
https://doi.org/10.1007/978-981-99-5085-0_24

methods claim to be post-quantum safe, neither guarantees absolute security. Shared information between sender and receiver must be secure while communicating with each other. The art of securing information with encryption and decryption techniques that let only access authorised users access it is called cryptography [3]. Communication takes place between Alice and Bob, who are referred to as the sender and receiver, respectively. Eve is the recipient's unauthorised access or extra involvement. In the traditional way, the first message is encrypted so that Eve cannot read it. Once encryption is done messages are shared with Bob from Alice with the key that is already known to both parties that are involved in the communication. There are chances, but at the same time, Eve also tries to read Alice's message and fetch our information. In that case, how we will secure keys? However, classically secure key exchange protocols are mathematically proven secure but there are ways through which Eve can interrupt them. As data is exponentially increasing [4] day-by-day classical protocols are somehow falling and compromising with computation and security. As an outcome, numerous hacking and security threats are coming into the picture. To address this issue there is an alternative for securing the key information using the properties of light way in quantum mechanics. According to the nature of light, we cannot create copies as well as tap because it follows no cloning and superposition laws. Further, it represents its unique property used as the basis of the QKD [5]. In summary, the sender and receiver communicate computational two levels called quantum bits or qubits that have been obtained in one of two completely at odds bases. Any eavesdropper initiative requires an assessment of the qubits, in which the eavesdropper should first select a single basis because quantum states cannot be cloned. As a result, inconsistencies in every original key occur when the intruder employs this same incorrect basis. Such inconsistencies, observed first by communicating entities, disclose eavesdropper efforts even before the key is generated as well as the incorporating string could be disposed of for data security. Furthermore, unlike traditional cryptographic algorithms, a delayed assessment of the transmission qubit isn't really possible. As a result, a secure manner of data transmission will continue to be protected in the future. In many research works, optical fibre is used for communication over a large distance and free space. BB84 and B92 are two-photon polarisation protocols in which light is used for secure communication [6]. These interesting principles of quantum mechanics have grabbed the attention of researchers worldwide. As a result with the aim of secure communication globally, research is breaking the new horizon and achieving good results every day.

2 Quantum Cryptography

These days, one of the really basic requirements in our everyday lives is privacy, and cryptography is a technique of securely communicating our confidential info over just a public network. There are two different types of cryptographic algorithms: symmetric or private-key cryptographic protocols and asymmetric or public-key encryption algorithms. To decrypt data (or decrypt) the plain text (or cipher-text), symmetric-key cryptographic algorithms are using a secret shared key, meanwhile, asymmetric key cryptography uses a public–private pair of keys for both encryption and decryption. One Time Pad (OTP) [7], Data Encryption Standard (DES) [8], Advanced Encryption Standard (AES) [9] and others are well-known symmetric key cryptography algorithms. Asymmetric key cryptographic [10] algorithms include Diffie–Hellman key exchange [11], ElGamal [12], RSA [13] and Elliptic curve cryptography [14], among others. This same requirement that the legitimate parties share a secret key prior to the exchange of information process presents a significant challenge for the symmetric key cryptosystem. Typically, each party creates the private key and securely transfers it to the second party. The question is currently how to share a private key in complete secrecy or without disclosing any information. Asymmetric key cryptographic algorithms can be applied to solve this issue traditionally key $k1$ and generate the secret key k for establishing communication with the other party, let's say, Bob. After that, Bob needs to encrypt k with key $k1$ and send it to Alice, who decrypts k using key $k2$, her private key. However, some mathematical hardness assumptions, such as the discrete log problem and integer factorisation problem, are the foundation for the confidentiality of the asymmetric key cryptosystem. However, Shor's algorithm makes the quantum computer a danger to asymmetric key cryptography because it can factor an integer in polynomial time. As a result, we can conclude that the key distribution problem cannot be solved conventionally without making some difficult mathematical assumptions.

3 Quantum Key Distribution

In accordance with the organisational concepts of quantum mechanics like the Heisenberg uncertainty principle [15] and quantum no-cloning theory [16], quantum cryptography offers absolute protection. Bennett and Brassard's BB84 QKD which they proposed in 1984, is the first quantum cryptographic protocol ever. The quantum conjugate coding concept put forth by Wiesner is the foundation for the BB84 QKD. It is conceivable for two or even more distant location users to create a shared secret key using quantum key distribution (QKD), and the security is solely reliant on the laws of quantum physics. As in the BB84 protocol, two individuals, Alice and Bob, create an arbitrary secret key using a pattern of high-energy photons arbitrarily provided in the diagonal basis (X-basis) and the rectilinear basis (Z-basis). Shor and Preskill provided a straightforward demonstration of the BB84 protocol's security, demonstrating the method's security. Another entangled state-based QKD protocol was put forth by Ekert in 1991. There have been numerous QKD protocol variations suggested by scientists up to this point, including BBM92, B92 and numerous others.

First, let's start by discussing QKD in greater depth. The concepts of quantum cryptography can be set in the late 1960s just before Stephen Wiesner first developed his concepts on conjugate coding, which was accepted for publication more than a couple of years later in 1983. Bennett et al. provided the first historical study that we cite. Wiesner's theories were the basis for Charles H. Bennett and Gill Brassard, who

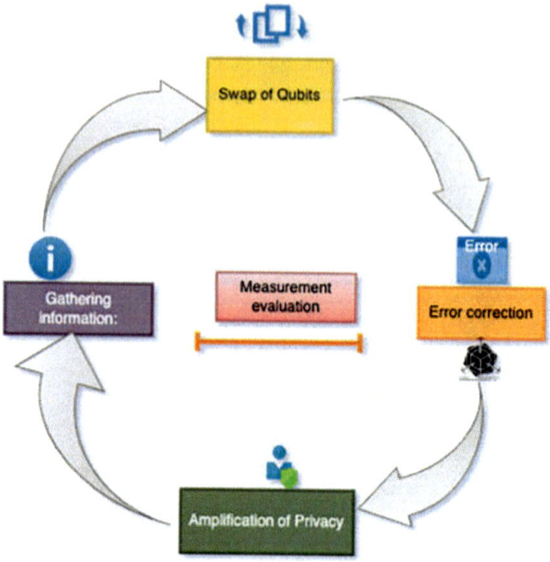

Fig. 1 BB84 protocol is further followed by several QKD protocols

in 1984 [17] proposed the BB84 protocol—the first real key distribution protocol—which employed the same quantum mechanical characteristics of photons to identify eavesdroppers' efforts. Through with there, innovative publication, this same study proposes using single photon polarisation to encrypt bits in various, chosen random bases as well as identify challenges by trying to compare a subset of outcomes. The same basic steps listed below, which were first presented in the BB84 protocol, are followed by several QKD protocols (Fig. 1).

- **Swap of Qubits**: Alice encrypts particles in one of at least two randomly chosen, non-commuting bases and transmits them to Bob, the receiving party, via a quantum channel. Bob then detects the photons on similarly randomly selected bases on his side. In comparison to entanglement-based schemes, this configuration is therefore known as a 'just prepare and measure setting'. The raw key is the bit stream that is transmitted.
- **Gathering information**: Unless Alice and Bob are verified (see below), those who can start sharing their lists of the bases they've both used over a traditional, public channel and only keep the measured data where both of them utilised the exact same basis (a process known as 'key sifting') should do so. They can keep the real measurement results to themselves. The sifted key, this same bit string achieved after the sifting step, must be deeply associated with in the absence of experimental flaws and adversarial eavesdropping (Eve).
- **Measurement evaluation**: The possibility required to listen could be discovered by making comparisons to a portion of the sifted key. These same rules of quantum systems state that every measurement made by such an unauthorised user just on transmission photons on the incorrect basis disturbs the quantum state and causes

inconsistencies in the sifted key that can be identified. The quantum bit error ratio (QBER), which also measures the likelihood that a bit will vary between Bob and Alice despite the fact that they utilised the same measurement basis, can be employed to quantify these errors. Be aware that quantum bit error rate in units of is a more commonly used phrase inside the research. Given that the QBER must be a probability, the key rate equations.

- **Error correction**: Inside a traditional post-processing stage, the same inconsistencies presented by that of the quantum channel, unsatisfactory state time to prepare, as well as the identification set-up must be corrected. As a result, a limited amount of information is leaked to the adversary, which must be rewarded in the final step.
- **Amplification of Privacy**: The confidentiality of a completed security code can be improved in some other traditional post-processing phases by using hashing algorithms that decrease key size, whereas improving security. These same remaining parts comprise the safe key that Bob and Alice share. The cryptographic key number is the rate of safe bits per time, and a favourable key rate indicates that protecting the public key is possible.

4 Principle Used for Secure Quantum Communication

No-cloning theorem makes it possible to detect the connectivity of eavesdropping. A single quantum cannot be duplicated, according to the theorem put forth in 1983 by W. H. Zurek and W. K. Wootters. The following equation holds for any quantum state $|a\rangle$, constant unitary matrix U and quantum cloning machines.

$$U(|a\rangle \otimes |0\rangle) = (|a\rangle \otimes |a\rangle) \tag{1}$$

In Eq. 2, replace

$$|a\rangle \otimes |b\rangle$$

$$U(|b\rangle \otimes |0\rangle) = (|b\rangle \otimes |b\rangle) \tag{2}$$

In Eq. 3, dot product of Eqs. 1 and 2 is shown

$$(\langle a/b\rangle) = (\langle a/b\rangle)^2$$

Hence,

$$\langle a/b\rangle \tag{3}$$

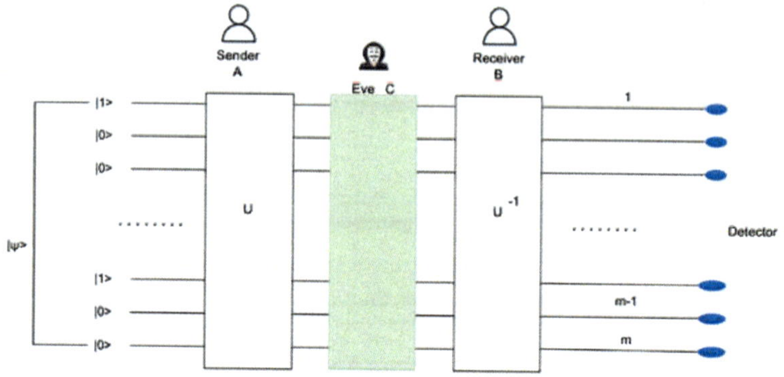

Fig. 2 Basic quantum theory

= either 0 or 1 will contradict the claim that there is no such thing as a quantum system. Accordingly, there are no quantum cloning devices where a third party cannot duplicate the original signal, and that would result in its collapse upon detection of the quantum. The possibility of variation between the issue that sent this same signal as well as the subject receiving it would allow them to detect third-party eavesdropping. Cryptography and quantum information exchange are referred to as reliable communication. Basic quantum theory is depicted in Fig. 2.

The very first quantum cryptographic protocol, also known as the BB84 protocol, was put forth by Bennett and Brassard in 1984 and allowed two individuals to share a secret key. Let the parties be Alice and Bob, who would like to exchange a private key. The following is the BB84 protocol:

Alice chooses the two 4n-bit strings a and b at random. She applies the subsequent method to create a finite series of 4n qubits Q from bit strings a and b.

For $1 \leq i \leq 4n$

If $a_i = 0$ and $b_i = 0$, she prepares $Q_i = |0\rangle$

If $a_i = 1$ and $b_i = 0$, she prepares $Q_i = |1\rangle$

If $a_i = 0$ and $b_i = 1$, she prepares $Q_i = |+\rangle$

If $a_i = 1$ and $b_i = 1$, she prepares $Q_i = |-\rangle$

Alice transmits Q to Bob, who randomly measures each qubit on the Z or X basis. He creates $4n$-bit string, and if the ith measured value is $|0\rangle$ or $|+\rangle$, $|1\rangle$ or $|-\rangle$, then $a'_i = 0(1)$.

Alice publicly declares b They reject the pieces of a and if Bob picks a different basis than Alice after a public debate on the choice of bases. It occurs with such a probability of $1/2$, therefore they have around 2n bit strings. If this is not the situation, the protocol is terminated. Alice selects n bits at random from the remaining 2n bits of a and informs Bob of the locations she has picked. Participants simultaneously compared the values of those n information bits and compute the margin of error to check for Eve's interference. If the error rate is not within acceptable limits, the protocol is terminated. Alice and Bob use a security enhancement and communication reconciliation technique to obtain an $m - bit(m \leq n)$ secret key as from left n bits.

5 Proposed Methodology

Shared information between sender and receiver must be secure while communicating with each other. The art of securing information with encryption and decryption techniques that let only access authorised users access is called cryptography. In the traditional way, the first message is encrypted so that Eve cannot read it. Once encryption is done messages are shared with Bob from Alice with the key that is already known to both parties that are involved in the communication. There are chances, but at the same time, Eve also tries to read Alice's message and fetch our information. In that case, how we will secure keys? However, classically secure key exchange protocols are mathematically proven secure but there are ways through which Eve can interrupt them. As data is exponentially increasing day by day classical protocols are somehow falling and compromising with computation and security. QKD is a modern security innovation that helps to solve security problems and defend against hackers. QKD is already inseparable from the BB84 procedure; hence, academics and researchers are continually working on it. Detailed BB84 protocol is discussed above. In the proposed protocol, we used the Qiskit simulator. In Fig. 3, a step-by-step discussion is shown. The purpose of this work is to analyse BB84 and polynomial-based BB84 protocol for secure communication. Alice starts with a random string of bits. Then random choices of basis are there for every individual bit. Again, a random selection of the basis is encoded from the Alice side. The message is shared from Alice to Bob. Bob also select randomly every qubit but results are not shared publicly. Only Bob is aware of the measurement results. Then basis is shared for every bit. If Bob is measuring the qubit from the same basis Alice used, they keep these bits for the generation of the secret key.

Therefore, the secret key is generated through the matched basis, other bits are discarded. Eventually, both parties share their keys, if their sample is the same their message is transmitted successfully. And in case the sample is not the same, interference from a third party is observed. There are scenarios, where we tried to change the secret key using polynomials. Because eavesdroppers action just on quantum connection interrupts the transmission qubit's state and causes errors inside the message being sent. In addition to eavesdropping, as the link transmission distance, the transmission signals also deteriorate, a phenomenon known as transmission loss.

Alice and Bob therefore first calculate the errors of their shared secret key (key sifting). The incorrect information bits are then corrected at the receiver, Bob, using the publicly verified communication link (key reconciliation). Any information shared over the communication page is taken to have been properly achieved by Eve, as well as the error-correcting process is known as key reconciliation. Eve can learn a little bit about the shared message because the information is connected to the private key.

We took 100 photons and Alice and Bob chooses their basis randomly. Key bit generation, error probability and information leakage have been taken as parameters of our analysis. Instead of normally revealing the bits for checking errors, in the case of polynomial-based BB84 we share a secret polynomial between Alice and Bob

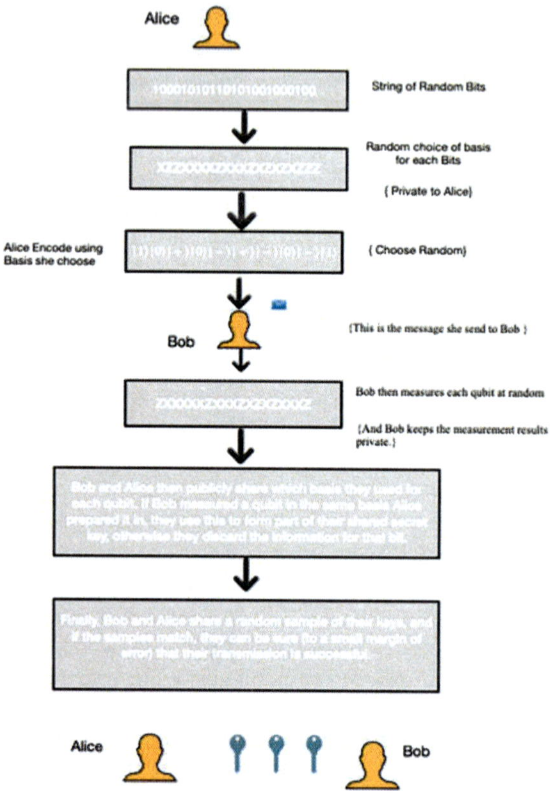

Fig. 3 Methodology of the work

in order to check errors. Once raw key bits are generated, we proceed with the key reconciliation process. Instead of CASCADE here also uses a shared polynomial for secretly correcting error bits. Any information shared over the communication page is taken to have been properly achieved by Eve. In this paper, we proposed a polynomial key reconciliation protocol that aims to transfer messages with zero (negligible or error bit) information leakage.

6 QKD and User Authentication Process Result and Discussion

Then basis is shared for every bit. If Bob is measuring the qubit from the same basis Alice used, an important fundamental duties of cryptography that can prevent an impersonation attack is authentication and authorisation verification. Ljunggren et al. introduced several QKD techniques with user identity verification via a trusted third party Eve in 2000. One of these systems is presented below, in which both Alice and Bob seek to communicate a secret key via quantum means. Eve and Alice (BOB) produce a secret key $K_A(K_B)$ using the BB84 protocol. Eve gives Alice (Bob) key K encrypted with the secret key $K_A(K_B)$ Eve individuals can exchange the secret message encoded with key K. Eve can hear the encrypted conversation since he appears to know the shared secret key K. The authors also recommend some QKD with authentication utilising entangled particles in the same study. Shi et al. [18] suggested a technique founded on entanglement state in 2001 that allows for the simultaneous realisation of QKD and quantum identification. Wei et al. [19], on the other hand, point out a flaw in Shi et al. [20], in which a hostile user might mimic a valid user without even being caught. They also provided an updated approach to avoid this flaw. Within this field, there are several other protocols inside this field, some of which are covered in [21].

7 Result and Discussion

Alice and Bob therefore first calculate the errors of their shared secret key (key sifting). The incorrect information bits are then corrected at the receiver, Bob, using the publicly verified communication link. Any information shared over the communication page is taken to have been properly achieved by Eve, as well as the error-correcting process is known as key reconciliation. In the modified BB84 protocol and existing BB84 protocol, the main difference is seen after key generation is completed. And instead of both parties, only Bob has revealed her key bits. CASCADE doesn't guarantee that all the error bits can be removed wherever polynomial-based error correction guarantees to remove every single error bit. CASCADE also reveals bits while removing the errors which shortens the key bits and further requires a privacy amplification step, whereas polynomial-based error correction doesn't require revealing the key bits. No information leakage is happening while error bits are removed secretly as compared to the Cascade algorithm which removes errors publicly (Fig. 4).

8 Challenges

In QKD protocols, the accessibility of the traditional authentication method platform has always been assumed indirectly. Regardless of whether a beginning privacy information is shared, authentication [22] must be repeated on a regular basis, which consumes a tiny proportion of a key, that is why efficient authentication protocols are required [23]. QKD can also be coupled with other post-quantum cryptography arrangements that support authentication [24]. Security certification is a challenge that is becoming more important even as advanced QKD systems mature. To ensure, the QKD system is functioning properly one could use a device-independent approach or another protocol [25] This includes certifying the randomness of the initial data,

Fig. 4 Analysis of existing versus proposed protocols

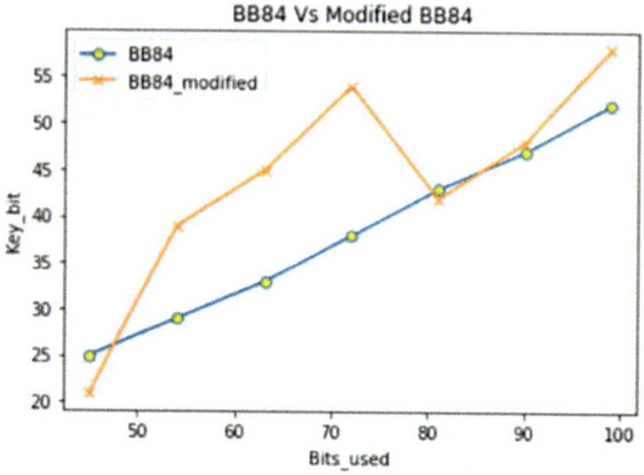

the same authenticity of a transmission medium, the purification of the generator and the accuracy of the post-processing procedure.

9 Conclusion

This paper briefly progresses with an overview of the fundamentals involved in the development of secure quantum communication. Detailed elaboration of Quantum Cryptography along with the Quantum Key Distribution protocol is covered. This article also discussed the depth of secure authentication. Our contributory works start with a simple execution of the existing BB84 protocol; then we found that there are ways through which we can secure secret keys. Therefore, a modified approach is developed in which an analysis of the existing work with the developed algorithm is described. Finally, the challenges of the present work are also discussed approach is developed in which an analysis of the existing work with the developed algorithm is described. Finally, the challenges of the present work are also discussed.

References

1. Rosenblatt ALFRED (2000) The code book: the evolution of secrecy from Mary Queen of Scots to quantum cryptography [Books]. IEEE Spectr 37(10):10–14
2. Bernstein DJ, Lange T (2017) Post-quantum cryptography. Nature 549(7671):188–194
3. Shor PW, Preskill J (2000) Simple proof of security of the BB84 quantum key distribution protocol. Phys Rev Lett 85(2):441
4. Gisin N, Ribordy G, Tittle W, Zbinden H (2002) Quantum cryptography. Rev Mod Phys 74(1):145
5. Bennett CH, Bessette F, Brassard G, Salvail L, Smolin (1992) Experimental quantum cryptography. J Cryptol 5(1):3–28

6. Bennett CH, Brassard G (1984) Proceedings of the IEEE international conference on computers, systems and signal processing
7. Vernam GS (1926) Cipher printing telegraph systems: for secret wire and radio telegraphic communications. J AIEE 45(2):109–115
8. Biryukov A, De Cannière C (2011) Data encryption standard (DES). In: Encyclopedia of cryptography and security, pp 295–301
9. Lasheras A, Canal R, Rodrıguez E, Cassano L (2020) Lightweight protection of cryptographic hardware accelerators against differential fault analysis. In: 2020 IEEE 26th international symposium on on-line testing and robust system design (IOLTS). IEEE, pp 1–6
10. Diffie W (1976) New direction in cryptography. IEEE Trans Inform Theory 22:472–492
11. ElGamal T (1985) A public key crypto-system and a signature scheme based on discrete logarithms. IEEE Trans Inf Theory 31(4):469–472
12. Rivest RL, Shamir A, Adleman L (1978) A method for obtaining digital signatures and public-key crypto systems. Commun ACM 21(2):120–126
13. Hankerson D, Menezes AJ, Vanstone S (2006) Guide to elliptic curve cryptography. Springer Science Business Media
14. Nikooghadam M, Zakerolhosseini A (2009) An efficient blind signature scheme based on the elliptic curve discrete logarithm problem. ISeCure 1(2)
15. Heisenberg W (1985)¨Uber den anschaulichen Inhalt der quantentheoretischen Kinematik und Mechanik. In: Original scientific papers wissenschaftliche originalarbeiten. Springer, Berlin, Heidelberg, pp 478–504
16. Wootters WK, Zurek WH (1982) A single quantum cannot be cloned. Nature 299(5886):802–803
17. Bennett CH, Brassard G (2020) Quantum cryptography: public key distribution and coin tossing. arXiv:2003.06557
18. Shi B-S, Li J, Liu J-M, Fan X-F, Guo G-C (2001) Quantum key distribution and quantum authentication based on entangled state. Phys Lett A 281(2–3):83–87
19. Wei T-S, Tsai C-W, Hwang T (2011) Comment on "quantum key distribution and quantum authentication based on entangled state." Int J Theor Phys 50(9):2703–2707
20. Zeng G, Wang X (1998) Quantum key distribution with authentication. quant-ph/9812022
21. Hwang T, Lee K-C, Li C-M (2007) Provably secure three-party authenticated quantum key distribution protocols. IEEE Trans Dependable Secure Comput 4(1):71–80
22. Guan D-J, Wang Y-J, Zhuang ES (2014) A practical protocol for three-party authenticated quantum key distribution. Quantum Inf Process 13(11):2355–2374
23. Sharma N, Ramachandran RK (2021) The emerging trends of quantum computing towards data security and key management. In: Archives of computational methods in engineering, pp1–14
24. Kiktenko EO, Malyshev AO, Gavreev MA, Bozhedarov AA, Pozhar NO, Anufriev MN, Fedorov AK (2020) Lightweight authentication for quantum key distribution. IEEE Trans Inf Theory 66(10):6354–6368
25. Sharma N, Saxena V (2022) An efficient polynomial based quantum key distribution approach for secure communication. In: 2022 8th international conference on signal processing and communication (ICSC). IEEE, pp 36–42

An Intelligent COVID Management System for Pre-schools

Vibha Gaur, Amartya Sinha, and Ravneet Kaur

Abstract Pre-schoolers were unable to adhere to the COVID-19 Protocols required to prevent the spread of the deadly coronavirus-caused disease. In public places like schools, a COVID management system can help enforce rules like wearing masks and keeping the appropriate distance between students. In the proposed system, machine learning algorithms and the Euclidean distance algorithm are used to recognize masks and figure out how far apart two faces are. The system is trained with the Convolutional Neural Network (CNN) algorithm, which is implemented with Python and TensorFlow library using a Raspberry Pi and camera.

Keywords Convolutional neural network (CNN) · Euclidean distance · TensorFlow · Face mask detection

1 Introduction

The COVID-19 outbreak has drastically altered students' lives worldwide. It affected public health, every sector of the economy, and social activities. The educational system was no different. According to information gathered by the United Nations Educational, Scientific, and Cultural Organization (UNESCO), school closings caused by COVID-19 affect about 290 million students [1]. This worldwide disruption in education is the most severe education crisis in human history. Closing schools to combat COVID-19 has raised concerns about its social impact. Closing schools was looked into as it reduces student social contact, which is a significant risk factor for virus transmission [2]. The world situation has gotten a lot better, and physical classrooms are back, but the coronavirus is still around, and its effects are uncertain. During the pandemic, it was mandatory for all citizens to wear a face mask in public as there were chances of getting infected. Before the pandemic, only people worried

V. Gaur · A. Sinha · R. Kaur (✉)
Acharya Narendra Dev College, University of Delhi, Delhi, India
e-mail: ravneetkaur@andc.du.ac.in

© The Author(s), under exclusive license to Springer Nature Singapore Pte Ltd. 2023
A. Mishra et al. (eds.), *Advances in IoT and Security with Computational Intelligence*, Lecture Notes in Networks and Systems 755,
https://doi.org/10.1007/978-981-99-5085-0_25

about their health wore masks to protect themselves from pollution. It was speculated that elementary school students could not adhere to COVID-19 regulations [3]. This led to the end of physical classroom instruction and a sudden switch to online instruction for the whole curriculum.

Online learning has changed the way traditional learning is done, but in the past year and a half, its popularity has been linked to a number of health problems for both students and teachers. Even a steady decline in student motivation has been observed. The vast majority of students turn off the camera and go about their daily routines. Online instruction significantly lowered students' attention spans because of visual distractions. The exponential increase in time spent in front of screens was making it harder for the brain to keep up with the flood of new information. There is a strong correlation between the amount of time spent in front of a screen and the frequency with which students complain of headaches. The ethics in the classroom have been severely compromised. Postures, a lack of routine, and attentiveness have all contributed to health risks. Additionally, constant sitting also led to weight concerns. Children have become overweight due to a lack of physical activity [4]. Therefore, school monitoring of children is necessary to counteract these issues. In these situations, the COVID face mask and social distance management system can be of great assistance.

The proposed COVID management system comprises a face mask detection model trained with the CNN (convolutional neural network) algorithm and the TensorFlow library. When the model is trained, it is put on a Raspberry Pi system along with a Python script (to find social distance and utilize the trained CNN model) with a camera (to get real-time video) and alarm system. For training, an image dataset is retrieved from Kaggle.com [5]. After training, the proposed system can be used in real time to find face masks. The CV2 library is used to retrieve images captured by live surveillance cameras. The Haar cascade frontal face model is used to identify faces in photographic images. With the help of the face points found by the Haar cascade frontal face model, social distance can be calculated using Euclidean distance.

2 Related Work

The World Health Organization (WHO) has mandated stringent regulations in response to the rapid global spread of the COVID-19 virus, which has devastating effects on both human health and economic growth. Face masks, social distancing, a shift to a virtual work culture, and other precautions are all recommended [6]. Among these norms, face mask detection is a cutting-edge tool that can help track how many people are covering their faces to limit the spread of disease and protect a large population. Many different kinds of software and programming languages are used to run the face mask detection technology. These include Python, Deep Learning, TensorFlow, Keras, OpenCV, and many others.

Students from the Indian Institute of Information Technology, Allahabad, and the VIT University in Chennai collaborated to propose a model that was developed by refining the InceptionV3, a deep learning model [7]. The model is based on Transfer Learning from InceptionV3 and outperforms the methods based on Simulated Masked Face Dataset (SMFD). However, it omitted many important factors that were not included in the pre-trained algorithm.

An intriguing SSDMNV2 model that uses OpenCV DNN and MobileNetV2 was presented [8], the latter of which places less strain on available hardware. Even though the model's accuracy of 92.64% is much higher than that of models like LeNet-5 and AlexNet, which were trained on the same dataset, it still needs to be improved before it can be used in real life.

An experiment with a hybrid model was presented that uses both traditional machine learning techniques and deep transfer learning [9]. Faces were classified using a decision tree, a support vector machine (SVM), and an ensemble algorithm, with feature extraction performed using a 50-layer deep CNN model called ResNet50. Another model for face mask detection to stop the spread of coronavirus was built using data from a variety of sources [10]. CNN assisted in the training of the model. Another group of researchers designed the multi-stage CNN architecture for face mask detection [11].

Many of these approaches have either relied on previously trained CNN architectures or improved upon previously trained CNN architectures through transfer learning. These solutions may provide accuracy by a small margin, but that is only possible with transfer learning on pre-trained models, which are trained on millions of datasets. But this work utilizes a self-created and trained CNN model for the COVID management system in pre-schools.

3 Research Methodology

Students and teachers have to go to their schools or universities, but at the same time, they should take care to prevent the transmission of the virus. But checking face masks and social distance manually is very tedious. There is an urgent need to develop a computerized system that can notify and announce in the classroom if students are wearing masks and keeping their distance from one another. Children in school often forget to wear masks. The proposed system keeps an eye on face masks and social distance rules for the good of all students. A Raspberry Pi with a camera module is used to make the proposed system compact and economical. The system is easy to set up and use. The proposed system is developed with Python 3.10 using the following steps.

3.1 Data Collection and Pre-processing

It entails gathering an image dataset of people wearing and not wearing masks. The dataset for training the mask detection model is obtained through Kaggle.com. The dataset is divided into three categories: "with mask", "without mask", and "incorrectly worn mask". Out of these, two categories, "with mask" and "without mask", are utilized in the proposed work to train the model. Each category contains about 3,000 images. The data is imported and pre-processed. Pre-processing includes reducing the size of the images and changing them into grayscale.

3.2 Training the CNN Model

CNN is beneficial for working on image datasets. For mask detection, a model must be trained. After importing the dataset and preparing the images, the next step is to make a CNN model and train it to identify face masks. The pre-processed images are reshaped into a NumPy array to work on the dataset that is split into training and testing groups. The CNN model was developed and trained using the TensorFlow library. TensorFlow is a popular machine learning library to train the model and does other image processing tasks. The TensorFlow library is used to build the sequential CNN model. It has Conv2D, MaxPooling2D, Flatten, Dense, and Dropout layers. The Conv2D layer with ReLU activation is used to make a convolution kernel with 200 filters (3×3). After that, MaxPooling2D down-sampled the input to 2×2. Again, a convolution kernel of 100 filters (3×3) is built using the Conv2D layer, while the activation function remains the same. The MaxPooling2D layer is used again for convolution. Dense layers are used to create a simple yet dense neural network. First, the neural network is composed of 100 and 50 units.

To avoid overfitting, dropout layers are also used at the rates of 0.50 and 0.30. The final dense layer is made up of two units. The Adam optimizer, along with categorical cross-entropy loss, is used for compilation due to its low cost. It can provide more performance at a lower training cost. Once the CNN model is developed, trained, and tested, it can detect masks. After obtaining satisfactory test results, the trained model is saved to disk for further application. The proposed framework for the application of the COVID management system is shown in Fig. 1, while Fig. 2 depicts the code snippet for building and training the CNN model. Table 1 summarizes the CNN model compiled for face mask detection.

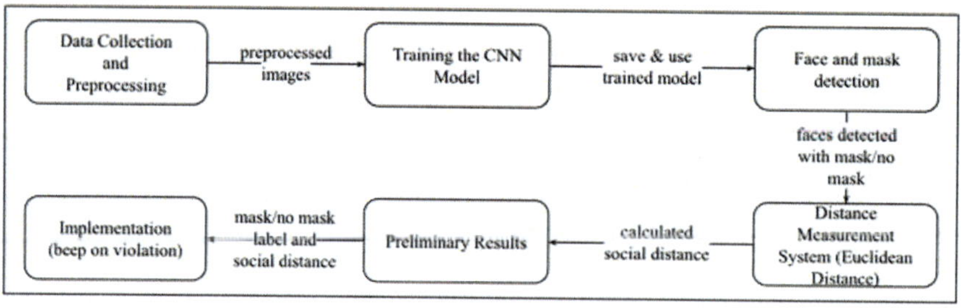

Fig. 1 Proposed framework for the application of COVID management system

```
model = tf.keras.Sequential([
        tf.keras.layers.Conv2D(200, (3, 3), padding='same',
            activation="relu", input_shape=data.shape[1:]),
        tf.keras.layers.MaxPooling2D((2, 2)),
        tf.keras.layers.Conv(2D100, (3, 3), padding='same',
            activation="relu"),
        tf.keras.layers.MaxPooling2D((2, 2)),
        tf.keras.layers.Flatten(),
        tf.keras.layers.Dense(100, activation="relu"),
        tf.keras.layers.Dropout(0.50),
        tf.keras.layers.Dense(50, activation="relu"),
        tf.keras.layers.Dropout(0.30),
        tf.keras.layers.Dense(2, activation="softmax")
   ])

model.compile(optimizer='adam', loss='categorical_crossentropy',
            metrics=['accuracy'])
train_data, test_data, train_target, test_target = train_test_split(
            data, target, test_size=0.1)

history = model.fit(train_data, train_target, epochs=20, callbacks=
            [checkpoint], validation_split=0.2)
```

Fig. 2 Code snippet of CNN model building and training

Table 1 Summary of compiled CNN model

Model: "sequential"		
Layer (type)	Output shape	Param#
conv2d (Conv2D)	(None, 100, 100, 200)	2000
max_pooling2d (MaxPooling2D)	(None, 50, 50, 200)	0
conv2d_1 (Conv2D)	(None, 50, 50, 100)	180,100
max_pooling2d_1 (MaxPooling2D)	(None, 25, 25, 100)	0
flatten (Flatten)	(None, 62,500)	0
dense (Dense)	(None, 100)	6,250,100
dropout (Dropout)	(None, 100)	0
dense_1 (Dense)	(None, 50)	5050
dropout_1 (Dropout)	(None, 50)	0
dense_2 (Dense)	(None, 2)	102
Total params: 6,437,352 Trainable params: 6,437,352 Non-trainable params: 0		

3.3 Face and Mask Detection

In this step, image frames are taken with a real-time camera using the CV2 library. The pre-trained deep learning Haar Cascade model detects faces in each captured frame [12]. Haar Cascade is a popular object detection algorithm used to identify faces in a given image or even from a real-time video source. Since this work uses real-time video sources to find faces, the Haar Cascade Frontal Face model works better than any other face detection model. The trained CNN model is then used to classify all of the faces that have been recognized as "with mask" or "without mask".

3.4 Distance Measurement System

The Euclidean distance is used to compute the distance between the two faces. The approach is simple. Each face's starting and ending points on the X and Y axes are determined, and these coordinates are used to compute the Euclidean distance using the Euclidean distance formula (1) [13]. Once the distance between two faces is calculated, it is compared with the minimum social distance allowed in public [14]. Its pseudo-code is explained in Fig. 3.

$$d_n(x, y) = \sqrt{\sum\nolimits_{i=1}^{n} (x_i - y_i)^2} \tag{1}$$

```
identify human body among other objects
find startX, startY, endX, endY axes for each body
store axes in a list

for axis1 in axes:
      for axis2 in axes:
            pass axis1 and axis2 in euclidean formula
            calculate social distancing

            if euclidean_distance <= 6 feet:
                  social distancing violation
                  activate buzzer
```

Fig. 3 Pseudo-code for social distance measurement

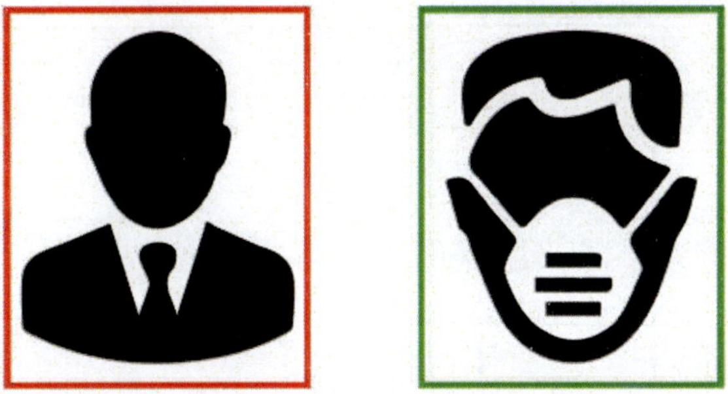

Fig. 4 Prototype of working model

3.5 Preliminary Results

The next step involves drawing a rectangle around the faces with the label "With mask" or "Without mask". The green box is used for "with mask", and the red box is drawn for "without mask", as shown in Fig. 4.

3.6 Implementation

The trained model and other software modules are installed on a system made up of a Raspberry Pi [15], a camera, and an alarm system [16]. The Python script for face mask detection and computing social distance detection will run on the Raspberry Pi.

Figure 5 shows the flowchart of the system on the Raspberry Pi. It shows the control flow over face mask detection and social distance violations. Once the model is trained and saved to disk, it is ready to be used for mask detection. Image frames are captured in real time using the live surveillance camera. Then the faces from these frames are extracted using the Haar Cascade model.

The trained CNN model is also loaded for recognizing masks on every face. A red box is drawn on the face, and the alarm system is activated if there is no mask. If the mask is worn, Euclidean distance is calculated between the two faces. Social distance violations will enable the alarm system. A green box is drawn on the face if social distance is also maintained while wearing a face mask.

4 Results and Discussion

The model achieved an accuracy of 99.01% with 20 epochs. Earlier, the trained models gained around 96% accuracy. However, the overfitting problem persisted, so additional Dropout layers were added. Not only did it fix the overfitting issue but also the accuracy also increased. The final model is achieved after 20 epochs because

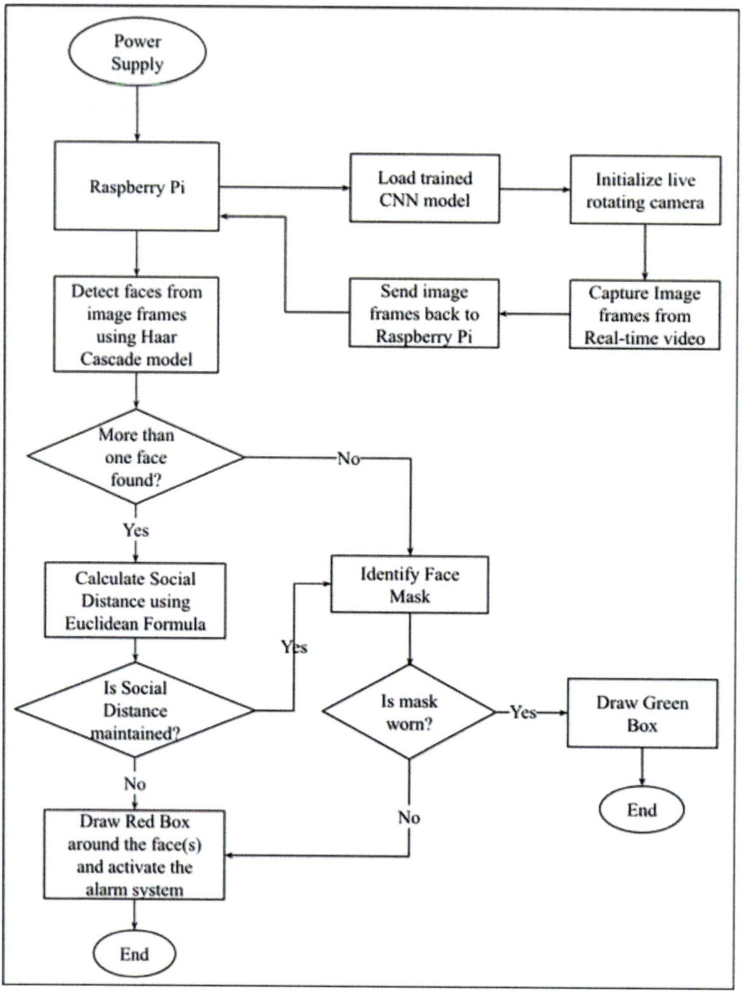

Fig. 5 Proposed flowchart for using the COVID management system in academic institutions

adding more epochs did not make a big difference. The training and testing graph is shown in Fig. 7.

A test sample of 300 images was taken. The false positives (Type I Error) are 11, and the false negatives (Type II Error) are 16, whereas the true positives are 284 and the true negatives are 288. Figure 6 shows the confusion matrix for the trained model.

Since the accuracy of the testing is 95.49%, the results from the confusion matrix are considered good. Table 2 shows the new model's accuracy and loss during training and testing. The model took 2 min and 22 s to train and was found to be 99.01%

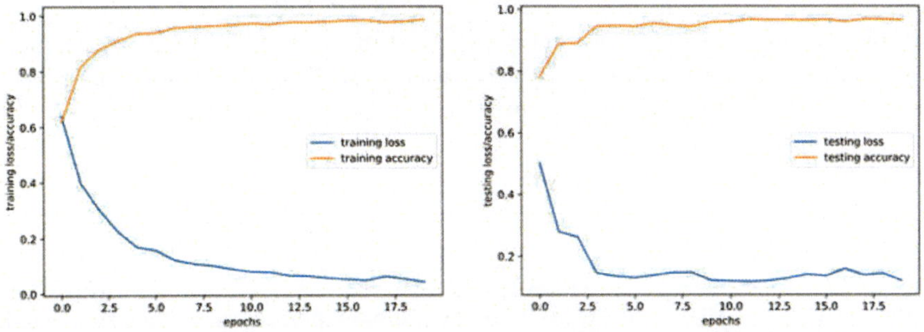

Fig. 6 Training and testing loss/accuracy graphs

Fig. 7 Confusion matrix (true labels versus predicted labels)

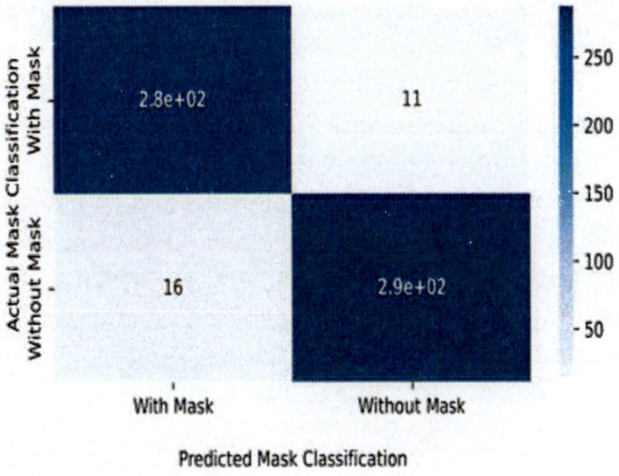

Table 2 Experimental results after model evaluation

Training accuracy	99.01%
Testing accuracy	95.49%
Training loss	0.0356
Testing loss	0.1144
Training time	142 s

accurate. Epochs are plotted on the X-axis, and training/testing accuracy and loss are plotted on the Y-axis.

5 Case Study

The proposed system was installed in a classroom with a capacity of 30 students. Initially, the proposed system was installed using the inbuilt camera of the laptop that could cover only a few students. Then the proposed system was implemented using the mobile phone's camera interconnected with the laptop. To utilize the mobile phone camera as a Live Video Capturing source, an IP Webcam application was used, which provided a URL to connect the phone's camera to the laptop on the same network. The number of students covered by the camera increased marginally, but the desired result was not obtained.

For better classroom coverage, the laptop was replaced by a Raspberry Pi with a camera module. The proposed system was installed on top of the blackboard in the classroom. But the camera did not cover the first row or capture the students sitting in the starting row. To overcome this issue, two camera modules were then connected to the Raspberry Pi 3 module. The first camera was connected using Camera Serial Interface (CSI), while the second camera module was connected using the USB port. It improved the coverage marginally. Even though some students were not covered, the proposed system could successfully identify the students who were not wearing masks. The proposed system recognized even the social distance obligation.

However, multiple systems can be installed for better coverage. To install multiple cameras, Multi-Camera Adapter Module can be used.

This implementation of the proposed system emphasizes the importance of taking into account a classroom's physical layout and the requirement for multiple cameras to ensure comprehensive coverage. It also emphasizes the importance of considering the use of various types of cameras and video-capturing devices to achieve the desired coverage. In addition, the proposed system can be useful for tracking whether or not students are following classroom safety protocols in the event of a pandemic of infectious diseases like COVID-19.

6 Conclusion

Over the past 2 years, the rapid global spread of the coronavirus disease COVID-19 has kept the entire world on edge. The most effective measures implemented to combat the spread of this virus involved the widespread use of face masks and the practice of keeping a safe distance from others while out in public. Effective methods for detecting the use of face masks and social distancing in public spaces were crucial in stopping the spread of the virus. Even though the world is getting over the damage done by the coronavirus, it is still important to take precautions and be ready to fight any new infectious diseases that may appear in the future. An intelligent system built on top of deep learning algorithms has been proposed to solve these problems.

Using Python, OpenCV, and TensorFlow, an intelligent computer vision system is developed using deep learning and conventional machine learning techniques. The work employs the CNN algorithm implemented in TensorFlow, coded in Python, to build the model to distinguish between students in pre-schools wearing masks and those who are not, with an average testing accuracy of 94.99%. The Euclidean formula also works well for determining how far apart students are on the social scale. Thus, the system can work efficiently to stop the transmission of the deadly coronavirus.

References

1. UNESCO 290 million students out of school due to COVID-19: UNESCO releases first global numbers and mobilizes response. https://www.unesco.org/en/articles/290-million-students-out-school-due-covid-19-unesco-releases-first-global-numbers-and-mobilizes. Last accessed 15 December 2022
2. Jackson C, Vynnycky E, Mangtani P (2016) The relationship between school holidays and transmission of influenza in England and Wales. Am J Epidemiol 184(9):644–651
3. Ahwireng D (2022) Confronting COVID-19 whilst elementary school students resume in-person learning. J Educ Learn 11(3):64–76
4. Idris F, Zulkipli IN, Abdul-Mumin KH et al (2021) Academic experiences, physical and mental health impact of COVID-19 pandemic on students and lecturers in health care education. BMC Med Educ 21:542
5. KAGGLE Face Mask Detection. https://www.kaggle.com/datasets/vijaykumar1799/face-mask-detection. Last accessed 30 December 2022
6. Clapham HE, Cook AR (2021) Face masks help control transmission of COVID-19. Lancet Digital Health 3(3):e136–e137
7. Jignesh Chowdary G, Punn NS, Sonbhadra SK, Agarwal S (2020) Face mask detection using transfer learning of InceptionV3. In: Bellatreche L, Goyal V, Fujita H, Mondal A, Reddy PK (eds) Big data analytics. BDA 2020. Lecture notes in computer science, vol 12581. Springer, Cham
8. Nagrath P, Jain R, Madan A, Arora R, Kataria P, Hemanth J (2021) SSDMNV2: a real time DNN-based face mask detection system using single shot multibox detector and MobileNetV2. Sustain Cities Soc 66:102692
9. Loey M, Manogaran G, Taha MHN, Khalifa NEM (2021) A hybrid deep transfer learning model with machine learning methods for face mask detection in the era of the COVID-19 pandemic. Measurement 167:108288
10. Shamrat FJM, Chakraborty S, Billah MM, Al Jubair M, Islam MS, Ranjan R (2021) Face mask detection using convolutional neural network (CNN) to reduce the spread of COVID-19. In: 5th international conference on trends in electronics and informatics (ICOEI). IEEE, pp 1231–1237
11. Chavda A, Dsouza J, Badgujar S, Damani A (2021) Multi-stage CNN architecture for face mask detection. In: 6th international conference for convergence in technology (I2CT). IEEE, pp 1–8
12. Cuimei L, Zhiliang Q, Nan J, Jianhua W (2018) Human face detection algorithm via Haar cascade classifier combined with three additional classifiers. In: 13th IEEE international conference on electronic measurement and instruments (ICEMI). IEEE, China, pp 483–487
13. D'Agostino M, Dardanoni V (2009) What's so special about Euclidean distance? Soc Choice Welf 33:211–233
14. Rezaei M, Azarmi M (2020) DeepSOCIAL: social distancing monitoring and infection risk assessment in COVID-19 pandemic. Appl Sci 10(21):7514
15. Sachdeva P, Katchii S (2014) A review paper on Raspberry Pi. INPRESSCO 4(6):3818–3819
16. Varsha B, Tiwari S, Chaudhari V, Patil V (2022) Face mask detection with alert system using Tensorflow, Keras and OpenCV. IJEAP 2(1):339–345

An Insight to Features Selection using ACO Algorithm

Priyanka and Anoop Kumar

Abstract The richness and diversity of network data provide a number of problems for data classification technology. The choice of features has always been a significant and challenging issue in categorization technology. Feature Selection (FS) method is the most important part of the data pre-processing steps in data modeling, as we know that all the data is not useful in final processing, so it is important to find out only the essential features as a subset from the superset of available features. Irrelevant and redundant features may add the overhead to the extraction of necessary information which becomes difficult due to the processing and storage needs. We may narrow down the available data to only the relevant information by using feature selection. The main aim of FS process is to reduce the number of features, remove irrelevant, noisy and redundant data from the dataset and provide better results and accuracy as compared to the complete set of features. In order to speed up processing, cut training time, prevent learning from noise (overfitting), reduce dimensionality, increase predictive accuracy and lower computational complexity, it is necessary to choose a good feature selection method based on the number of features investigated for sample classification. The goal of this work is to identify the most widely used feature selection optimization strategies that deal with improving classification performance in terms of accuracy and efficiency. This paper highlights various methods of FS and additionally suggests a feature selection method based on ACO.

Keywords Classification · Feature selection · Optimization · Dimensionality reduction · ACO

1 Introduction

The feature can be identified as an attribute, property, dimension or characteristic of anything which defines it. Feature extraction is the process of conversion of given input data into a subset of essential features. Some predictive modeling problems

Priyanka (✉) · A. Kumar
Banasthali Vidyapith, Tonk, Rajasthan, India
e-mail: priyankarattan85@gmail.com

© The Author(s), under exclusive license to Springer Nature Singapore Pte Ltd. 2023

A. Mishra et al. (eds.), *Advances in IoT and Security with Computational Intelligence*, Lecture Notes in Networks and Systems 755,
https://doi.org/10.1007/978-981-99-5085-0_26

involve a large number of variables, which can slow down model development and training and finally require a large amount of system memory. In addition, some models may perform poorly if the target variable contains unrelated input variables. It starts with the initial set of data that is readily available and develops the features, which are also known as borrowed values. These features are meant to be descriptive and non-redundant, and they will make the subsequent learning and classification processes simpler. It will speed up the learning and generalization phases of the learning process by reducing the amount of redundant data in the dataset, which will minimize the quantity of data needed to create the model and require less machine work. When the given input in the algorithm is very large for handling and it is suspicious of being redundant, then there is a need to convert it into a decreased set of features. Undoubtedly, feature selection refers to the choosing of a restricted subset of the extracted characteristics. If the desired features are able to extract the necessary information from the provided set of data, the task can be accomplished successfully by accepting the reduced data rather than the entire set of offered data. Basically, it is connected to dimension reduction [1]. One method for reducing classifier calculation errors in the selection of important attributes and the elimination of redundant, irrelevant and noisy attributes from the original dataset is feature selection (FS) [2]. Data is continuously created and expands exponentially over time in real-world applications [3]. For instance, doing wet lab experiments to get a complete feature set in bioinformatics may be expensive and feature storage may be impractical. The classification of tweets in social media [4] and the classification of images in video surveillance [5, 6] are two other examples. It might not be possible for such applications to be fully feature-complete before the selection process begins.

Wrapper, filter, and embedding models are the three main groups into which feature selection techniques are typically divided [7–10]. The learning-based classifiers are used in the wrapper feature selection approach to choose a subset of features from the available feature superset. Wrapper methods may choose acceptable features, but they may not be practical for streaming features because we must retrain the classifier each time a new feature is added in order to evaluate the model's performance. The computational cost of filter-based feature selection approaches, on the other hand, which evaluate the performance of each feature separately, is lower; nonetheless, it's likely that the weak features may perform better when combined [5, 11, 12]. As a result, filter-based approaches might potentially not offer a comprehensive representation of streaming features.

2 Related Work

Any unsupervised setting makes feature selection more difficult since label information extraction takes more time [13]. This paper discussed an unsupervised filter-based feature selection method where the features are prioritized according to their applicability to the target class using the Ant Colony Optimization (ACO) process, with the feature search space being represented as a graph. In order to discover the best subset of features iteratively rather than utilizing the learning algorithm, another filter-based unsupervised feature selection method employing ACO is provided in [14]. The closeness between the features is used to calculate the feature relevance, which reduces redundancy. Unsupervised feature selection is done using a streaming technique for social media [14, 15]. In order to discover which properties are closely related, this strategy makes use of the connection information. Users who share same hobby, like football and cricket, frequently exchange knowledge and vocabulary. So, when describing a group, words are used as features [16]. This paper discusses a method for feature choice that is based on group output.

In order to reduce computation time and memory requirements while improving performance, a variant of ACO is developed in [17], which travels only a directed path rather than the full graph. To accurately categorize email messages from the data stream, an ACO-based method is discussed in [18]. An approach to feature selection that uses clustering-based ACO [19] and divides features into groups over the whole search space has been recommended for improvement in [20]. ACO is then utilized for the subset of optimal features. Pheromone initialization in the updated technique is based on evaluating the value of each feature to the target class. Additionally, by updating the evaluation function with multiple discriminant analyses, the redundant and irrelevant features are given low likelihood. According to [21], the wrapper technique is then used for further evaluation of the candidate feature set (CFS) and guidance of the integration process after instance-based learning is originally employed to determine the CFS.

Feature selection (FS) approach is essentially used to create search space problem prediction accuracy. There are many different search methods, such as involuntary hybrid exploration algorithms, heuristic search, probabilistic search and thorough search. Algorithms are using meta-heuristics to identify the best answer, you must search the features much more quickly because doing so takes too much time [2].

Because the feature selection problem is multi-objective in nature, optimizing feature subsets with regard to a single criterion is insufficient [22]. Therefore, a multi-objective genetic algorithm will be used to optimize selected subsets in order to incorporate several feature selection criteria. The suggested technique may identify a variety of ideal feature subsets that are evenly distributed over the entire feature space and guarantee higher classification accuracy. Regarding prediction accuracy, non-redundancy and cardinality of feature subsets, the various strategies satisfy fair levels of optimality. These systems provide users a variety of feature subset selections.

Rough sets and the bat algorithm are the foundation of the novel selection method that is suggested. When using the Bat Algorithm (BA), which is a feature selection method, bats fly in the feature subset space and find the optimal feature combinations [23]. The BA just needs basic and straightforward mathematical operations, not complex ones like crossover and mutation. It has a low computational cost in terms of memory and runtime. Fitness balances the classification performance and reduction size by taking into account both the classification accuracy and the number of chosen features. The used rough set-based fitness function guarantees improved classification results while maintaining a small feature size. By using a purely structural approach, rough set theory offers an arithmetic tool to identify data dependencies and minimize the number of features in the dataset. Producing all conceivable reductions and selecting one with a negligibly low cardinality is the entire method for detecting nominal reductions.

The FACO technique, also known as an improved FS algorithm, is suggested. It combines the ACO algorithm with FS. The security of the network is now endangered by additional network attacks, such as APT and DDoS attacks, as a result of the exponential growth in network data and feature sets. In the anomalous sector of data discovery, a classification technique is widely used to more quickly find network anomalies. Due to massive redundant and irrelevant features in the available dataset, the efficiency of anomaly detection classifiers will be affected. Therefore, the ACO method accepts to search the optimal subset of features and selects the redundant features independently of the classifier in order to improve the performance of classification for classifiers. This effectively reduces the complexity time of algorithms for classification and improves classification accuracy.

3 Feature Selection Methods

1. **Filter method**: In this method, features are dropped based on their relation to the output or how they are correlating to the output. With this approach, features are chosen based on statistical metrics. The pre-processing step of this method selects the subset of features and is independent of the learning algorithm.
2. **Wrapper method**: In this case, the data is subdivided and used to train a model. We can add or remove features and retrain the model based on the model's output. We will select our ultimate set of features based on the model's output. Using this method, selecting features is accomplished by approaching it as a search problem and by constructing, examining and contrasting alternative combinations.
3. **Embedded or intrinsic or hybrid method**: It combines the qualities of both filter and wrapper methods to create the best subset. This model will train and checks the accuracy of different subsets and select the best among them. By taking into account feature interaction and low computational cost, it combines the benefits of both filter and wrapper approaches. These are quick processing techniques that are comparable to the filter method but are more accurate.

4 Proposed Methodology Using Ant Colony Algorithm

There are various random search techniques or heuristic search techniques to solve the problem of feature selection. The proposed algorithms which remove the irrelevant and duplicate features from the feature space and extract the useful subset of key features are:-The random forest algorithm [24], genetic algorithm [25], relief [26] and ACO algorithm [27].

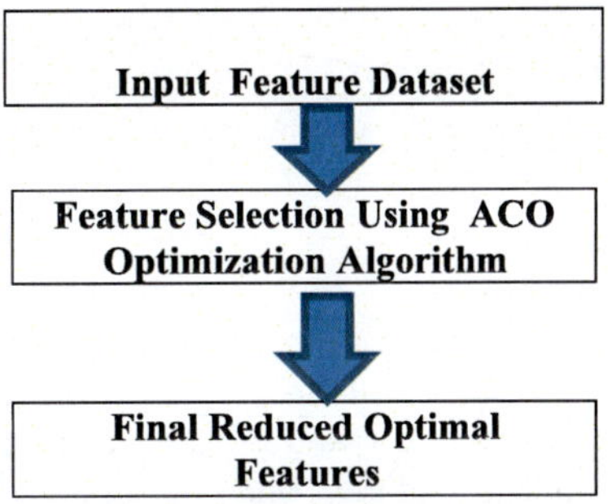

Fig. 1 Flowchart of proposed work

Currently, ACO is gaining popularity as a new approach to the FS [17, 18, 20]. Experimental results of available literatures illustrated that ACOs are better than the conventional approaches, such as sequential search and the genetic algorithm [18]. It was first presented by Dorigo and colleagues in the early 1990s as a meta-heuristic optimization technique based on ant behaviour. ACO uses Swarm Intelligence, a fresh method of using artificial intelligence to solve problems (SI). SI refers to the property of a system in which local interactions between uninformed people and their surroundings lead to the emergence of coherent functioning global patterns. Other insects, like bees and ants, also live in colonies. When working alone, people can only act in simple ways, but cooperative work in a colony is an example of complicated activity [19]. The ACO algorithm uses ant colonies as a model for social behaviour. Ants don't have eyes, but they may use chemical signals they leave behind to find the shortest route from a food source to their nest [21]. Ants release a chemical hormone known as a pheromone along their trail when searching for food and their movement direction depends on the amount of the pheromone present in the area. This method is particularly attractive for FS, as there seems to be no heuristic that can lead the search to the optimal minimal subset every time [17]. The main aim of this paper is to propose an algorithm for the optimal FS for the given dataset. ACO will help in generating features using initialization of ants and finally achieve the optimal features as the final outcome. The flow chart of the proposed methodology is shown in Fig. 1.

4.1 Graphical and Mathematical Representation

The entire directed graph with n characteristics serves as the representation for the ACO algorithm. The tabuk list, which is the set for each ant, will keep track of the locations that the ant has visited at first. It is anticipated that m ants will be distributed at random among n feature nodes. The pheromone concentration $ij(0)$ is initialized to 0 on both sides at the same time. Depending on the pheromone intensity on each side, the ant chooses the subsequent node. According to [28], the likelihood that the ant would switch from characteristic i to characteristic j throughout iterations t is given as $pk_{ij}(t)$:

$$P_{ij}^{K}(t) = \begin{cases} \frac{T_{ij}^{\alpha}(t)\eta_{ij}^{\beta}(t)}{\sum_{s\in tabu_k} T_{is}^{\alpha}(t)\eta_{is}^{\beta}(t)}, & j \in tabu_k \\ 0, & \text{else} \end{cases} \quad (1)$$

where η_{ij} is heuristic information, which is generally $\frac{1}{d_{ij}}$; dij is the Euclidean distance between two features $\tau_{ij}(t)$ is the pheromone concentration along the route that connects features i and j throughout the course of t iterations. Using the information heuristic factor and the expecting heuristic factor, respectively, the values of heuristic information and pheromone concentration are distributed. When the ant changes the information concentration along each path, the traversal is complete.

$$T_{ij} \leftarrow (1-p)T_{ij} + p\sum_{k=1}^{m} \Delta T_{ij}^{k} \quad (2)$$

where p represents the weighting factor (0p1). The path between features i and j during the traversal's pheromone increment, or k_{ij}, is stated as follows:

$$\Delta T_{ij}^{k} = \begin{cases} \frac{Q}{L_k}, & (i, j) \in pathofk \\ 0, & \text{else} \end{cases} \quad (3)$$

4.2 *Proposed ACO-Based Feature Selection Algorithm*

Algorithm 1 ACO-based feature selection algorithm

1. Begin
2. Include Feature Dataset
3. Initialize all the parameters
4. Generate Ants
5. Initialize a population of ants with random positions
6. Evaluate each ant $i = 1, 2, 3, \ldots, n$
7. Repeat evaluation of ants Until selection of feature else stop the evaluation and generate selected subsets
8. Evaluate the selected subsets
9. Evaluate the termination/stopping criteria
10. Continue until Termination criteria is satisfied, update pheromone, Generate new ants and repeat step 6~9 else stop and return best subset
11. Return Reduced Features
12. End

The procedure begins by creating a large number of ants, each of which starts with a single random characteristic, before being randomly put on the graph, as shown in Fig. 2. An alternative is to have the number of ants on the graph match the number of features in the data collection, with each ant beginning the construction of its path at a distinct feature. From these starting places, they probabilistically traverse nodes and features until a traversal halting requirement is met. The created subsets are gathered and evaluated. If an optimal subset is found or the algorithm has been run a preset number of times, the procedure pauses and outputs the best feature subset found. In the absence of any of these events, the pheromones are renewed, new ants are created and the cycle is repeated.

5 Conclusion

We have attempted to give an introduction to feature selection strategies in this work. The literature on feature selection approaches is extremely extensive and covers a wide range of applications. According to the review of the literature, all the researchers presented feature selection strategies to address real-world challenges (feature selection difficulties) by putting forth fresh ideas (based on a single algorithm or a hybrid algorithm), and the outcomes were widely accepted. This paper illustrates the significance of feature selection in classifications. Most classification tasks require the supervised learning approach, which aids in developing prediction models. Reduce these redundant characteristics to create a decent classifier. The effectiveness of learning models is impacted by large amounts of irrelevant and duplicated features. In order to find the best feasible combination of subsets of features that helps to reduce computing complexity, optimization techniques are applied. The extensive search of feature subsets from the training dataset can be optimized with the aid of specific meta-heuristic methodologies to discover the best subsets among the search parameters.

6 Future Work

In the future, the proposed method for feature selection can be applied to the IoT attack datasets like NSL-KDD, DS20S traffic traces, KDD Cup 99, UNSW_NB15 Dataset, etc. To verify the performance of the ACO feature selection algorithm, any of the above dataset can be used as the experimental data to simulate the algorithm on the Python or MATLAB2014a platform.

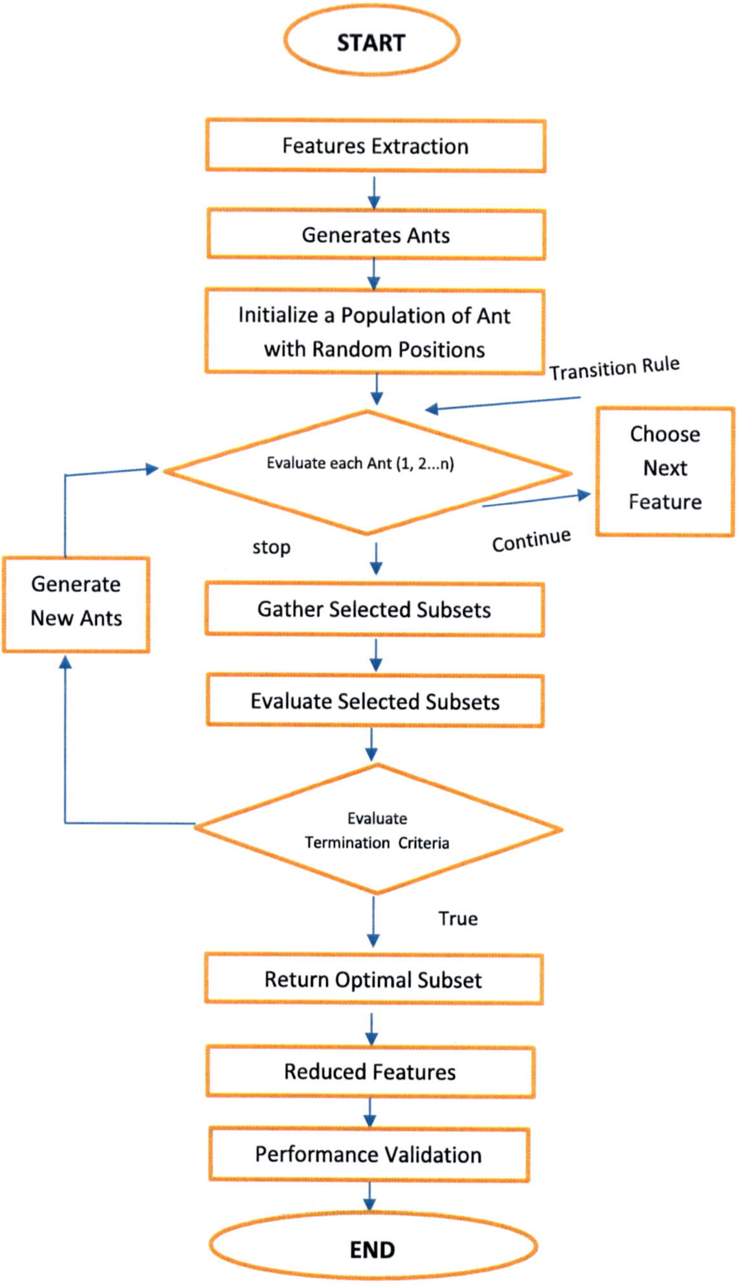

Fig. 2 ACO feature selection process

References

1. Deng L, Yu D et al (2014) Deep learning: methods and applications. In: Foundations and Trends® in signal processing, vol 7, no 3–4, pp 197–387
2. Liu H, Motoda H (2012) Feature selection for knowledge discovery and data mining. Springer Science & Business Media, vol 454
3. Larabi S, Sainte M, Nada A (2020) Firefly algorithm based feature selection for Arabic text classification. J King Saud Univ 32(3):320–328
4. Jovic A, Brkic K, Bogunovic N (2015) A review of feature selection methods with applications. In: Proceedings of the IEEE international conference on information and communication technology, electronics and microelectronics, Opatija, Croatia
5. Tahir SF, Cavallaro A (2014) Cost-effective features for reidentification in camera networks. IEEE Trans Circuits Syst Video Technol 24(8):1362–1374
6. Xindong Wu X, Kui Yu K, Wei Ding W, Hao Wang H, Xingquan Zhu X (2013) Online feature selection with streaming features. IEEE Trans Pattern Anal Mach Intell 35(5):1178–1192
7. Nemati S, Basiri ME, Ghasem-Aghaee N, Aghdam MH (2009) A novel ACO–GA hybrid algorithm of feature selection in protein function prediction. Expert Syst Appl 36(10):12086–12094
8. Guyon I, Gunn S, Nikravesh M, Zadeh LA (eds) (2009) Feature extraction: foundations and applications, vol 207. Springer
9. Abualigah LMQ (2019) Feature selection and enhanced Krill Herd algorithm for text document clustering. In: Studies in computational intelligence
10. Abualigah LM, Khader AT, Al-Betar MA, Alyasseri ZAA, Alomari OA, Hanandeh ES (2017) Feature selection with β-hill climbing search for text clustering application. In: 2017 Palestinian international conference on information and communication technology (PICICT). IEEE, pp 22–27
11. Tang J, Alelyani S, Liu H (2014) Feature selection for classification: a review. In: Algorithms and applications, vols 37–70
12. Li J, Cheng K, Wang S et al (2016) A feature selection: a data perspective. J ACM Comput Surv 55(6):94
13. Tabakhi S, Moradi P, Akhlaghian F (2014) An unsupervised feature selection algorithm based on ant colony optimization. Eng Appl Artif Intell 32(1):112–123
14. Li J, Hu X, Tang J, Liu H (2015) Unsupervised streaming feature selection in social media. In: Proceedings of the ACM international conference on information and knowledge management, Melbourne, Australia
15. Roy A (2015) Automated online feature selection and learning from high-dimensional streaming data using an ensemble of Kohonen neurons. In: Proceedings of the IEEE international joint conference on neural networks, Killarney, Ireland
16. Wang J, Zhao Z-Q, Hu X, Cheung Y, Wang M, Wu X (2013) Online group feature selection. In: Proceedings of the international conference on artificial intelligence, Beijing, China
17. Aghdam MH, Ghasem-Aghaee N, Basiri ME (2008) Application of ant colony optimization for feature selection in text categorization. In: Proceedings of the IEEE congress on evolutionary computation (CEC'08), pp 2872–2878
18. Lee KJ, Joo J, Yang J, Honavar V (2006) Experimental comparison of feature subset selection using GA and ACO algorithm. In: Advanced data mining and applications. LNCS, vol 4093. Springer, pp 465–472
19. Engelbrecht AP (2005) Fundamentals of computational swarm intelligence. Wiley, London
20. Rijsbergen CJ (1979) Information retrieval, 2nd edn. Butterworth, London, UK
21. Bonabeau E, Dorigo M, Theraulaz G (1999) Swarm Intelligence: from natural to artificial systems. Oxford University Press, New York
22. Spolaor N, Lorena AC, Lee HD (2010) Use of multi objective genetic algorithms in feature selection. In: 2010 Eleventh Brazilian symposium on neural networks

23. Nakamura RYM, Pereira LAM, Costa KA, Rodrigues D, Papa JP, Yang X-S (2012) BBA: a binary bat algorithm for feature selection. In: 2012 XXV SIBGRAPI conference on graphics, patterns and images.
24. Adnan MN (2015) Improving the random forest algorithm by randomly varying the size of the bootstrap samples. In: Proceedings under IEEE international conference on information reuse and integration (IRI), Aug. 2015, pp 303–308
25. Whitley D, Rana S, Heckendorn RB (2015) The island model genetic algorithm: on separability, population size and convergence. J Comput Inf Technol 7(1):33–47
26. Huang H, Ding C, Huang H, Zhao H (2012) Multi-label relieff and f-statistic feature selections for image annotation. In Proceedings of the IEEE conference on computer vision and pattern recognition, pp 2352–2359
27. Musa R, Arnaout JP, Jung H (2010) Ant colony optimization algorithm to solve for the transportation problem of cross-docking network. Comput Ind Eng 59(1):85–92
28. Peng H et al (2018) An improved feature selection algorithm based on ant colony optimization. IEEE, pp 69203–69209

Analysis and Prediction of Sea Ice Extent Using Statistical and Deep Learning Approach

Ramakrishna Pinninti, Nirmallya Dey, S. K. Abdul Alim, and Pankaj Pratap Singh

Abstract Census of the Southern Hemisphere of ice using the Recurrent Neural Networks (RNN) predicting the flow of ice is important in the oceans model of interoperability and safe operation of warships in cooler climates. In this study, we are investigating the power of RNN and the model of logistic growth in a few movement predictions in the coming days based solely on NSIDC data. We collect the largest website of the dataset and provide the total extent for each day for the entire time period (1978–2019). RNN and logistic growth model are used to calculate the movement of the extent sea ice between consecutive days from the dataset. Our model approaches a root mean square error (RMSE) of 0.58 and 0.71 for Recurrent Neural Networks and logistic growth model, respectively.

Keywords Recurrent neural networks · Logistic growth model · Decay cycle · Growth model · Cyclic model

1 Introduction

Knowledge of navigation in tropical and subtropical regions is vital for the safe operation of naval vessels and modeling energy and mass interchange between space and sea. Therefore, timely forecasting of the future of the extent sea ice is important in planning for safe shipping, oil and gas exploration, and fisheries, and may be the answer to developing maritime models. Despite its significance, the challenging task of predicting future movements is yet to be considered. The few studies presented rely heavily on numerical models that require large amounts of data from multiple sources, including surface, ice, water currents, rheology data, collision data, etc., and other sources of uncertainty.

In contrast, in this study we aimed to analyze the prediction of sea ice extent using statistical and deep learning approaches. It is based solely on the observance of the

R. Pinninti · N. Dey · S. K. Abdul Alim · P. P. Singh (✉)
Central Institute of Technology, Kokrajhar, Assam, India
e-mail: pankajp.singh@cit.ac.in

© The Author(s), under exclusive license to Springer Nature Singapore Pte Ltd. 2023
A. Mishra et al. (eds.), *Advances in IoT and Security with Computational Intelligence*, Lecture Notes in Networks and Systems 755,
https://doi.org/10.1007/978-981-99-5085-0_27

dataset used for this study which was originally released by the National Snow and Ice Data Center (NSIDC). This is the open scope for predicting southern sea ice extent based solely on the dataset. Predicting motion is still a daunting task. Recent in-depth learning methods are used to predict the extent of sea ice based on a large numerical database from past conditions.

1.1 Logistic Model Theory

Contrary to exponential models, logistic models are quite effective at computing usable, real-world data. Due to this reason, exponential growth rates remain constant regardless of population size and these exponential kinds of models are also useful in the near term. Consider the population P, which is growing exponentially as an example. For exponential progress, (t) = a (1 + r) t (where 'a' is the beginning amount and 'r' is the growth rate), r is the rate of expansion. Population progress rate with regard to time is expressed as dp/dt.

Displayed equations are centered and set on a separate line:

$$dP/dt \propto P \Rightarrow dP/dt = kP \tag{1}$$

If 'k' is positive in this case then it represents the growth; if the 'k' value is negative, it shows the decay. Now, in an ideal world, this would function; unfortunately, as is clear, population number has no bearing on it; only the initial quantity, growth rate, and time do. Realistically, most populations grow rapidly before plateauing after a certain point.

Till up to a carrying limit K (or decrease toward K if they ever exceed K), dP/dt thus equals kP, if P is very smaller than dP. Otherwise, $dt < 0$, if $P > K$. The following phrase contains the two presumptions mentioned above: $dP/dt = (1 - PK)$. This is a representation of the logistic differential equation's growth rate. The logistic progress model of a population over time t is as follows for constants a, b, and c:

$$(t) = c1 + ae - bt \tag{2}$$

Additionally, the growth rate is proportional to P anytime P/K approaches zero (if P K), whereas the growth rate is negative (i.e., decay rate) if P > K. The slope or steepness of the function is represented by the constant b in the logistic growth model. If the slope is negative, we have an appropriate decay model, and the differential equation changes to a negative value, thereby indicating the decay rate: $(t) = c1$.

1.2 Recurrent Neural Networks

Recurrent neural networks (RNNs) are able to learn features and long-term dependence from consecutive data and time series. RNNs have a number of non-linear units where at least one single link between units constitutes a targeted cycle. A well-trained RNN can model any flexible system; however, training RNNs suffer the most from long-term dependency learning problems. In this paper, we present the RNN research and the new development of new entrants and experts in the field.

2 Related Work

In terms of sea level ice forecasting, from the literature review, we have noted that different authors suggest different types of methods for making their models. The authors used GKMI for the selection of the attribute and applied it to marine activity [1]. Ice is divided using a combination of data from many different sensors. They modeled the login function using the Python system to create the precision material model and also information collected from multiple hours of research [2]. After that, check their database and find data following the growth and decay cycle. They were able to predict sea ice-driven data nicely. In this paper, they proposed Long Short-Term Memory (LSTM) incorporating a monthly sea forecasting method [3]. This article introduces ensemble conformalized quantile retreat, a potential innovative prediction technique (EnCQR) [4]. EnCQR generates Prediction intervals (PIs), ideal for non-fixed and heteroscedastic time series data, and gradually validates them. It may be applied to any predictive model. This paper discussed the risk prediction of flood severity in the Assam region using space-based observations [5].

By incorporating ML algorithms and causal analysis into Model prediction, one of the functions offered the idea of short-term sea ice forecast [6]. For a full explanation of how well ML algorithms perform on short-term offshore, two tests were created for predictability: (1) SIC Prediction in Kara Strait (single site: 71N, 58E) at various lead times. (2) SIC forecasts made at various lead times in the B-K sea (region: 70–80N, 20–80E). Here, another work describes the steps necessary to employ the standard feature selection method first presented [7]. Now, the HSERNN measurement method's design is checked, which was suggested. It is noted that they were looking into the effectiveness of Recurrent Neural Networks (RNNs) for the prediction [8, 9]. Here, they suggested an elaborate LSM learning network to forecast ice migration in the ocean. We combine a convolutional neural network (CNN) and recurrent neural network (RNN) analysis of both spatial and temporal variations in sea level with an independent, accurate lifting approach to predict sea-level disturbance throughout the year, taking into account [8–10].

3 Proposed System

We have collected the targeted datasets. After data cleaning and data curation, the initial datasets have been pre-processed. Then the pre-processed datasets have been split into two parts training and testing. Both the training and testing datasets are implemented using the logistic growths model and Recurrent Neural Network where we have trained our models. The model's performance is assessed using the average squared error, which serves as the loss function for least squares regression, to determine the extent of the difference between an observation's predicted value and its actual value. This difference is calculated using the Root Mean Square Error (RMSE).

3.1 Datasets

In this study, the ice dataset is used which was collected from the National Snow and Ice Data Centre (NSIDC). It supports research on frozen terrains like glaciers and climatic interactions that make up the environment around the Earth. In addition to managing and distributing scientific data, NSIDC also develops data access tools, offers user support, carries out scientific investigations, and informs the public about the cryosphere. The dataset contains each day's cumulative extent over the whole time period (1978–2015) [11, 12]. Seven variables are present (Table 1).

3.2 A Framework Description

In Fig. 1, an RNN-based model is shown. The proposed framework is based on a logistic growth model as shown in Fig. 2.

Table 1 Description of the input features and values

Input features	Values
Year	1978–2019 years
Month	1–12 months
Day	1–31 days
Extent: unit is 10^6 km^2	Values are continued
Missing: unit is 10^6 km^2	Values are continued
Source: source data product	Data reference, source

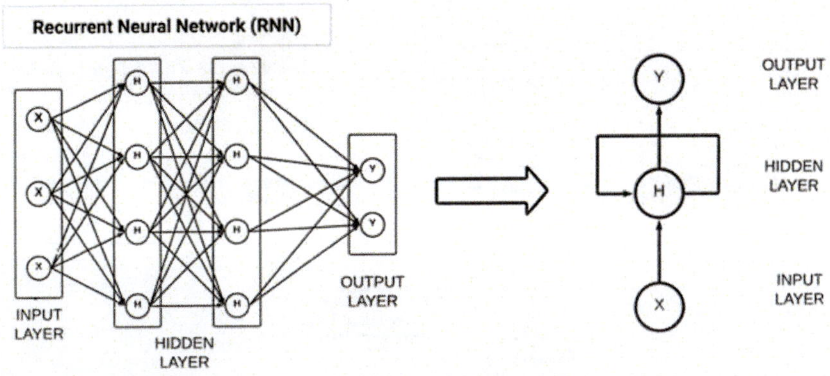

Fig. 1 A designed framework based on RNN model

3.3 *Tech Stack*

In this experiment, Google Colaboratory 2 with Python 3.1 environment is used to train and test the model with the system configuration of AMD RyzenTM 4-core 3.6 GHz CPU and 8 GB RAM. SciPy is a Python library that is available for free and open source and is used for technical and scientific computing [13]. A complete open-source machine learning platform is called TensorFlow. Researchers can advance the state of the art in ML, thanks to its extensive, adaptable ecosystem of tools, libraries, and community resources, while developers can simply create and deploy ML-powered apps [2].

4 Results and Analysis

Initially, pre-processing is applied on the loaded dataset, and thereafter 'Datapoints' as an additional attribute is also added for making easy tracing. To better comprehend the data, data is visualized (see Fig. 3). For instance, we should check to see if the data displays the logistic s-curve pattern. We observed a cycle of logistic improvement and decline over time. To test this further, we must refine and train our values for a, b, and c in the function:

$$\frac{C}{\left(1+\left(axe^{-bt}\right)\right)} \tag{3}$$

We will withhold some of our most current data so that we can test our function later. We also need to design two separate models, one for growth and the other for decay, because the data shows a cycle of development and decay. Since the growth cycle comes to an end at the last point in our data, we simulate the growth cycle first.

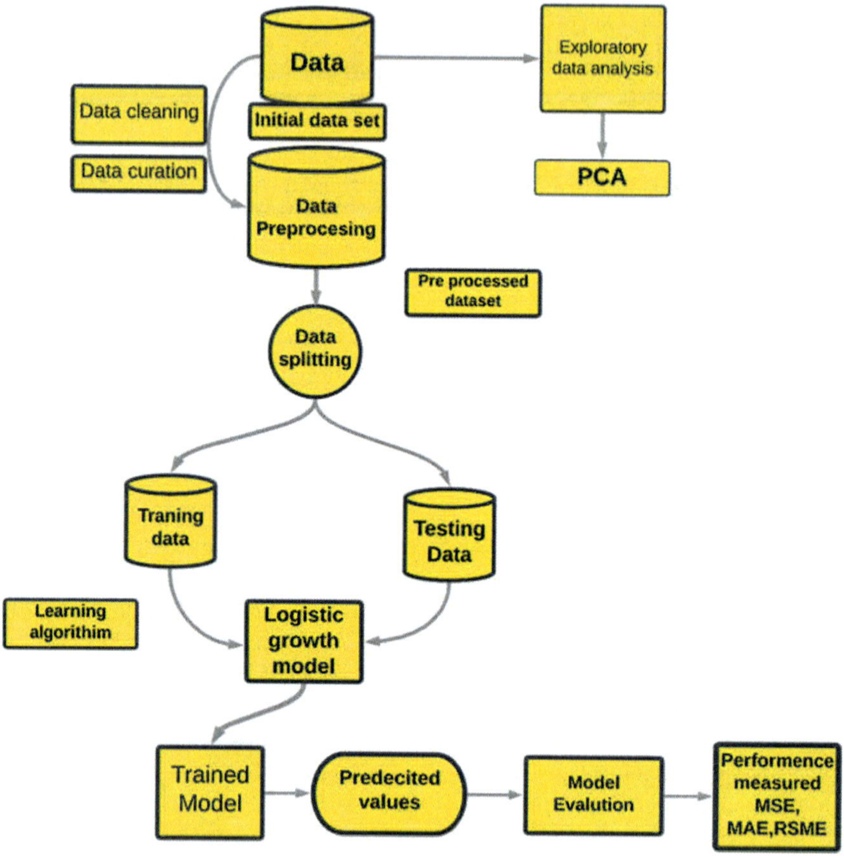

Fig. 2 A designed framework based on logistic growth model

We need a rough idea of how lengthy the cycles of development and decay are in order to accomplish this.

4.1 *RNN-Based Regression and Prediction*

To make it simpler to track, we are adding a "Data Points" column to the dataset. Here, we select To = 7000, which means that the RNN will only be trained with the first 7000 data points before being allowed to forecast the long-term trend (next > 13,177 data points, approximately) as shown in Fig. 4. Considering the quantity of test points, that isn't a lot of training data, is it? To better understand the data, we need to visualize it as our next step (Fig. 5).

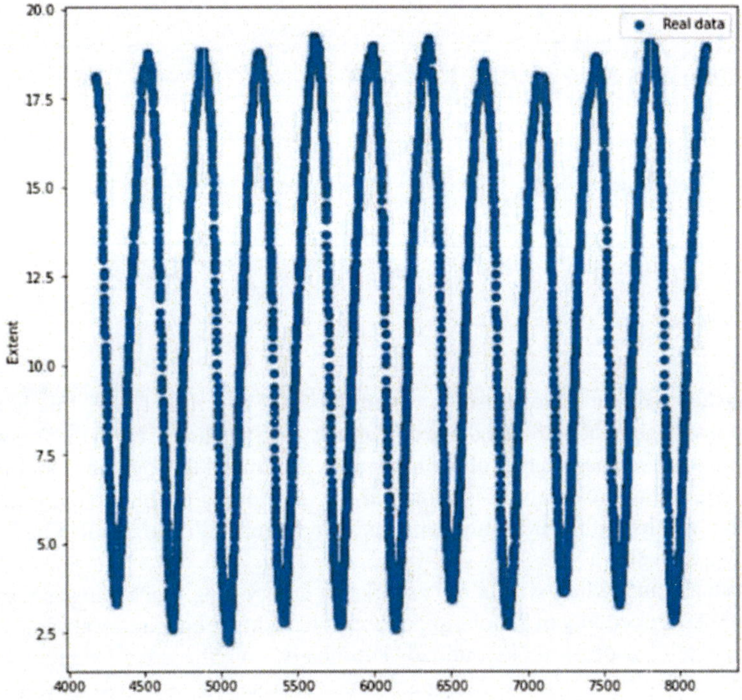

Fig. 3 Real dataset

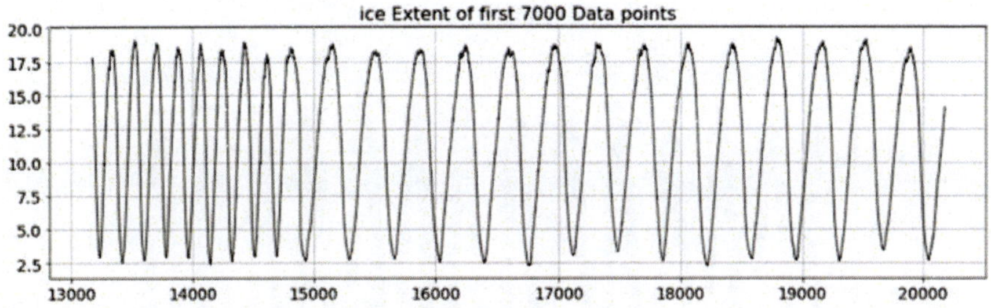

Fig. 4 Data points

4.2 Logistic-Based Regression and Prediction

When constructing the model, we took into account the user's capacity to supply not just a future date but also a former date or an exact date in contrast to the date of our most recent observation (2005–09–21). With this model, we now make an effort to predict our holdout data, which were dates following our most recent observation, and we contrast the results with the actual data. Despite some discrepancies between

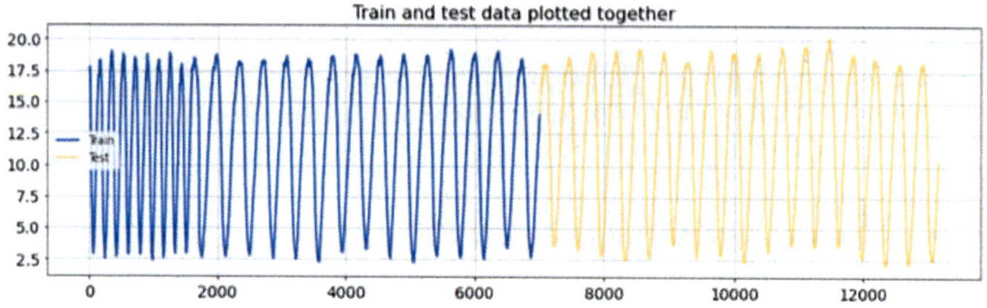

Fig. 5 Plotting of the train and test dataset

our logistic model and the real data, given the 5000-day time frame, this is to be expected, and, generally, the model is still a very good predictor. Now, the following trends are observed here in the following figures. Plotting of the logistic model-based data points and the real dataset is shown in Fig. 6. Plotting of the ground truth and prediction results with small embedding and large embedding are shown in Figs. 7 and 8, respectively.

Too small embedding size is not useful, but a very long embedding is also not effective. An embedding of 8 looks good for this data. More epochs are always better. A batch size of 8 or 16 looks optimal. Ultimately, an exhaustive hyperparameter tuning is needed for the best overall performance. Also, the predictive power is not well-defined as we are judging the quality of the prediction mostly visually here, but a numerical metric (or a few of them) would be a better approach.

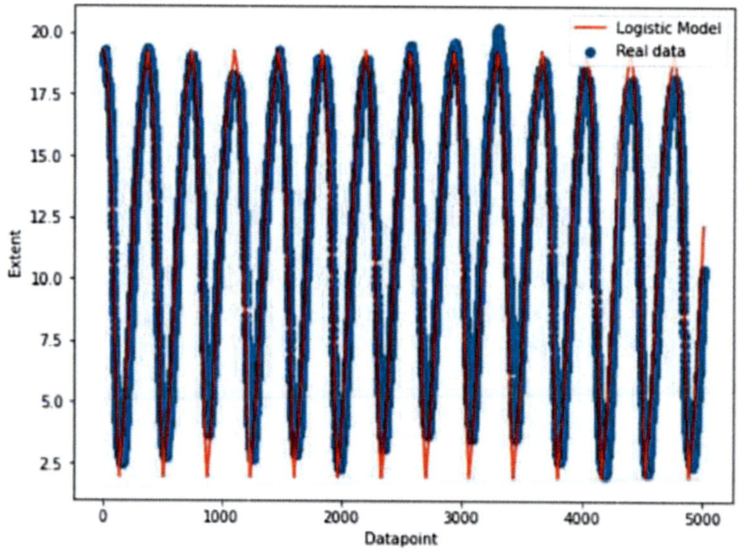

Fig. 6 Plotting of the logistic model-based data points and real dataset

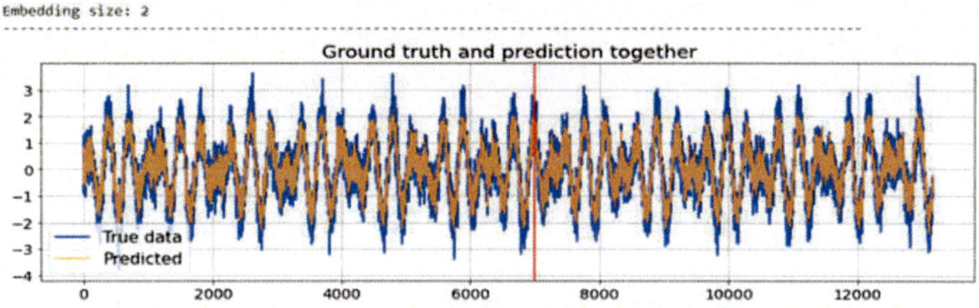

Fig. 7 Plotting of the ground truth and prediction results with small embedding

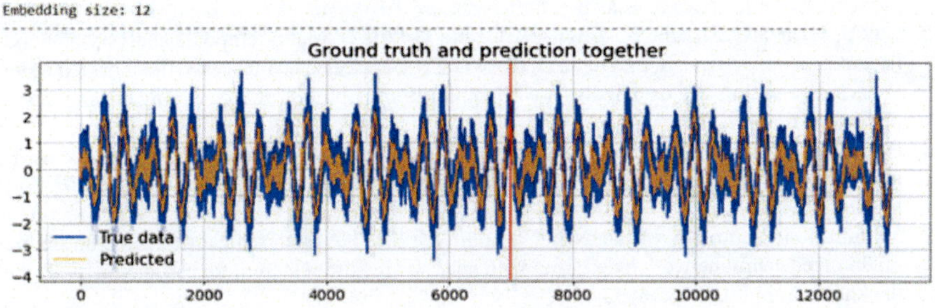

Fig. 8 Plotting of the ground truth and prediction results with large embedding

5 Conclusion

In this paper, analysis and prediction are done using a logistic growth model and RNN-based model. The NSIDC information is utilized for predicting sea ice extent. This created model is giving better prediction by reducing RMSE. During the implementation, the Southern Hemisphere portion of datasets is trained and tested with the two approaches such as statistical and deep learning. This model is successfully given the output for the prediction of sea ice extent. Regression is used for the classification of the abovementioned data; therefore RMSE is evaluated with a minimum computational resource. There are still some challenges in this context with previous methods.

This proposed work may help in predicting the land extent of any particular type of natural disaster. This paperwork is based on Arctic Ocean sea ice. In the same manner, Antarctic Ocean sea ice can also be checked for its prediction.

References

1. Khachatrian E, Chlaily S, Eltoft T, Dierking W, Dinessen F, Marinoni A (2021) Automatic selection of relevant attributes for multi-sensor remote sensing analysis: a case study on sea ice classification. IEEE J Select Top Appl Earth Obs Remote Sens 14:9025–9037
2. Sharma BA, Owens G, Asthana G, Mishra P (2022) Time series forecasting of southern Hemisphere's sea ice extent using the logistic model. SSRN Electron J. https://doi.org/10.2139/ssrn.4103583
3. Ali S, Huang Y, Huang X, Wang J (2021) Sea ice forecasting using attention-based ensemble LSTM. In: Tackling climate change with machine learning workshop at the international conference on machine learning, ICML, San Diego, CA
4. Jensen V, Bianchi FM, Anfinsen SN (2022) Ensemble conformalized quantile regression for probabilistic time series forecasting. https://doi.org/10.48550/arXiv.2202.08756
5. Roy S, Ojah SK, Nishant N, Singh PP, Chutia D (2022) Spatio-temporal analysis of flood hazard zonation in Assam. In: Gupta D, Goswami RS, Banerjee S, Tanveer M, Pachori RB (eds) Pattern recognition and data analysis with applications, LNEE, vol 888. Springer, Singapore, pp 521–531
6. Petrou ZI, Tian Y (2017) Prediction of sea ice motion with recurrent neural networks. In: IEEE international geoscience and remote sensing symposium (IGARSS), pp 5422–5425. IEEE, Fort Worth, TX, USA
7. Mozaffari A, Scott KA, Azad NL, Chenouri S (2017) A hierarchical selective ensemble randomized neural network hybridized with heuristic feature selection for estimation of seaice thickness. Appl Intell 46:16–33
8. Viatkin D, Zhuro D, Zakharov M, Malysheva S (2021) Calculation of northern hemisphere sea ice area using recurrent neural networks. IOP Conf Ser Earth Environ Sci 937:042094
9. Braakmann-Folgmann A, Roscher R, Wenzel S, Uebbing B, Kusche J (2017) Sea level anomaly prediction using recurrent neural networks. In: Proceedings of the 2017 conference on Big Data from space, pp 297–300. Publications Office of the European Union, Luxembourg, Toulouse, France
10. Li M, Zhang R, Liu K (2021) Machine learning incorporated with causal analysis for short-term prediction of sea ice. Front Mar Sci 8:649378
11. http://nsidc.org/data/nsidc-0051.html. Last accessed 21 September 2016
12. https://www.kaggle.com/datasets/nsidcorg/daily-sea-ice-extentdata?select=seaice.csv
13. scipy.optimize.Bounds-SciPy v1.8.0 Manual (n.d.) SciPy documentation

Advancements in e-Governance Initiatives: Digitalizing Healthcare in India

Sudakshina Das, Vrinda Garg, and Jerush John Joseph

Abstract In order to improve the quality of service delivery to the public, to encourage interactive communications between government and citizens or government and business, and to address development challenges in any given society, information and electronic governance is the sophisticated fusion of a wide range of information and communication technologies with non-technological measures and resources. Digital technology advancements over the past ten years have made it possible to quickly advance data gathering, analysis, display, and application for bettering health outcomes. Digital health is the study and practice of all facets of using digital technologies to improve one's health, from conception through implementation. Digital health strategies seek to improve the data that is already accessible and encourage its usage in decision-making. Digital patient records that are updated in real-time are known as electronic health records (EHRs). An electronic health record (EHR) is a detailed account of someone's general health. Electronic health records (EHRs) make it easier to make better healthcare decisions, track a patient's clinical development, and deliver evidence-based care. This concept paper is based on secondary data that was collected from a variety of national and international periodicals, official records, and public and private websites. This paper presents a review of advancements for scaling digital health within India's overall preparedness for pandemics and the use of contact tracing applications in measuring response efforts to counter the impact of the pandemic. The paper provides information about the government of India's EHR implementation and initiatives taken toward the establishment of a system of e-governance. The document also covers the advantages of keeping EHR for improved outreach and health care. Further, this paper discusses

S. Das · V. Garg · J. J. Joseph (✉)
School of Commerce, Finance and Accountancy, CHRIST (Deemed to Be University), Bengaluru, India
e-mail: jerush.john@christuniversity.in

S. Das
e-mail: sudakshina.das@commerce.christuniversity.in

V. Garg
e-mail: vrinda.garg@commerce.christuniversity.in

© The Author(s), under exclusive license to Springer Nature Singapore Pte Ltd. 2023

A. Mishra et al. (eds.), *Advances in IoT and Security with Computational Intelligence*, Lecture Notes in Networks and Systems 755,
https://doi.org/10.1007/978-981-99-5085-0_28

in depth the effectiveness of using contact tracing applications in enhancing digital health.

Keywords e-Governance · ICT digital health · EHR · Contact tracing applications

1 Introduction

Information and communication technology (ICT) has made it possible for people to communicate more effectively, retrieve data more quickly, and utilize information more effectively. ICT is mostly used in e-Government, which is the delivery of government services to the public online. There are a number of issues with implementing e-Government activities in developing nations like India where literacy levels are extremely low and the majority of the population lives below the poverty line. People are not even aware of the benefits of e-Governance activities and do not use information and communication technologies to a great extent. Low and middle-income nations (LMICs) have shifted more and more from paper-based to digital information systems during the past ten years, and they are now using new technologies to gather data and implement policies. As digital technologies and data tools are used more frequently to assist global health security, nations can now prevent, detect, and respond to outbreaks with greater accuracy and speed. The fusion of digital technologies and new data models with health systems, or "digital health," when maximized, can assist guarantee that the required data and information are accessible in the proper locations, at the appropriate times, and to the appropriate audiences [1, 2]. With a 90% increase in the digital adoption index between 2014 and 2017, India is one of the few nations that have seen one of the fastest growing digital economies in recent years. To increase quality and accessibility, the Indian healthcare industry has embraced the digital revolution. India's digital healthcare industry, which was valued at 116.61 billion dollars in 2018, is anticipated to grow at a CAGR of 27.41% to reach 485.43 billion dollars by 2024 [3].

By 2024, it is predicted that the mHealth category would control 40% of the market, as more people are projected to utilize health and fitness applications to track and monitor everyday activities. This will drive market expansion. The telehealth section then comes next. Patients will be more likely to adopt mHealth services and solutions, such as healthcare apps and wearable technology, if accessibility and convenience are improved [4]. The relevance of digital technology and health is driving capital investments. Health care received 53% of angel investments in 2019 overall. A 50% increase in volume (21 deals compared to 14 deals in 2019) and a 14% rise in value (US$ 452 million versus US$ 398 million in the first quarter of 2019) were recorded in the healthcare industry in the first quarter of 2020. Curefit (US$ 111.5 million), Iora Health Inc. (US$ 126 million), and HealthCare Global Enterprise Ltd. (US$ 119.7 million) were a few of the noteworthy deals [5]. India has made a number of wise investments in a range of digital tools and methodologies over the past five years. It has worked to increase the ability of local partners to use these resources. Future predictions for the global digital health sector predict a sharp increase on both the supply and demand sides of the ecosystem. India will benefit from a focus on building a technology-enabled healthcare ecosystem as the backbone to address many current difficulties. Intelligent patient monitoring, supply chain issues, clinical inefficiency, claims resolution, and patient safety are all problems that can be solved with the help of a health IoT, AI, 3D printing, and robotics ecosystem [2, 6, 7]. While start-ups have demonstrated some early success, they now need to scale their operations and prove how ICT solutions can be used in the healthcare system. A bigger push must be made to scale up digital activities within the healthcare sector in order to attract more investments in the health start-up ecosystem [8].

2 Literature Review

Through telemedicine, there are two ways either occurring in real-time, i.e. video conferences, or asynchronously, as in the storing and forwarding of data from monitoring home-based measurements of body weight, blood glucose, blood pressure, and other bodily variables [9]. The use of telemedicine for monitoring cardiovascular risk factors seems to be particularly interesting. This comprises health software to enhance the way of life and adherence to medication through supervision, instruction, psychological assistance, and interactive motivational elements. They have the capacity to lower the burden of illness, lower the risk of complications, hospitalizations, recurring occurrences, and early mortality, as well as enhance quality of life. Maintaining the primary focus on patients' unique needs while avoiding being overwhelmed by the incredibly rapid advancements in technology and informatics and continuing to carefully evaluate the research supporting the practice will be a significant challenge for those involved in these new technologies. Digital Surveillance increases the possibility of rights abuses, particularly against those who are historically marginalized [10]. In this digitalization era, the purpose of surveillance has expanded because of security risks and a system of social management and control that modifies behavior. Due to the possibility of being denied access to public services and locations if people refuse to submit their personal information, this has raised the data burden on average citizens. It is, hence, noticed that without adequate protections, surveillance can be used as a means of repression and exclusion [11]. The concept of telemedicine and its deliverables are expected to be delivering high-quality, reasonably priced health care from the Indian perspective. Whether telemedicine is a success in providing adequate healthcare services or not, and the gap it covers have been discussed in their report. Anything cannot be implemented without any certain rules or policies to uplift its effect, hence, policy majors taken has also been taken into consideration. At last, the challenges and barriers faced by telemedicine in India have been thoroughly discussed. Various types of digital technologies are utilized for health promotion and their socio-political consequences [12].

Many digital health promotion efforts, it is argued, place an emphasis on the individual's responsibility for their health rather than taking into account the social, cultural, and political implications of using digital technology. Individuals' digital data privacy and security are currently not sufficiently secured. The increasing use of commercial health promotion apps, platforms, self-tracking gadgets, smart environments, and objects, as well as the use of digital data on people's health and wellbeing-related behaviors and biometrics for profit, move the field of health promotion into uncharted political territory. The human rights in digital health technologies include the right to privacy, right to non-discrimination as per digital literacy, right to health, and right to benefit from scientific progress [13] including web applications, sensors, etc. The focus is also on regional data protection frameworks which would highlight the list of data subjects' favorable rights. In order to effectively respond to epidemics, timely and reliable information is required to estimate disease loads, monitor newly emerging outbreaks, and support disease prevention and control activities. The fusion of digital technologies and new data models with health systems, or "digital health," when maximized, can assist guarantee that the required data and information are accessible in the proper locations, at the appropriate times, and to the appropriate audiences. But at the same time, there is a need to study the impact and effectiveness of maintaining EHR and the prudent use of the same.

3 Key Themes

Over the past ten years, low- and middle-income countries (LMICs) have increasingly switched from paper-based to digital information systems, and they now use new technologies to collect data. Digital technology advances throughout the previous five years have allowed for quick collection progress, utilizing data for analysis, presentation, and improvement of health results. Nations may now prevent, detect, and respond to epidemics with better accuracy and speed because of the increased use of digital technologies and data tools to support global health security. This study contends that the growing digitalization of health care could have serious consequences for the right to health, despite legitimate concerns about data protection and the right to privacy.

This research paper mainly focuses on the following key areas:

- Effectiveness of these digital health applications in monitoring emerging outbreaks
- E-Governance through the usage of contract tracing implementation
- Persistent challenges in e-Governance of health care.

The effectiveness can be measured by the features of these healthcare medicare apps which help in personal self-tracking devices for health and fitness which could reduce the possibility of getting infected by any external disease. These devices generate comprehensive data that can be easily shared with others via social media platforms or with healthcare or public health specialists that are keeping an eye on people's biometrics and health-related behaviors. In order to measure the effectiveness, contact tracing applications are the crucial elements in order to establish contact

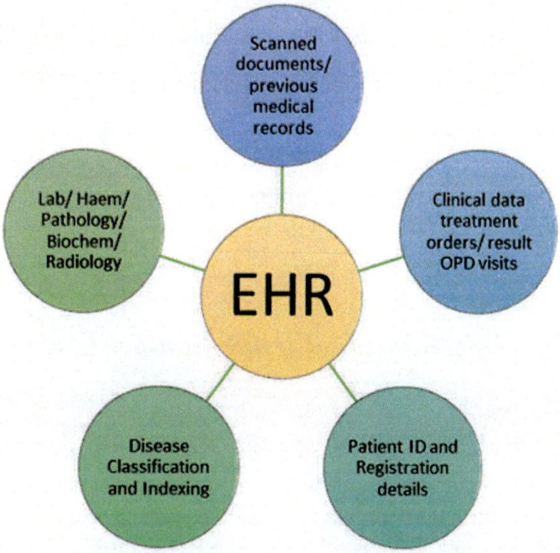

Fig. 1 A simple electronic health record system

between a user and a patient who has an infection [14]. Data breaches infringe on a person's right to privacy and damage public confidence in the healthcare system. Individuals have a right to know what personal information is kept in databases and why. Additionally, they have the right to request that any files that contain inaccurate personal information or that "have been gathered or processed in violation of the provisions of the law" be corrected or deleted. Hence, the risk of data breaches rises as technology develops and health systems become more intricate [15]. In order to prevent these issues, the Indian government has passed an act "Digital Health Information in Healthcare Security Act" which primarily focuses on the protection of the users' data being used without their knowledge. Digitization of health helps the users to self-track their health and wellbeing and can work on it accordingly, and also they have the facility to share their health status through social media. There is a great advancement in them as they now also can be seen related to digital gaming technologies. Sensor-based technologies are becoming more prevalent in both home and urban settings. Numerous self-tracking and self-care tools feature digital sensors to detect location, physical movement, and biometric information [16] (Fig. 1).

4 Discussion

"An Electronic Health Record (EHR) is defined as a collection of various records that get generated during any clinical encounter or events" [17]. "The electronic health record includes all information contained in a traditional health record including a patient's health profile, behavioral and environmental information. As well as content the EHR also includes the dimension of time, which allows for the inclusion of information across multiple episodes and providers, which will ultimately evolve into a lifetime record" [18]. An electronic health record, or EHR, is a real-time, singular longitudinal record of a single person's complete personal health information, including medical details like history, physical exam, diagnosis, lab findings, allergies, information on immunizations, therapy, etc. in a digital format. Over the course of his or her lifespan, healthcare practitioners submit the data electronically.

4.1 Effectiveness of Using Contact Tracing Applications

Comparing digital contact tracing to labor-intensive manual approaches, efficiency benefits are clear. However, there is a semblance of agreement that using digital contact tracing alone as a form of governance is worthless. It must be utilized by a sizable enough percentage of people. A model to determine the necessary minimum user percentage was created by the Big Data Institute at Oxford University. At least 60% of people must download the app in order to stop the virus's spread. This does not imply that any lower rate will not be advantageous. If 20% of people download the contact-tracing program, the propagation of the infection will be inhibited. For instance, TraceTogether has only been downloaded by 19% of Singaporeans. 6% of the populace has so far downloaded the Turkish information and quarantine tracing program. The government must present strong justifications for why they are reliable and helpful in order to attract more people (Table 1).

The primary prerequisite for the vast majority of produced apps is owning a smartphone. Even in industrialized nations like the United States or Germany, the penetration rate for smartphones is over 80%. Only South Korea has a smartphone adoption rate of above 90%. Thus, contact tracing apps will be inaccessible to millions of people. Unsurprisingly, the most susceptible group of people are typically those without smartphones. People over 65 are, for example, less likely to use a phone that supports the apps. Philadelphia-based data management business Microshare has created a low-cost contact tracing system using Bluetooth-enabled bracelets. The business claims that the Universal Contact Tracing technology was created for use in factories, warehouses, and other sites where the use of smartphones is restricted or outlawed. Citizens will need to have faith in their governments. Likewise, government agencies will have to have faith in app users. They will have to have faith in the fact

Table 1 Model of contact tracing application

Level of risk	Excessive data collection	Limited data collection
High risk of mission creep and/or hacking	Centralized GPS-location data	Centralized Bluetooth data
Low risk of mission creep and/or hacking	Decentralized GPS-location data	Decentralized Bluetooth data

that people will disclose their symptoms at the very beginning of the infection, if not at all. People will stop using the app if there are too many instances of false negatives.

4.2 *Initiatives Taken by the Government of India*

- In September 2013, the Ministry of Health and Family Welfare (MoH&FW) released the first set of Electronic Health Record (EHR) standards for India. The Ministry of Health and Family Welfare (MoH&FW) established the EMR Standards Committee, which issued recommendations for these [19]. The document offered suggestions for creating a standardized method for healthcare practitioners to create and maintain EHRs. These guidelines were updated, and December 2016 saw their notification.
- The Ministry of Health and Family Welfare (MoH&FW) proposed creating the National e-Health Authority of India (NeHA) in 2015 with the aim of establishing the e-Health ecosystem in India. To "set down data management, privacy and security rules, guidelines and health records of patients in compliance with statutory provisions" is one of the goals of NeHA, according to the Ministry of Health and Family Welfare of the Government of India.
- As a legislative organization for the promotion or acceptance of e-Health standards, the Ministry of Health and Family Welfare (MoH&FW) has proposed a draft act for parliament named Digital Health Information in Healthcare Security (DISHA). According to the Ministry of Health and Family, the Digital Health Information in Healthcare Security Act (2018) is "an Act to provide for establishment of National and State eHealth Authorities and Health Information Exchanges; to standardise and regulate the processes related to collection, storage, transmission, and use of digital health data; to ensure reliability, data privacy, confidentiality, and security of digital health data; and to address such other matters related and incidental thereto".
- The National Institution for Transforming India (NITI Aayog) has suggested the "National Health Stack," a forward-thinking digital architecture, with the goal of developing digital health data for all Indian people by the year 2022.

4.3 *Persistent Challenges Present in e-Governance of Health Care*

Digitization does not have a significant impact on Health Outcomes

Countries have gradually shifted from paper-based to digital information systems over the past ten years, and by utilizing the corresponding data, they have developed new skills and insights. To digitally gather and visualize better data, such as client medical records, supply chain updates, and the location of healthcare workers, a variety of open-source and for-profit software applications have been created. However, not all data digitization initiatives have produced better data for decision-making. Furthermore, despite significant development in new digital tools and methods, capacity-building among local and regional players has not kept pace. Without a major increase in training budgets, the adoption of digital tools would be delayed and they could be misused. Additionally, efforts will be hindered even in environments.

When it is most necessary to collaborate, there are barriers to data exchange

The governance frameworks and regulations that support the technologies, tools, and users are just as crucial as the tools and technologies themselves. Effective digital and data governance places enabling frameworks to allow the use of data to inform the decisions and actions necessary for epidemic response at all levels of the health system. This supports a dynamic policy and regulatory environment. National digital health plans assist in ensuring that tested and expandable technologies are deployed with the guidance of competent national health authorities and that they enhance rather than displace the capabilities of current health information systems. An effective epidemic surveillance and control system must include governance of information flows, including data ownership, access, standards, privacy, security, and sharing. For instance, in the area of data sharing, the governance surrounding data sharing frequently acts as a roadblock to the efficient use of data and digital health tools in the context of epidemic planning and response. At its core, the regulation of digital data is a political and capacity issue rather than a technical one. Aligning and streamlining decision-making processes is a fantastic place to start, but the health workers who will be at the center of digital and data governance systems must also have the knowledge, skills, and power to lead. combining this capacity-building with the establishment of solid digital health governance rules, procedures, and institutions through legislative and executive authorities. In order to connect the existing health, data, and digital infrastructures and to create structures that can iterate and adapt to address future disease trends and public health needs, the Indian government should collaborate with partners from other governments.

Private sector contributions are not adequately leveraged

The private sector has played a significant role in the development of digital tools, but consistent industrial support for global health goals is underused. Private corporations contribute specialized knowledge to the creation of digital health advances, but they frequently lack the access to user interfaces and develop an understanding of how the public health systems operates and thereby creating the best solutions. Without more effective cooperation between the private sector and governments, funders, public health authorities, and other global health stakeholders, their capacity, in-kind contributions, and resources cannot be fully utilized. Furthermore, the opportunity to develop digital health innovation is being missed due to the ineffective engagement of industry players. The Indian IT sector has enormous untapped potential for participation in preparedness and response initiatives. Indian businesses have high expectations for data security and privacy, and many have overcome these obstacles in a country with complicated and shifting norms and regulations.

5 Conclusion

The Government of India has the chance to dramatically increase worldwide capacity to prevent, identify, and respond to threats to global health security by utilizing digital health tools. This can be accomplished by putting serious effort into creating interoperable digital health information systems, improving data governance, increasing capacity, and making efforts to improve country coordination and leadership, like through EHR. The Indian government has a chance to demonstrate what is possible when harnessing digital tools to prevent, identify, and respond to infectious disease epidemics by making coordinated and strategic investments across ministries and programs. Through the involvement of public and commercial sector partners both domestically and internationally, a more organized and focused effort will improve leadership in global health security and stimulate breakthroughs in digital health. While enhancing global capabilities to manage health security challenges wherever they may arise, creating this enabling environment would help safeguard India's health security and fully realize the potential of digital health innovation.

References

1. Eruchalu C et al. (2021) The expanding digital divide: digital health access inequities during the COVID-19 Pandemic in New York City. J Urban Health 98(2):183–186. Available: https://doi.org/10.1007/s11524-020-00508-9
2. World Health Organization (2021) Global strategy on digital health 2020–2025. World Health Organization
3. USAID (2017) Global Health News - Digital Health. [Online]. Available: www.usaid.gov/global-health/global-health-newsletter/digital-health. Accessed 19 September 2022
4. Chandran R (2020) Privacy concerns as India pushes digital health plan, ID. Thomson Reuters Foundation, 19 February 2020. [Online]. Available: https://www.reuters.com/article/indiahealth-tech-idUKL8N2G536U. Accessed 19 September 2022
5. Sharma A (2022) India's progress in digital healthcare ecosystem: evolution of policy, regulations and future. BW Business world, 24 September 2022. [Online]. Available: www.businessworld.in/article/India-s-Progress-In-Digital-Healthcare-Ecosystem-Evolution-Of-Policy-Regulations-And-Future/24-06-2022-433956/. Accessed 19 September 2022
6. Dhingra D, Dabas A (2020) Global strategy on digital health. Indian Pediatrics 57(4):356–358, Available: https://doi.org/10.1007/s13312-020-1789-7. Accessed 20 September 2022
7. Labrique A et al. (2018) Best practices in scaling digital health in low and middle income countries. Globalization Health 14(1). Available: https://doi.org/10.1186/s12992-018-0424-z. Accessed 20 September 2022
8. O'Neill PH, Ryan-Mosley T, Johnson B (2020) A flood of coronavirus apps are tracking us. Now it's time to keep track of them," MIT Technology, 7 May

2020. [Online]. Available: https://www.technologyreview.com/2020/05/07/1000961/launching-mittr-covid-tracing-tracker/. Accessed 19 September 2022
9. Saner H (2019) Digital health implementation: How to overcome the barriers?. Europ J Preventive Cardiol 26(11):1164–1165. Available: https://doi.org/10.1177/2047487319848222. Accessed 20 September 2022
10. Mahapatra S (2021) Digital surveillance and the threat to civil liberties in India. Hamburg: german institute for global and area studies (GIGA)—Leibniz-Institut für Globale und Regionale Studien, Institut für Asien-Studien
11. Kustwar R, Ray S (2020) eHealth and telemedicine in india: an overview on the health care need of the people. J Multidiscip Res Healthcare 6(2):25–36. Available: https://doi.org/10.15415/jmrh.2020.62004. Accessed 20 September 2022
12. Lupton D (2015) Health promotion in the digital era. Oxford University Press, March 2015. [Online]. Available: https://www.jstor.org/stable/48518369. Accessed 19 September 2022
13. Sun N (2020) Human rights and digital health technologies. The President and Fellows of Harvard College, December 2020. [Online]. Available: https://www.jstor.org/stable/27039994. Accessed 19 September 2022
14. Lupton D (2014) Quantified sex: a critical analysis of sexual and reproductive self-tracking using apps. Cult Health Sexuality 17(4):440–453. Available: https://doi.org/10.1080/13691058.2014.920528. Accessed 23 September 2022
15. Sahi M et al. (2022) Privacy preservation in e-healthcare environments: state of the art and future directions. IEEE Access 6:464–478. Available: https://doi.org/10.1109/access.2017.2767561. Accessed 22 September 2022
16. Umashankar G, Abinaya P, Premkumar J, Sudhakar T, Krishnakumar S (2022) Evolution of electronic health records. Internet of medical Things (IoMT), pp 143–160. Available: https://doi.org/10.1002/9781119769200.ch7. Accessed 28 September 2022
17. Schneider EC (2014) Promoting patient safety through effective health information technology risk management. RAND Corporation. Available: https://www.jstor.org/stable/https://doi.org/10.7249/j.ctt14bs3z5. Accessed 19 September 2022
18. Bhavaraju S (2018) From subconscious to conscious to artificial intelligence: a focus on electronic health records. Neurology India, 17 September 2018. [Online]
19. Ozair F, Jamshed N, Sharma A, Aggarwal P (2015) Ethical issues in electronic health records: a general overview. Perspect Clinical Res 6(2):73–76. Available: https://doi.org/10.4103/2229-3485.153997. Accessed 19 September 2022

Mobile Application for Leaf Disease Classification

Surendiran Balasubramanian, Santhosh Rajaram, S. Selvabharathy, Vallidevi Krishnamurthy, V. Sakthivel, and Aditi Ramprasad

Abstract The study uses a machine learning model to predict various classes of leaf diseases. As a front end for this model, the study also aims to develop a mobile application for the Android and iOS platforms. An application that classifies leaf diseases informs the user whether a leaf has a disease or not. The currently used techniques are expensive and time-consuming. Both low-class farmers and many urban populations have had issues with this. Deep learning techniques have made it possible to solve this issue more affordably than with naive approaches. The image is classified using the Convolutional Neural Network (CNN) architecture, and a model with good accuracy is saved for use in mobile applications. Users of the mobile application can learn which disease the leaf or plant has by uploading just one photo, which was done for user-friendliness. These use cases have a close connection to the training set of data. Generalizability can be attained by selecting the appropriate data with care, gathering more of it, and using it for training. Using neural networks to classify an image is the primary goal of the study.

Keywords Classification · MobileNet · Flutter · Mobile application development · Deep neural network

S. Balasubramanian (✉) · S. Rajaram · S. Selvabharathy
National Institute of Technology Puducherry, Karaikal, India
e-mail: surendiran@gmail.com

V. Krishnamurthy
Vellore Institute of Technology, Chennai, Tamil Nadu, India
e-mail: vallidevi.k@vit.ac.in

V. Sakthivel
Konkuk University, Seoul, South Korea

A. Ramprasad
Sri Sivasubramania Nadar College of Engineering, Kalavakkam, Chennai, India

© The Author(s), under exclusive license to Springer Nature Singapore Pte Ltd. 2023
A. Mishra et al. (eds.), *Advances in IoT and Security with Computational Intelligence*,
Lecture Notes in Networks and Systems 755,
https://doi.org/10.1007/978-981-99-5085-0_29

1 Introduction

For classifying a plant disease, there are numerous chemical and biological testing facilities and procedures. These methods, however, are expensive and time-consuming. These techniques would be outmoded for someone with a roof garden, and they wouldn't continue to take care of the plants. With the help of the designed app, a user can identify a plant's issue from a simple photograph on his phone. One such deep learning-based software solution has been presented. A type of lightweight deep neural network called MobileNetV2 is the deep learning method used. The next-generation deep neural network, known as MobileNetV2, is used in embedded and mobile devices. These MobileNets are used in many mobile-based classification applications, such as mobile phone face and fingerprint matching. This application can function without a server and is designed to take up little space on mobile devices. The image is classified in offline mode using MobileNetV2 and the application is updated by simply swapping out a Tflite file.

2 Related Work

There has been rising interest in building small and efficient neural networks in leaf disease detection and classification, e.g. [1–5]. Deep learning models have been trained on public datasets that contain diseased and healthy plant leaves collected under controlled conditions. Deep learning architectures like AlexNet, GoogLeNet, LeNet and MobileNet are employed to get high rates of accuracy close to 99%. Methods such as particle swarm optimization are used to optimize the weights during backpropagation. MobileNet is used in a wide range of embedded and mobile device applications, and the results are effective in a wide range of applications like object detection, and fine-grain classification. LeNet works well under challenging conditions such as illumination, complex background, different resolution, size, pose, and orientations of real scene images. The k-means clustering technique provides efficient results in the segmentation of RGB images [6] which is used widely in leaf segmentation research. A standard technique is followed across studies [7–9] where the green pixels are masked and removed using a specific threshold value followed by a segmentation process and passed through a classifier. They employ image acquisition, image pre-processing, image segmentation, feature extraction and classification with an accuracy close to 90%. The above technique was optimized [9] by removing pixels with zeros, red, green and blue values on the boundaries of the infected cluster.

The choice of the dataset can also be based on colour or texture properties. Techniques with colour as the context are used across leaves from beets, vines and grapes to produce accurate results. A CNN on an unmanned aerial vehicle [10] is used to identify diseases in grapefruits after the initial colour classification. Colour imaging and texture feature analysis [11] are demonstrated and could be used for classifying citrus peel diseases. An automated method to classify [12] by using high-resolution multispectral stereo images and further generating 3D models of leaves with them before passing the data through a Bayes classifier.

Surveys on leaf detection and classification [13, 14] discuss and analyse various methods and their drawbacks on Monocot and Dicot family plants. Different deep learning and classification techniques [14] like Artificial Neural Networks (ANN), Back Propagation Neural Networks (BPN), Fuzzy Logic and Support Vector Machines (SVM) are put forth and applied only when they meet certain feature requirements. Techniques of segmentation and feature extraction like thresholding, region growth, partition clustering, watershed and edge detection have also been elucidated. Gray-Level Co-Occurrence Matrix(GLCM) and CNN features are used for plant leaf disease identification and classified using SVM classifier [15].

3 Methodology

Plant diseases are currently a major concern for farmers. As a result of not knowing the type of disease, farmers frequently are unsure of which pesticide or insecticide is required to treat a specific diseased plant. This leads to the application of incorrect pesticides, which harms the plants and reduces plant yield. A system that quickly recognizes some common diseases that affect tomato plants by simply looking at the plant's leaves as a way to solve this issue has been developed. However, the road to victory is often marred with hurdles, and in order to filter out those unforeseen obstacles various steps like understanding the dataset and data pre-processing are carried out in an effective and organized manner.

3.1 Dataset

The dataset that was used for this project was found on various websites. The dataset includes images of various disease classes. The current dataset in use is made up of images from various image classes combined. A total of 87,000 images have been collected which were split into training and testing datasets in a ratio of 4:1. Images from 38 labels are used in the data to classify leaf diseases. Only the class labels of the various leaf diseases are represented by the labels in the training set. It does not give the model the information it needs to know what to do in the absence of a leaf. Therefore, a 1500-image dataset was collected with images from environmental surroundings to give the model the necessary information for a class without leaves.

3.2 Data Pre-processing

Reading the image and converting it to a NumPy array organizes the data. To prevent the MobileNetV2 model from overfitting, feature scaling and adjustments like shearing, zooming and focus are used. The information is then transformed to fit the image's dimensions and the MobileNet's specifications. The model is saved after being trained on the data.

3.3 Architecture

When the MobileNetV2 model has been trained, predictions can be made using it. Currently, 80% of the leaf images from all the classes are used as training input. A random number is used to select the neural network's initial weight.

In order to determine whether the predictions made using the current weights are accurate, the MobileNetV2 model would first check them. If not, backpropagation would then occur, leading to a weight adjustment. This procedure will continue until the weights are stabilized during the neural network's training.

After the MobileNetV2 model's training phase, testing is carried out. Finding out how well the trained MobileNetV2 model performs on actual data is the goal of testing. A dataset is typically split into a training and testing set in a ratio of 4:1, though this division is not required. The output of the test datasets is then predicted using the trained MobileNetV2 model, and the output is compared with the labels of the dataset to determine accuracy.

3.4 Model and Integration

The MobileNetV2 model was created using the Keras library. A CNN-based model has been employed. After pre-processing the image, a $224 \times 224 \times 3$ matrix is obtained and used as the input by the MobileNetV2.

Up until the final matrix dimensions of 8×8 and 256 kernels, which are intended to reduce the matrix size by half and increase the number of kernels by twice, the values are converted into features. The exact opposite occurs from the following layer, and eventually a $128 \times 128 \times 1$ matrix is obtained that can be used to build the wave later. These layers are concatenated with the preceding layers of the same size in order to maintain the original values. A softmax activation function is reached at which point an output layer can be designed in accordance with the classes. 38 output nodes, both from the 38 classes and those that do not belong to them, make up the output layer.

For the integration of the model, a package called tflite_flutter has been used; it is a Flutter plugin that provides a flexible and fast solution for accessing the TensorFlow

Lite interpreter and performing inference. The API is similar to the TFLite Java and Swift APIs. It directly binds to TFLite C API making it efficient (low-latency). It offers acceleration support using NNAPI, GPU delegates on Android, Metal and CoreML delegates on iOS, and XNNPack delegates on Desktop platforms.

3.5 Training and Predicting

The MobileNetV2 model is saved as a Tflite file after training so that it can be later integrated with the mobile application. The MobileNetV2 model is then run against the Test data to determine how well it performs. The MobileNetV2 model's accuracy and confusion matrix is then located. The parameters can be tweaked and the MobileNetV2 model can be retrained till the desired MobileNetV2 model is found. There is an accuracy of 91% on the testing dataset. The MobileNetV2 model is then saved as a Tflite file for later use. The saved MobileNetV2 model is then used to integrate the android application with the MobileNetV2 model using Firebase's ML kit. The mobile image that was saved is stored as bytes to be processed by the mlkit model.

Once the MobileNetV2 model has been trained, it can be saved and used in the future. The MobileNetV2 model is then imported, and a single image is provided as input to predict an image. The probability that an image belongs to a particular class will be predicted by the MobileNetV2 model after it has been trained on the training set. After determining the maximum probability, the corresponding class that an image belongs to is printed.

3.6 Activation Function

The activation function transforms the weighted sum of a node's inputs to activate another node or serves as an output for that input. Due to vanishing gradient issues, sigmoid and hyperbolic tangent functions cannot be applied to neural networks with many layers. In order to solve the vanishing gradient problem and enable the MobileNetV2 model to train and learn more quickly while also performing better in predictions, we used the ReLU activation function. ReLU is a linear activation function that, if the input is positive, outputs the input directly; otherwise, it returns zero.

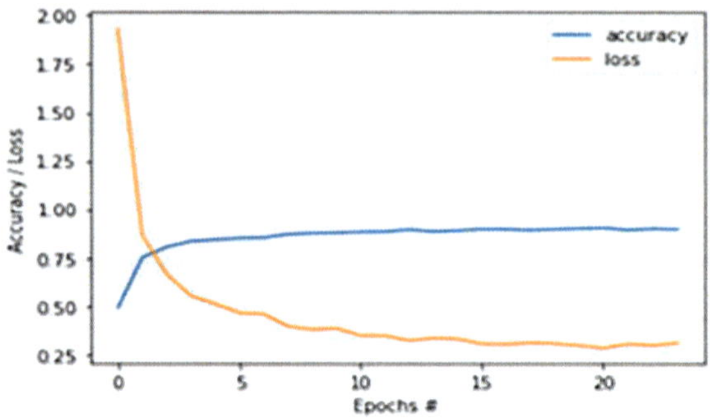

Fig. 1 Graph of MobileNetV2 on different epochs

4 Results and Discussion

Figure 1 depicts how the MobileNetV2 model converged almost 50 epochs later. It shows the MobileNetV2 model's accuracy and training loss. The MobileNetV2 model has a training accuracy of about 90%.

5 Summary of Model

A neural network developed by Google is called MobileNet. Depending on the requirements, the MobileNetV2 model can be constructed around it. The MobileNetV2 model was built with a MobileNet in the middle with customized layers to satisfy the classification needs. An input layer was added to the MobileNetV2 to input the image of size 224 × 224 × 3 and dense layers to output the 38 labels of leaf and one class label to predict the non-leaf images. A summary of the model built is shown in Fig. 2.

5.1 Testing the Dataset

Before training the MobileNetV2 model, the dataset is split into training and test sets in a 4:1 ratio. The MobileNetV2 model is used on the test set after training it on the training set to determine the prediction's accuracy. To create test data with the appropriate image aspects, the Testgen generator from the Keras library was used.

The MobileNetV2 model was applied to the batches after we divided the images into 32 image batches. On the test dataset, an accuracy of about 91% was obtained which can be seen in Fig. 2. The confusion matrix on the test dataset can be seen in Fig. 3.

```
Model: "model"
_________________________________________________________________
Layer (type)                 Output Shape              Param #
=================================================================
input_2 (InputLayer)         [(None, 224, 224, 3)]     0
_________________________________________________________________
mobilenetv2_1.00_224 (Functi (None, 7, 7, 1280)        2257984
_________________________________________________________________
global_average_pooling2d (Gl (None, 1280)              0
_________________________________________________________________
dropout (Dropout)            (None, 1280)              0
_________________________________________________________________
dense (Dense)                (None, 38)                48678
=================================================================
Total params: 2,306,662
Trainable params: 48,678
Non-trainable params: 2,257,984
_________________________________________________________________

550/550 [==============================] - 122s 221ms/step - loss: 0.2524 - accurac
y: 0.9198

[0.25242072343826294, 0.9198155999183655]
```

Fig. 2 MobileNet model summary with its performance on test data

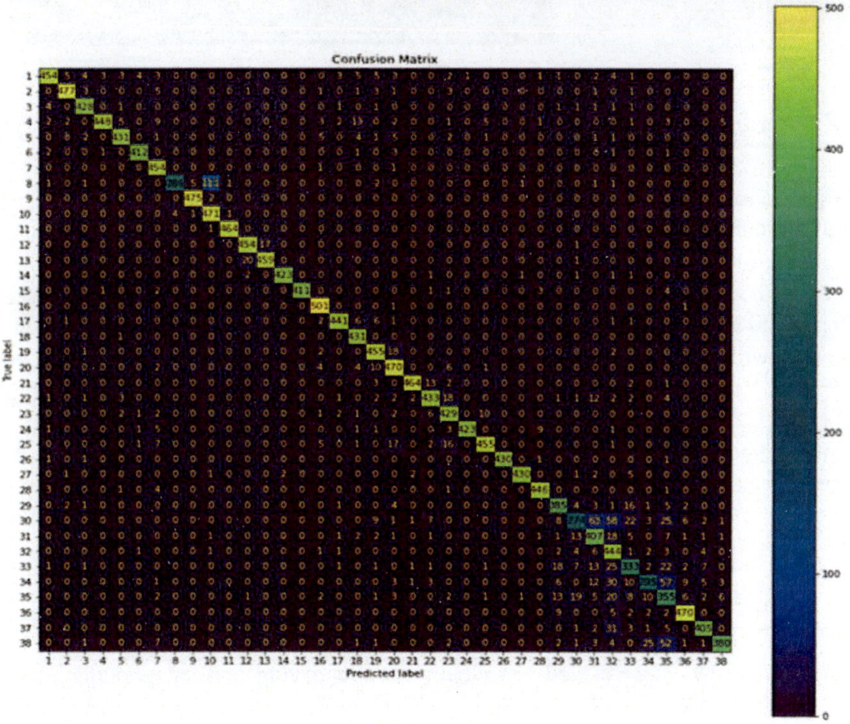

Fig. 3 Confusion matrix

Fig. 4 App leaf disease image gallery

5.2 *Android Application*

Using the Flutter SDK, an android application was created. MobileNetV2 model is used for training and testing. Here, in this mobile app, the user could upload an image or choose one from the app gallery shown in Fig. 4 on the app's basic page.

Now, the user can choose from the list of the diseased leaf images here or the user can upload a captured leaf image from the mobile phone. Based on user selection, the user will have the following image preview page shown in the mobile app as in Fig. 5. Now the user can use the "predict" option to predict the disease from the mobile app.

If the predict button is pressed, the app switches to the result page shown in Fig. 6, where the name of the leaf disease is displayed to the user through the mobile app.

6 Conclusion and Future Enhancement

There are many sophisticated techniques for identifying and categorizing plant diseases using sick plant leaves. But as of now, there isn't a practical commercial solution that can be used to identify illnesses. In our research, images of healthy and diseased plant leaves are used to use MobileNetV2 to identify plant diseases. The standard Kaggle dataset with 72,000 images was used, all of which were taken in

Fig. 5 Leaf image preview page—mobile application

Fig. 6 Predicted leaf disease—result page in the mobile application

lab settings, to train and test the MobileNetV2 model. This dataset consists of 38 distinct classes of images of 14 different species' healthy and diseased leaves.

The application was first built to serve as the front end for the MobileNetV2 model. With coloured images, it required 565 s/epoch. The deep learning MobileNetV2 model that was implemented has better loss and accurate prediction capabilities. Compared to other machine learning methods, the MobileNetV2 model required much less time to train. The MobileNetV2 architecture is a deep convolutional neural network that has been optimized for mobile use by minimizing the number of parameters and operations. This paved way to create a programme that incorporates the MobileNetV2 model for image classification and predicts the classification of leaf diseases.

References

1. Howard AG, Zhu M, Chen B, Kalenichenko D, Wang W, Weyand T, ... Adam H (2017) Mobilenets: efficient convolutional neural networks for mobile vision applications. arXiv preprint arXiv:1704.04861
2. Amara J, Bouaziz B, Algergawy A (2017) A deep learning-based approach for banana leaf diseases classification. In: BTW (workshops), vol 266, pp 79–88
3. Chanda M, Biswas M (2019) Plant disease identification and classification using back-propagation neural network with particle swarm optimization. In: 2019 3rd international conference on trends in electronics and informatics (ICOEI), pp 1029–1036. IEEE
4. Mohanty SP, Hughes DP, Salathé M (2016) Using deep learning for image-based plant disease detection. Front Plant Sci 7:1419
5. Ferentinos KP (2018) Deep learning models for plant disease detection and diagnosis. Comput Electron Agric 145:311–318
6. Al Bashish D, Braik M, Bani-Ahmad S (2011) Detection and classification of leaf diseases using K-means-based segmentation and. Inf Technol J 10(2):267–275
7. Arivazhagan S, Shebiah RN, Ananthi S, Varthini SV (2013) Detection of unhealthy region of plant leaves and classification of plant leaf diseases using texture features. Agric Eng Int CIGR J 15(1):211–217
8. Dhaware CG, Wanjale KH (2017) A modern approach for plant leaf disease classification which depends on leaf image processing. In: 2017 international conference on computer communication and informatics (ICCCI), pp 1–4. IEEE
9. Al-Hiary H, Bani-Ahmad S, Reyalat M, Braik M, Alrahamneh Z (2011) Fast and accurate detection and classification of plant diseases. Int J Comput Appl 17(1):31–38
10. Kerkech M, Hafiane A, Canals R (2018) Deep leaning approach with colorimetric spaces and vegetation indices for vine diseases detection in UAV images. Comput Electron Agric 155:237–243
11. Kim DG, Burks TF, Qin J, Bulanon DM (2009) Classification of grapefruit peel diseases using color texture feature analysis. Int J Agric Biological Eng 2(3):41–50
12. Bauer SD, Korč F, Förstner W (2011) The potential of automatic methods of classification to identify leaf diseases from multispectral images. Precision Agric 12(3):361–377
13. Akhtar A, Khanum A, Khan SA, Shaukat A (2013) Automated plant disease analysis (APDA): performance comparison of machine learning techniques. In: 2013 11th International conference on frontiers of information technology, pp 60–65. IEEE
14. Patil S, Chandavale A (2015) A survey on methods of plant disease detection. Int J Sci Res 4(2):1392–1396

15. Leong KK, Tze LL (2020) Plant leaf diseases identification using convolutional neural network with treatment handling system. In: 2020 IEEE international conference on automatic control and intelligent systems (I2CACIS), pp 39–44. IEEE

An Energy-Saving Approach for Routing in Wireless Sensor Networks with ML-Based Faulty Node Detection

Nibedita Priyadarsini Mohapatra and Manjushree Nayak

Abstract Wireless Sensor Networks (WSNs) touched the peak in research groups for unique challenges with the concept of quadruple any, i.e. anyone, anything, anywhere and anytime. Advancements in new techniques in the field of wireless communication open the path for creating tiny economic sensor nodes with limited power and provide a wide range of functions. The routing technique in sensor networks is controlled by the network layer. As we know invested power and energy is a vital factor to be discussed; investigation tells that radio communication consumes a large amount of energy. Saving energy is thus a main constraint to be taken care of. The current study focuses on making an economic energy model which strengthens the lifespan of the network as well as recharges itself with an energy harvesting strategy. Thus, we could avoid costly battery removal procedures for most of the applications such as sensor nodes deployed in dense and hazardous environments. Here, we try to present how the economic model works in the context of economic routing ability (EcoSEP) compared with the Improved LEACH model (basic LEACH modified with two-tiered grid approach) and implementation is done using MATLAB. Here, the author also focuses on the harvesting power method. We use unsupervised machine learning approach for faulty node detection which is based on the data model and node status function. We implement this function using Python programming.

Keywords Economic routing · Wireless sensor networks · Harvesting power · Unsupervised learning · Status function

1 Introduction

The cordless sensing environment received a considerable amount of attention in recent years for its broad application range, which begins from small academic applications to complex activities in military applications, then from health care, traffic

N. P. Mohapatra (✉) · M. Nayak
NIST Institute of Science and Technology (Autonomous), Berhampur, Odisha, India
e-mail: mohapatranibedita10@gmail.com

© The Author(s), under exclusive license to Springer Nature Singapore Pte Ltd. 2023

A. Mishra et al. (eds.), *Advances in IoT and Security with Computational Intelligence*, Lecture Notes in Networks and Systems 755,
https://doi.org/10.1007/978-981-99-5085-0_30

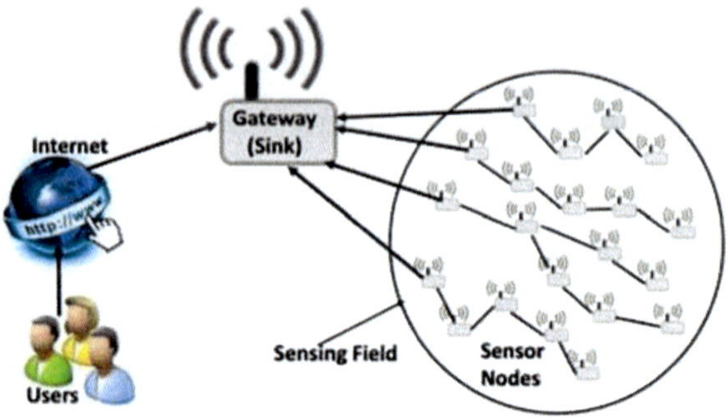

Fig. 1 Wireless sensor network architecture

control, environment monitoring, structural monitoring, etc. It consists of a huge number of tiny sensor nodes and a base station (BS); we may call it sink(s). The duty of these sensors is to collect the information and forward it to the BS or Sink using radio energy. Here, the limitation is power and computational ability. WSNs can be used in various applications, such as biomedical, military, academic and environmental applications. Finding the best route and reserving the route is not an easy job at all, due to energy limitation unancipated positional values of the sensor nodes make uncertain changes [1–7]. Saving energy is the prime issue for plotting the routing methodology in WSNs. Cluster-based routing or graded routing is very popular in WSNs. In the graded approach, nodes with higher energy can be used for the communication process with sink, while nodes with low energy can be used to do the duty of sensing the environment with propinquity of the target; LEACH, PEGASIS, TEEN and APTEEN are some of the hierarchical routing protocols (Fig. 1).

Applications of WSN could be classified into two broad categories: tracking and monitoring [8]. The probable functions include environment, health, home, traffic surveillance, military sensing, air traffic control, industrial, manufacturing automation, and other commercial areas. In the field of WSN, the model targets improving the lifespan that provides overall power saving and conveying collected data to the base station in an efficient manner. Several routing methodologies have been put forward in order to work out issues of routing in WSNs. Deployment, energy consumption and security are the most vital factors its impact on routing techniques cannot be disregarded. Thus, time demands us to give significant grip on plotting energy-saving sensor nodes and technologies that could support different functions. This paper gives a layout of the cordless sensor environment and is assembled in the following order: The second segment provides a short proposal on economic routing. Segment three elucidates the fundamental aspect of economic graded approach methods. Segment four contrasts SePSEP, EcoSEP and LEACH. Segment five unveils the experimental results. Segment six exposes the thought of harvesting power in cordless sensor environments (PH-WSN). Ultimately segment seven gives concluding remarks on the paper (Fig. 2).

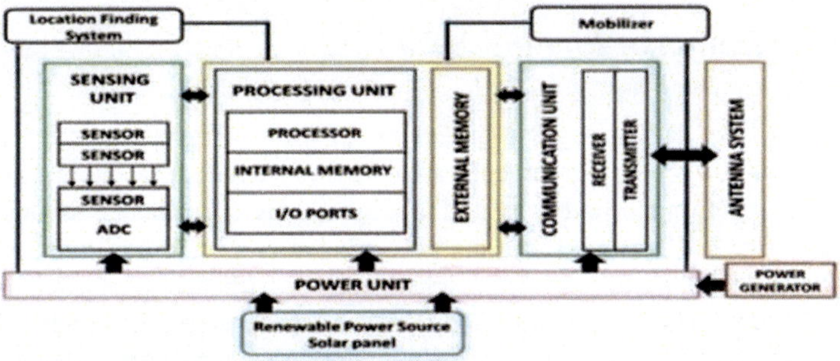

Fig. 2 Wireless sensor network structure

2 Economic Routing in WSN

Cordless sensor nodes are mostly installed in grim physical places and are an easy target for malfunctioning of the system [8–11]. The communication process is less energy-consuming. Here, limited distance transmission among nodes resulted in confined interlinks in the network. Different components of a sensor network must be utilized. The full duplex conveying process provides directionless pictorial representation, and unbalanced interaction resulted in the controlled graph. Providing the most effective logical replica for cordless sensor nodes and actuators is impossible work to be done as different variants of the electric eye are available, and differ in their work process and technological viewpoint. As everyone knows, cordless sensor field has promising future potential. Usability starts from war ground to secure complex applications; these sensor nodes could be used to monitor factual changes and analyze environmental elements over a longer period of time. We implement two-tiered grid networks for effective node placement. This can be illustrated by the following scenario. Where a clique of cordless sensors is randomly installed in a given section, a few of them do the passing of information to the base station duty, provided that each cordless sensing node is capped by at least one pass-on node (node responsible for conveying information to the base station). The goal is to optimize to pass on the node population. Thus, make the net field k-connected (predefined value of k is 2) (Fig. 3).

Fig. 3 Classification of routing protocols in WSN

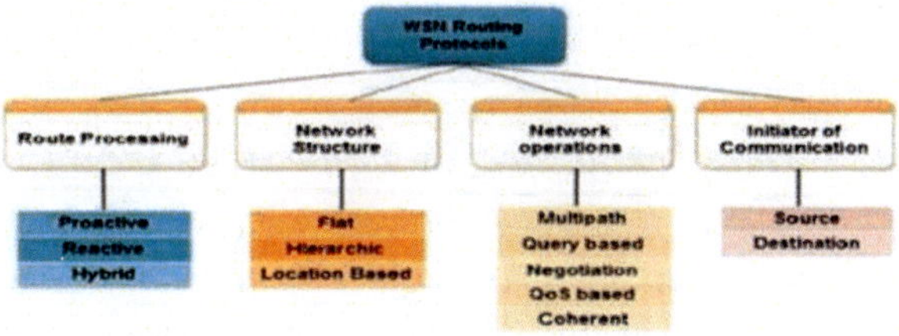

2.1 *Classification of Routing Methods*

In [10–12], a brief layout regarding ranking of routing agreements in cordless environment guiding path agreements are classified depending on the net field shape, communication methodology and topological style. Authentic strategies are described in Fig. 3. Here, we focus on Network Structure and how the nodes are placed efficiently, so we followed two schemes as described in Sects. 2.2 and 2.3. Sensor nodes have power constraints and transmission capacity; they totally depend on energy awareness for every layer of the networking protocol suite. The main role of the network layer is to find ways for energy-saving methods to create and relay data from sensor nodes to sink with reliability constraints in order to improve the lifetime of the network. Selection of routing techniques is a vital issue in WSNs. All routing methods have the most common targets [13] like net field security, updating, performance and durability; strengthening net field life; minimizing net field traffic, power saving; reductions in communication hold-up time with minimal information loss.

2.2 *Node Positioning*

Approach MPP-1 is used for the cordless net field of sensor nodes and pass-on nodes, in which sensor and pass-on nodes must be linked. Its clue depends on two basic concepts. In the polynomial-time approximation technique (PTAM), we must identify the smaller number of positional value points where, for any given error $\in$, the ratio of key value is not greater than $1+\in$ and $\in$ is constant. The structure must then be built using the Steiner point finding method (STP-MSP), and positional value points must be optimized. In the first stage, for any given error $\in$ we use the MPP-1 method to find a group of pass-on nodes that capped all sensor nodes. We have to keep in mind that these pass-on nodes must not be linked if the distance among them is more than d. To link these pass-on nodes, several pass-on nodes are required. In the second phase, the Steiner point finding method, using the STP-MSP algorithm pass-on nodes requirement is minimized to a greater extent. As a result of this, the net field environment is created with sensor nodes or sensing nodes, and pass-on nodes linked in the given net field environment (Fig. 4).

2.3 *Node Positioning in Two-Tiered Grid Environment*

Here, we present the node positioning technique used for updating the previous single-tiered net field. MPP-2 approach is utilized for making the net field two-tiered and pass-on node value increased here because MPP-2 requires the network be two-tiered, linking each sensing node with at least two pass-on nodes. We must traverse

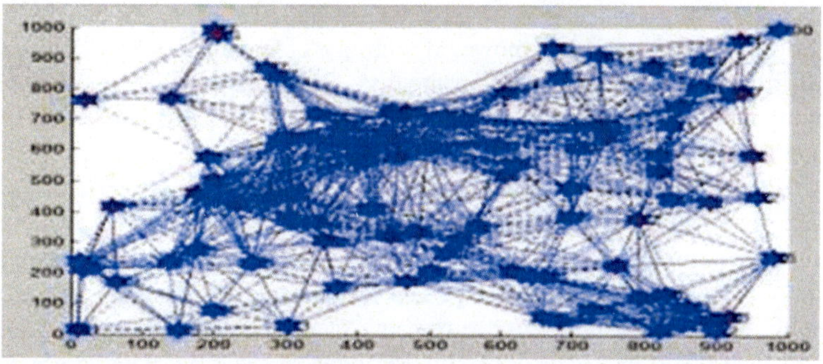

Fig. 4 Single tiered cordless sensor net field

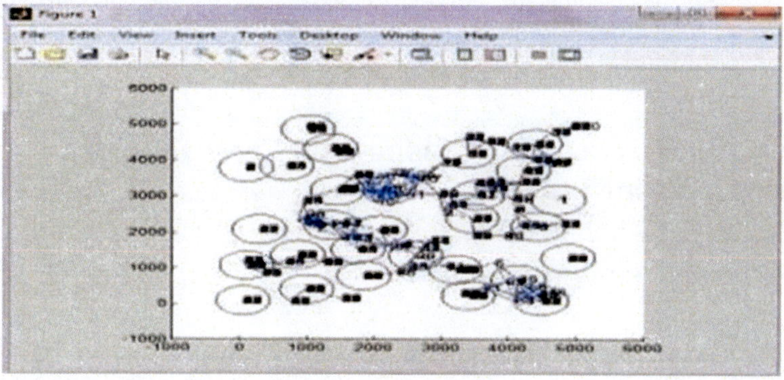

Fig. 5 Pass-on node positioning cordless sensor net field

all the sensing nodes present in the net field. In this scenario one pass-on node is used for conveying information if sensing nodes and a second pass-on node is utilized as backup service [12]. In Fig. 5, the circle represents the transmission range of the pass-on node group, and the transmitting radius value is r; if sensing node is outside the range of r, it is not linked to any pass-on node. Here, we represent this feature as two-tiered grid net field protocols.

2.4 Group Architecture

At first, we assemble the sensing nodes in the field environment in two different groups, each group of sensing nodes guided by a leader node known as group head (GH). The grouping method provides an efficient power budget and minimizes net field scalability. X presented as a route metric: guiding sensing node path schemes, select the next sensing node by aiming the shortest distance with nearby sensing hop to it. Multipath guiding method maintains the power budget overall as the single path guiding method results in power draining problem in less duration, so it is better to follow the multipath guiding method, make the net field more stable by minimizing the failure recovery time; inserting some more pass-on nodes gives the assurance that net field range is increasing; at the same time we have to optimize the number of pass-on nodes, enhancing the communication area. X pass-on node positioning less number of nodes should be added. Pass-on node positioning: the effective positioning of the nodes through uniform distribution or by adding a few pass-on nodes provides better coverage and space and avoids trigger hot spots. X Sink mobility: nodes located close to the sink face quick power draining because almost all data traffic is conveyed by these nodes to the sink. This stress can be stabilized by using a flexible sink which gathers node information by moving in the sensor net field. The performance of the power-efficient guiding method is evaluated using constraints like net field life duration, average power dissipation, limited power consumption, total number of sensing nodes alive, etc.

3 Economic Graded Path Guiding with Error Detection Using Machine Learning

3.1 Economic Graded Routing

In WSNs, a noble concept is the graded structure. A leader or group head (GH) has the highest priority under this approach, which divides the entire net field into smaller groups, with sensing nodes having a comparable group priority value given the same service. The grouping method reduces the net field load, finds some pattern in data values, summing up them, which results in economic sensor model GHs are take the responsibility for collecting and data fusion with some decisive capability, after the data aggregation task completed, group head convey it to the sink. The key job of the graded path guider method is tactfully handling the power constraint of sensing nodes but multi-ho technique within a group and also do data fusion and data aggregation. It results in minimal data traffic, grouping properties refer to group size, intra-group transmission and inter-grouping communication. The GHs can be static or could be equipped with a mobility facility. Net field may be homogeneous or it may be heterogeneous. The election procedure of GH has a solid impact on the grouping performance. Grouping algorithms can be centralized or distributed. Each technique has its own protocols to elect the group head. There are several graded methods such as LEACH, LEACH-C, SEP, TEEN, APTEEN, PEGASI and HEED [12–14] of which this paper compares the variants of LEACH and SEP methods in the next section.

3.2 Unsupervised Learning

Unsupervised learning is a type of machine learning where a model mainly works on its own to discover new and unknown patterns in untagged input data during the testing prediction phase. It is executed without any involvement of users and the model is trained without providing information for correct or incorrect results [16]. In contrast to supervised learning, unsupervised learning is done by presenting only features of different objects to a ML model without specifying their labels. After some repeating in epochs, the model could begin to differentiate the features and sort them in its own way. Thus, the model will eventually recognize correctly an unknown target without the need for a specific label for it.

4 Differentiation of LEACH and EcoSEP Methods

4.1 Short Energy Flexible Grouping Graded Technique

It is a self-assembling single hop graded method that clarifies that nodes make groups on their own and are linked with one of them behaving as GH. If architecture is a fixed one then it will degrade its power rapidly. The rotation of the group head provides a way to tackle the power-draining issue, thereby enhancing the life of sensing nodes in the net field. GHs are chosen at stipulated times with certain probability constraints and then they broadcast their status to non-group leaders. Non-GH nodes choose those GHs that need minimal transmission power and decide to which group they should belong. The short energy flexible grouping graded method consists of two phases such as setup and a steady state phase. In the setup stage, the group is assembled and choosing of the group head takes place. Data communication is carried out in the steady state phase. Data fusion is followed before the data transmission process to sink, thus minimizing power dissipation. The short energy flexible grouping graded method makes WSN alterable and sturdy [13].

4.2 Short Energy Flexible Graded-Centralized Method

It is one variant of basic LEACH; here, sink takes the group formation decision. Every node in the net field is provided with the facility of finding the energy requirement. Group members convey this information to BS with a positional value of them. Sink elects GHs based on the power constraint and distance between the sensing node and Sink. Initially, Sink does the average power calculation. If the sensing node receiving signal strength value is above the average power value, then the node is selected as GH for the ongoing round. This approach also follows up the two basic phases (setup and steady state phase) as the LEACH protocol [12].

Table 1 Comparison of variants of SEP protocol

Protocol	Scalability	Energy efficiency
SEP	Good	Good
EcoSEP	Very High	High
PEGASIS	Good	Good
LEACH	Average	Good

4.3 Stable Election Protocol (SEP)

We investigate the diversification among nodes with respect to power; in the cordless sensor net field, the groups are based on the graded approach. In such net fields, some of the nodes become group heads, aggregate the information of their group members and convey it to the sink. The percentage of the sensor population is provided with additional energy called advance nodes. It shows the main difference sensing node and the advanced node. All nodes are uniformly installed in the net field environment, and they are static. All the coordinate values are previously known. When the first sensing node loses all its power and becomes dead, the net field becomes unstable. The basic old grouping method assumes all nodes are equipped with the same amount of power. It cannot give the effective heterogeneity utilization. We put forward Economic Stable Election Protocol (EcoSEp), a heterogeneous-power saving protocol to improve the time interval before the death of the first node (we refer to it as the stability period), which is crucial for several applications where the feedback from the sensor network must be reliable. SEP is based on weighted election probabilities of each node to become cluster head according to the remaining energy in each node. By employing simulation results, we demonstrate that SEP consistently outperforms the stability period obtained using the most recent clustering algorithms (and that the average throughput is higher). SSEP takes the advantage of both LEACH and SEP. Here, CH selection is based on the classical approach and heterogeneity in the energy level of sensor nodes related to SEP. We wind up by analyzing the effectiveness of our EcoSEP method by providing greater stability to the network and improving the lifespan of the network and saving overall energy too. Higher values of energy constraints will be managed by the stability factor and performance of advance nodes (Table 1).

4.4 Fault Detection Model

In this paper, we try to detect erroneous sensor data; faulty sensor nodes will remain active and can communicate with the sink (base station), but their sensed data is abnormal [15].

Define G(V, E) as the communication graph for wireless sensor networks. Here, V represents the set of sensor nodes in the networks. E is the edge set between nodes.

For better understandability, some related symbols and definitions are presented as follows:

Node represented as Vj and its coordinates are (Xi, Yi) and distance between two nodes presented as $d(Vi, Vj)$.

$$V = \{V_1, V_2, V_3, V_4 \ldots V_n\}V_1 \tag{1}$$

$$d(V_{i,}V_j) = \sqrt{(X_i - X_j)^n + (Y_i - Yj)^n} \tag{2}$$

Edge set between nodes is represented by E and communication radius by R.

$$Neighbour(V_i) = V_j \in V : (V_i, V_j \in E) \tag{3}$$

Sensor node sensed data V_i is displayed as α_i, and the following equation is used to convey the data of neighbor node V_j:

$$\beta_i = \{\beta_{i1}, \beta_{i2}, \ldots \beta_{ik-1}, \beta_{ik}\} \tag{4}$$

Historical perceptional data average is given by E_α calculated using the following formula:

$$E_{\alpha_i} = 1/m \sum_{i=1}^{m} \alpha_i \tag{5}$$

Neighbor's sensed perceptional data average is given by the following formula:

$$E_{\beta_i} = 1/k \sum_{j=1}^{k} \beta_{ij} \tag{6}$$

4.5 Data Model

$$|\boldsymbol{\alpha}_i - \boldsymbol{\beta}_{ii}| < \theta, \theta$$

Model 1: Here we follow the Neighbor D-value Data Model. The above mathematical formula represent theta as a Threshold value for a given model; it may vary depending on different practical application:

$$b_i = \omega \frac{\alpha_i - E_{\alpha_i}}{\sigma_1} + (1 - \omega) \frac{\beta_{ij} - E_{\beta_i}}{\sigma_2} \tag{7}$$

$$\sigma_1 = \sqrt{\frac{\sum_{i=1}^{m} (\alpha_i - E_{\alpha_i})^2}{m - 1}} \quad (8)$$

$$\sigma_2 = \sqrt{\frac{\sum_{j=1}^{k} \beta_{ij} - E_{\beta_i}{}^2}{k - 1}} \quad (9)$$

ω is the collaboration coefficient between historical node Vi and its reliable neighbor sensor nodes. Data value range [0,1] [17] $\omega = 1$ means detection based on self-historical data where $\omega = 0$; it reflects that process depends on neighbor's data and ignores self-data values.

4.6 Status Estimation and Faulty Node Detection Algorithm

Step-1: For all sensor nodes present in the network, we consider about the following steps.

Step-2: If the difference of self-data and neighbor's data is less than the threshold value then the comparison function is set to zero else one.

Step-3: If similar data, then the node is not a faulty node.

Step-4: If faulty node, then its state function is set to 1 else 0.

Step-5: Reliable neighbor selection depending upon threshold value θ and status function.

Step-6: Normal node has decision capability and faulty node does not have the decision capability. Node status is decided by its collaboration coefficient (ω) and volatility factor (τ) (Figs. 6, 7, 8, and 9).

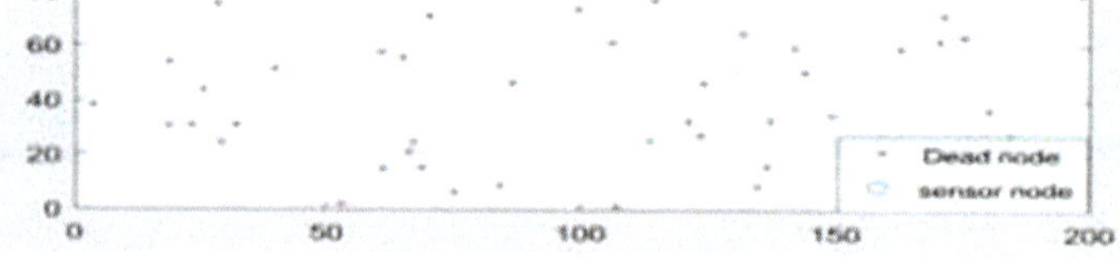

Fig. 6 LEACH (100 nodes, 1000 rounds)

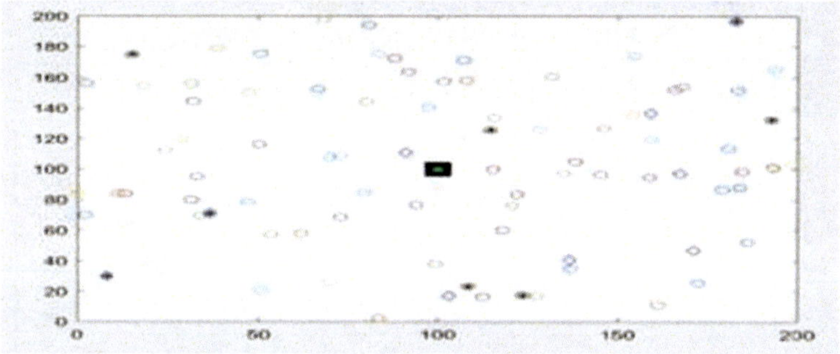

Fig. 7 SEP (100 nodes, 1000 rounds)

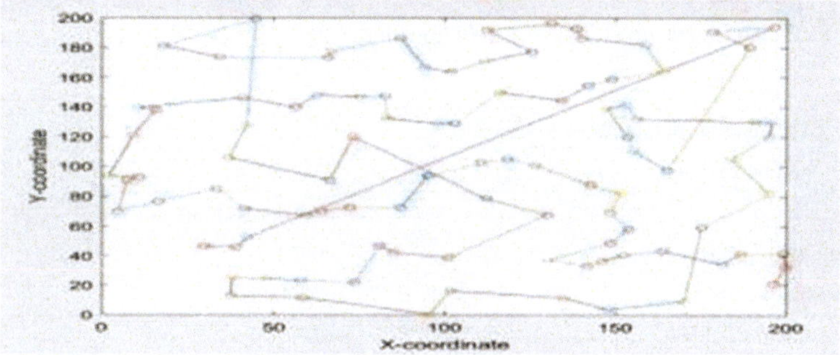

Fig. 8 Pegasis (100 nodes, 1000 rounds)

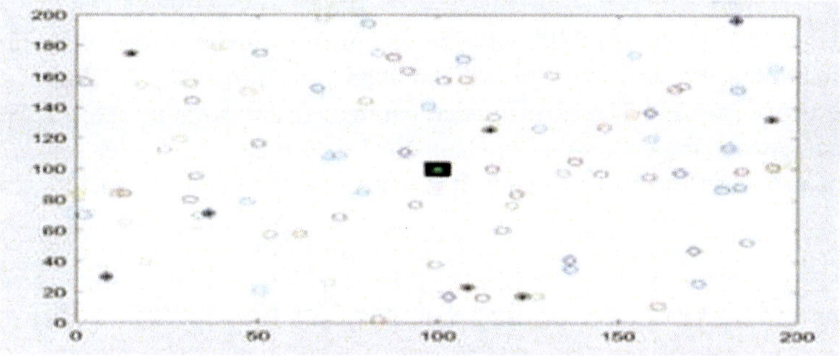

Fig. 9 EcoSEP (100 nodes, 1000 rounds)

5 Experimental Results

Using MATLAB R2016a and the following experimental parameter values, LEACH and PEGASIS were matched: The number of active nodes (node status) per round, the amount of energy left in the net field, and the number of expired nodes per round

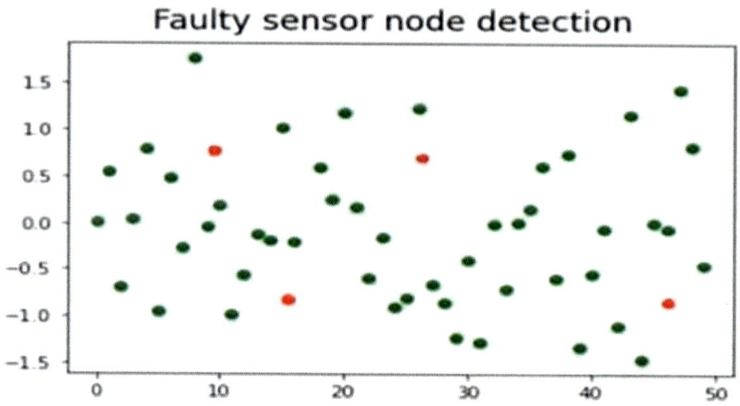

Fig. 10 Faulty node detection (50 nodes, 1000 rounds)

Table 2 Experimental parameters

Network parameters	Value
Network size	200×200, $50 \times 15\ m^2$
Initial energy of nodes	0.7 J
Packet size	2000bits
Number of nodes considered per simulation	200
Number of rounds taken per simulation	1000
Transceiver idle state energy consumption	50nj/bit

determine how long the net field will last. On the assessment of the above factors, it was found that EcoSEP provides a 75% improvement over LEACH and a 60% improvement over SEP. The experimental parameters are given in Table 1. We use the same network parameter and mathematical data model for faulty sensor node detection using Python programming. It is represented in Fig. 10 (Table 2).

6 Power Harvesting Cordless Sensor Net Field (PH-WSN)

Power constraint or power limitation is a dominant factor in the plotting of WSNs. It has a limitless lifetime when it is not confined by a limited power source. The power harvesting method takes advantage of renewable energy sources such as solar, wind and water which are being harvested to make the net field efficient. PHWSNs convert ambient power from the net field environment into electrical power to recharge the sensing nodes. The prime goals of PH-WSNs are to accelerate the performance in terms of maximum throughput and minimize hold-on time. Efficient power harvesting technology and saving energy are keys to improving net field lifespan [15].

7 Concluding Remarks

This paper presents the layout of cordless sensor field path guiding techniques and the graded structure and gives a quick review regarding power harvesting cordless sensor field. Design of networks must target keeping sensing nodes alive for a long duration to satisfy the application need and meet the adaptability constraint. Graded method is examined to be the finest to come up with adaptability, and improved lifespan of the net field. We analyze four methodologies LEACH, PEGASIS, SEP and EcoSEP by utilizing MATLAB, and our assessments furnished that EcoSEP outperforms LEACH, PEGASIS and SEP. Sensing nodes lose their energy quickly and cannot deal with the requirement unless they regained their energy. We use a machine learning approach for faulty node detection and get the desired result. For this implementation, we use Python programming.

References

1. Akkaya K, Younis M (2005) A survey of routing protocols in wireless sensor networks. Elsevier Ad Hoc Netw J 3/3:325–349
2. Liu H, Nayak A, Stojmenovic I (2009) Fault tolerance algorithms /protocols in Wireless sensor networks. Department of Computer Science, Hong Kong Baptist University, Hong Kong. https://doi.org/10.1007/978-1-84882-218-410_c. Springer, London Limited
3. Al-Karak JN, Kamal AE (2004) Routing techniques in wireless sensor network: a survey. IEEE Wirel Commun 11:6–28
4. Liu H, Wan P-J, Xiaohuajia (2005) Fault-tolerant relay node placement in wireless sensor networks, L. Wang (Ed.): COCOON, LNCS 3595, pp 230–239
5. Ming Lu Y (1999) Multipath routing algorithms for wireless sensor networks. B. EngEcole-Polytechnique de Montrel
6. Zheng J, Jamalipour A (2009) Wireless sensor networks: a networking perspective. IEEE Press, New Jersey, pp 20–21
7. Valada A, Kohanbash D, Kantor G (2010) Design and development of wireless sensor network system for precision agriculture. Carnegie Mellon University, Pittsburgh, Pennsylvania, Robotics Institute, p 19
8. Lagemann A, Nolte J, Weyer C, Turau V, Applying self- stabilization to wireless sensor networks
9. Blumenthal J, Handy M, Golatowski F, HseMarc T, Dirk (2003) Wireless sensor networks—new challenges in software engineering. emerging technologies and factory automation. Proceedings. ETFA '03. IEEE Conference, vol 1, 16–19
10. Liao C, Hu S (2009) Polynomial time approximation schemes for minimum disk cover problems, © Springer Science + Business Media, LLC
11. Gupta G, Younis M (2003) Load-Balanced clustering of wireless sensor networks. Proceeding of IEEE ICC'2003, pp 1848–1852

12. Blumenthal J, Handy M, Golatowski F, HseMarc T, Dirk (2003) Wireless sensor networks—new challenges in software engineering. emerging technologies and factory automation, 2003. Proceedings. ETFA '03. IEEE Conference, vol 1
13. Pantazis N, Nikolidakis SA, Vergados DD (2013) Energy efficient routing protocols in wireless sensor networks: a survey. Commun Surv Tutorials IEEE 15(2):551–591
14. Sen F, Bing Q, Liangrui T (2011) An improved energy-efficient PEGASIS-based protocol in wireless sensor networks. Fuzzy systems and knowledge discovery (FSKD), 2011 eighth international conference on, vol 4. IEEE
15. IBM, Supervised vs. unsupervised learning: what's the difference?, https://www.ibm.com/cloud/blog/supervised-vsunsupervised-learning. Accessed 25 May 2021

Risk Prediction in Life Insurance Industry Using Machine Learning Techniques—A Review

Prasanta Baruah and Pankaj Pratap Singh

Abstract Technological advancement has resulted in producing a large amount of unprocessed data. Data can be collected, processed, analyzed, and stored rather inexpensively. This capability has enabled to make innovations in banking, insurance, financial transactions, health care, communications, Internet, and e-commerce. Risk management is an integral part of the insurance industry as the risk levels determine the premiums of the insurance policies. Usually, insurance companies claim higher premiums from the insurance policy holders having higher risk factors. The higher the accuracy of risk evaluation of an applicant for a policy, the better is the accuracy in the pricing of the premium. Risk classification of customers in insurance companies plays an important role. Risk classification is based on the grouping of customers on the basis of their risk levels calculated by using machine learning algorithms on historical data. Evaluation of risk and calculation of premium are the function of underwriters.

Keywords Risk assessment · Logistic regression · Predictive modeling · Geographic Information System · Machine Learning · Life insurance business

1 Introduction

Revolutionary big data technologies enable insurance firms to collect, process, analyze, and accomplish data with increased efficiency [1, 2]. As a result, it spreads across various insurance industry sectors such as customer analytics, claims analysis, marketing analytics, product development, risk assessment, underwriting analysis, fraud detection, and additional coverage [3, 4]. Telematics is a classic case of using big data analytics massively executed and is changing the mode of auto insurer's value of the premiums of single drivers. Single life insurance groups are still based

P. Baruah · P. P. Singh (✉)
Central Institute of Technology Kokrajhar, Assam, India
e-mail: pankajp.singh@cit.ac.in

© The Author(s), under exclusive license to Springer Nature Singapore Pte Ltd. 2023

A. Mishra et al. (eds.), *Advances in IoT and Security with Computational Intelligence*, Lecture Notes in Networks and Systems 755,
https://doi.org/10.1007/978-981-99-5085-0_31

on the predictable formulas based on the mortality rates and accordingly, life insurance policy premiums are estimated. Life insurance businesses have started to carry out predictive analytics to uplift their business efficacy. Even though it is still lacking behind in the extensive level of research for predictive analytics but it can enhance the life insurance sector. Researchers have focused on data mining techniques to detect fraud among insurance companies, and such critical problems can provide significant loss [5–7]. Such kinds of predictive analytical methods are certainly helpful to the underwriters to estimate the appropriate premiums for the different risk levels of the customers without being stuck on adverse selection. For over 20 years, Property and Casualty (P&C) insurers have been utilizing predictive analytics, primarily for assessing the likelihood of recovery in disability claims. Meanwhile, in life insurance, the use of predictive analytics primarily focuses on modeling the mortality rates of candidates to enhance underwriting verdicts and improve business profitability [8].

In any life insurance business organization, the risk profiles of every individual applicant are comprehensively examined by underwriters. The main task of the underwriters is to finalize the status of the life risk of each applicant based on various factors, and this risk should be at the evaluation level. After evaluating the risks, the premiums are accurately evolved for deriving the business smoothly. The term "risk classification" is commonly used among insurance firms, referring to the process of grouping customers based on their probable level of risk, which is determined from their historical data [9]. For a long period, companies of the life insurance business have been trusting the customary actuarial formulas and mortality tables to approximate life expectancy and formulate underwriting rules, although the unoriginal techniques took lots of time and may be varied till several days and were also too much costly. Hence, it is indeed necessary to find ways to create the underwriting procedure faster and more reasonably. Predictive analytics have been demonstrated to streamline the underwriting process and enhance the decision-making process [10, 11]. Still, broad research has not been accompanied to this extent. The main purpose of this kind of research is to relate predictive modeling for classifying the risk level which is based on the historical data in the life insurance industry. It also endorsed the best suitable model to consider risk and deliver solutions to improve underwriting practices.

The utilization of credit and additional notching models marks a subtle change in actuarial practices. This change encompasses two related aspects. Firstly, credit data is behavior-based and does not have a direct correlation with insurance losses, unlike most traditional rating variables. Instead, it is believed to act as a proxy for unobservable, latent variables such as "risk-taking disposition" or "cautious personality" that are not accounted for by customary insurance rating factors. This leads to the consideration of other external sources of facts, such as environmental data, household, lifestyle, purchasing, and social networks which can be valuable in making actuarial estimates. Secondly, the usage of credit and other notching models marks an early sample of the expansion of predictive prototypes in insurance. It is a natural progression for actuaries to adopt advanced analytical and predictive methods to find more effective solutions to conventional actuarial issues, such as loss reserve setting, mortality estimation, and classification rate-making. However, with the rising of actuaries and insurance analytics, professionals have to be more dependent on predictive modeling methods to enhance business processes that have conventionally been the domain of human experts [12].

Life insurance shows a vital role in the financial protection of millions of households. The US life insurance industry manages trillions of dollars in benefits collectively. Although there are many different types of insurance agreements, a common aspect is the evaluation of individual mortality risk through the underwriting process. This has traditionally been done through manual means, using human judgment and point-based systems that consider each risk factor individually. These methods, while adequate in the industry, are rudimentary and prone to inconsistencies. As a result, traditional underwriting restricts the ability of insurers to accurately assess risk from data and offer cost-effective products. The presence of extensive historical datasets presents an opportunity for data mining to revolutionize underwriting in the life insurance sector. MassMutual, a prominent insurance and financial services firm, has amassed a dataset of nearly one million applicants, covering a 15-year span and containing health, behavioral, and financial information. Combining this data with advancements in data mining techniques and survival modeling enables accurate estimation of mortality risk [13].

1.1 Relevant Use of Machine Learning in Life Insurance Business

Predicting the risks of the applicants in the life insurance industry is highly relevant since the predictive models developed by using machine learning algorithms provide more accurate results [14]. The more accurate result does help to make better decisions in the business procedures. With the use of predictive modeling, machine learning provides solutions to market segmentation and then market basket analysis for answering the business-related questions with a higher level of accuracy. Several data mining techniques are utilized in the development of the insurance industry. These are classification, prediction, regression, association rule mining, and summarization. A robust algorithm can acquire essential knowledge for predicting risk in the life insurance business by utilizing a hybrid and advanced technique.

1.2 Predictive Modeling and Its Related Features

A successful analytical model must have the capability to highlight some key aspects of the business issue. The objective of predictive models is to identify patterns in the data that are nontrivial, previously unknown, potentially beneficial, and capable of being acted upon. Hence, the analysis of predictive modeling features and predictive models is very important.

Predictive Modeling Features. In order to make an enhanced decision in risk prediction, it is essential to extract useful information from the existing dataset. In the business world, the patterns discovered in historical and transactional data are used in predictive models to uncover risks and opportunities. The models capture the interplay between various factors, enabling the evaluation of risk or potential linked to a specific set of circumstances, thereby informing the decision-making process. Features associated with predictive modeling are as follows.

Data analysis and manipulation. Tools for data analysis create new datasets, modify, combine, categorize, merge, and filter existing datasets.

Visualization. The visualization features include interactive graphics and reports.

Statistics. Statistical tools are used to produce and confirm the relationship between the variables in the data. Statistics from various statistical software can be incorporated into certain solutions.

Hypothesis testing. Construction of models, evaluation, and selection of the appropriate model.

Predictive Modeling. According to Gartner, "Predictive modeling is a widely utilized statistical method for forecasting future behavior. Predictive modeling solutions utilize data-mining technology to analyze existing and historical data, and develop a model to anticipate future results". Predictive modeling involves the examination of vast datasets to uncover meaningful relationships and make inferences, with the aim of improving the prediction of future events. Utilizing statistical methods to differentiate organized patterns from random fluctuations, this information is transformed into business rules, contributing to enhancing decision-making. Essentially, it is a self-restraint that actuaries have been practicing for a significant period. One of the earliest instances of statistical analysis influencing business verdicts is the use of mortality tables for pricing pensions and life insurance policies, a practice that dates back to the effort of John Graunt and Edmund Halley in the seventeenth century [15]. The following steps are associated with predictive modeling (see Fig. 1).

2 Related Work

Many algorithms have been developed to estimate the risk of applicants of life insurance organizations. To confine the range of the literature review, only pioneering approaches based on data mining techniques have been examined. Hutagaol et al. (2020) did a study to observe how machine learning can be beneficial for risk calculation of the applicants in life insurance businesses. The study found that the procedure using the Random Forest has the highest precision in comparison to the Support Vector Machine (SVM) algorithm and Naive Bayesian algorithm [16]. Mustika et al.

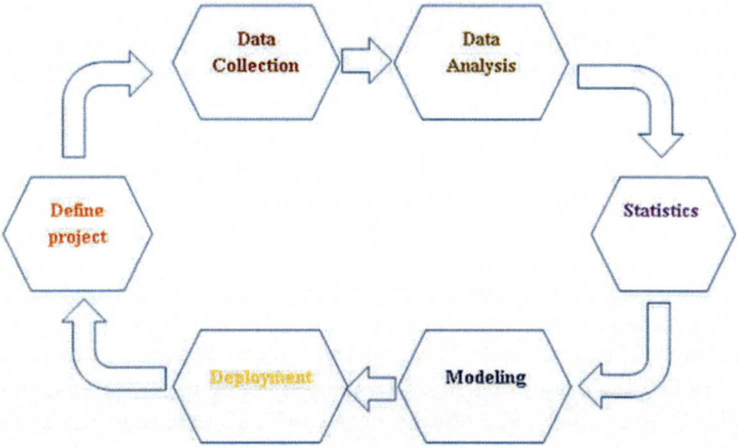

Fig. 1 Steps associated with predictive modeling

(2019) applied Extreme Gradient tree boosting (XGBoost), a model based on a decision tree to calculate the risk level of customers of life insurance businesses. The results of the study illustrate that the accuracy of the XGBoost model is very good compared to the Bayesian ridge models, decision tree, and random forest [17].

Jain et al. (2019) proposed a solution known as ensemble learning to solve the problem of assessing risk associated with a life insurance policy applicant. The experiment was carried out on a real-world dataset having 128 attributes using two ensembles, namely artificial neural network and gradient boosting algorithm XGBoost to study the risk value associated with a policy applicant [18]. Boodhun et al. (2018) applied a few algorithms from machine learning to estimate the level of risk of the applicants of life insurance industries. The algorithms implemented were Artificial Neural Network, Multiple Linear Regression, REPTree, and Random Tree classifiers. The data dimension was reduced by feature selection techniques and feature extraction, namely Correlation-Based Feature Selection (CFS) and Principal Components Analysis (PCA) [19]. Biddle et al. (2018) applied Logistic Regression, XGBoost, and Recursive Feature Elimination techniques of data mining to automate some parts of the underwriting process. The objective is to reduce the burden on the underwriters. To predict the outcomes of the underwriting process, the algorithms Logistic Regression, XGBoost, and Recursive Feature Elimination were implemented. The experiment was conducted on a dataset delivered by a leading Australian life insurance company and shows that their early stage results were promising [20].

Dwivedi et al. (2020) applied machine learning classification methods to assess the risk of the customers in the life insurance industries. Among the algorithms Multiple Linear Regression, Random Tree, Artificial Neural Network, and Random Forest, the last one produced the most accurate risk prediction of the candidates compared to the other algorithms [12]. Prabhu et al. (2019) used both structured and unstructured data to implement a new Convolutional Neural Network-Based Multimodal

Disease Risk Prediction (CNN-MDRP) algorithm. The prediction accuracy of this approach is quite high along with a faster convergence speed than the CNN-Based Unimodal Disease Risk Prediction (CNN-UDRP) model [21]. Franch-Pardo et al. (2020) reviewed various scientific reports on the geospatial and spatial-statistical exploration of the geographical sides of the 2019 coronavirus disease (COVID-19) pandemic [22]. For disease mapping, a number of themes were identified such as web-based mapping, spatiotemporal analysis, data mining, health, environmental variables, and social geography.

The knowledge of spatiotemporal dynamics is helpful for the mitigation of the COVID-19 pandemic, since it aids to make clear the extent and consequence of the pandemic and can help in planning, decision-making, and taking a course of action in the COVID-19-affected community. Health geography is crucial for having the knowledge of how public health officials interact with affected people as well as the first responders to improve the estimations of the spread of disease and the possibility of new outbreaks. Saran et al. discusses the increasing relevancy of geospatial technologies such as geographic information system (GIS) in the public health domain, particularly for infectious disease surveillance and modeling strategies [23]. Infectious diseases are possible to surveil using geospatial technologies. This study is categorized into four broad classes to emphasize the handling of a large physical area and time-based problems in public health decision-making and planning policies: (1) Using spatial and temporal information of diseases, it is possible to map the diseases and understand their behavior across geography; (2) Involving the citizen in such a way that they voluntarily provide their health and disease data. These data can be used to assess the critical situation such as how the disease is spreading and can be prevented from spreading into the neighborhood; (3) Analyzing health-related behavior in a scientific way with the help of geospatial methodologies; (4) Development of a capacity building program. Table 1 shows the Comparative analysis of the various proposed machine learning approaches by several researchers and also the best performance among all for risk-level analysis in life insurance companies.

3 Scope of Deep Learning and GIS

Post COVID-19 shows that people's health problem is increasing as well as the complexity of the health problem. With the increase in health problems, the advancement of health systems is also happening rapidly. This is because of the fact that the size of health data is increasing rapidly in numerous formats. There are many sources from which these data originate such as patient record history in the hospital, mobile devices, and digital records. Nowadays, health data analysis is becoming easier due to the advent of big data and thereby enhancing the risk prediction of life insurance businesses in innovative ways. Deep learning is suitable in complex applications where machine learning algorithms do not produce expected results. Some of the methods such as Bayesian fusions and neural networks are suitable examples of deep learning methods for extracting data and making logical inferences. Combining

Table 1 Comparison of machine learning methods of various research

Authors	Objective	Methods used	Best method based on performance
Hutagaol et al. [16]	Examined risk level using machine learning of customers in life insurance companies	Random Forest, Support Vector Machine (SVM), and Naive Bayesian algorithm	Random Forest has highest precision in comparison to the SVM algorithm and Naive Bayesian algorithm
Mustika et al. [17]	Applied a machine learning model to predict the risk level of applicants in life insurance	Extreme Gradient tree boosting (XGBoost), Decision tree, Random forest, and Bayesian ridge models	The results of the study show that the accuracy of the XGBoost model is very good compared to the other algorithms
Jain et al. [18]	Proposed ensemble learning method for assessing risk associated with a policy applicant	Artificial neural network and gradient boosting algorithm XGBoost	XGBoost provided the best result
Boodhun et al. [19]	Implemented machine learning algorithms to predict the risk level of applicants	Multiple Linear Regression (MLR), Artificial Neural Network, REPTree and Random Tree classifiers	REPTree performed better with the Correlation-Based Feature Selection (CFS) whereas MLR showed the best performance using PCA method
Biddle et al. (2018) [20]	To automate the underwriting process	Logistic Regression, XGBoost, and Recursive Feature Elimination	XGBoost is the most ideal one giving better accuracy
Dwivedi et al. [12]	To predict the risk level of candidates using machine learning algorithms	Artificial Neural Network, Multiple Linear Regression, Random Tree, and Random Forest	Random Forest came out to be most efficient one
Prabhu et al [21]	Proposed a new Convolutional Neural Network for disease risk prediction	CNN-MDRP and CNN-UDRP	CNN-MDRP produced better result

information fusion paradigms with deep learning and applying it to big health data, it has been possible to produce not only more comprehensive but also more reliable risk predictions of applicants.

Due the data extraction and reasonable inferencing capabilities, deep learning techniques are becoming very popular in the assessment of health risks in a more reliable way. Deep learning has the power of generating high-level information from low-level data with the help of iterative inferencing. Moreover, some of the methods such as convolutional neural network belonging to deep learning have the capability

Table 2 Application of GIS-based geospatial technologies

Author	Objective	Applications	Relevance in risk prediction
Franch-Pardo et al. [22]	Reviewed the significance of geospatial and spatial-statistical analysis in 2019 coronavirus disease (COVID-19) pandemic	To identify highly infected locations of COVID-19 pandemic across the world thereby enabling to predict the neighboring areas likely to be infected by the disease as well as can help in decision-making, planning, taking course of action for curing patients and prevention in spreading the COVID-19 virus	To identify the areas or locations where most of the inhabitants were infected by COVID-19 disease. This will enable to estimate the risk of the applicants of these locations more appropriately
Saran et al. [23]	To discuss the relevance of geographic information system in infectious disease surveillance and modelling	Suitable for applications to the public health decision-making and planning policies as the behavior of diseases across geography can be understood by its spatial and temporal information	Mapping of the diseases across geography is possible and thereby it will help in calculating the risk of the applicants of the geography

to improve the accuracy of analysis in bulky training datasets. This specialism has great merit in this era of big data [24]. Recently, Geographic Information System (GIS) has shown how effectively and efficiently data can be created, managed as well as analyzed. In addition to that GIS can map all types of data. Through GIS, data are connected to a map indicating the location of data as well as a description of data in the map. Thus, GIS allows the mapping of data of all types and making the analysis easier than the traditional approaches of machine learning methods. With the aid of GIS, the understanding of the patterns and their relationship in terms of geographic context has become easier for users. Moreover, GIS provides more accurate results and helps users to take quick and better decisions even in complex situations. Although there are various examples of GIS applications, some of the recently used applications are mentioned below in Table 2.

4 Conclusion

This paper covers a detailed review of the risk prediction factor of the applicants in life insurance industries. In a large dataset, the elimination of unnecessary data is quite essential. Therefore, data preprocessing will have an important role to remove the noisy or irrelevant data or the outliers to accomplish the target dataset. Data preprocessing also includes some strategies required to deal with the inconsistencies in the dataset. Such a dataset will provide a better facility to analyze the data. The cases of inconsistencies can be solved by transforming specific variables so that analysis and interpretation become easier. This paperwork also provides a deep insight into the several algorithms for achieving the task of risk prediction in life insurance businesses.

References

1. Sivarajah U, Kamal M, Irani Z, Weerakkody V (2017) Critical analysis of big data challenges and analytical methods. J Bus Res 70:263–286
2. Joly Y, Burton H, Irani Z, Knoppers B, Feze I, Dent T, Pashayan N, Chowdhury S, Foulkes W, Hall A, Hamet P, Kirwan N, Macdonald A, Simard J, Hoyweghen I (2014) Life Insurance: genomicsStratification and risk classification. Eur J Hum Genet 22:575–579
3. Umamaheswari K, Janakiraman D (2014) Role of data mining in Insurance Industry. Int J Adv Comput Technol 3:961–966
4. Fan C, Wang W (2017) A comparison of underwriting decision making between telematics-enabled UBI and traditional auto insurance. Adv Manag Appl Econ 7:17–30
5. Goleiji L, Tarokh M (2015) Identification of influential features and fraud detection in the Insurance Industry using the data mining techniques (Case study: automobile's body insurance). Majlesi J Multimed Process 4:1–5
6. Joudaki H, Rashidian A, Minaei-Bidgoli B, Mahmoodi M, Geraili B, Nasiri M, Arab M (2016) Improving fraud and abuse detection in general physician claims: a data mining study. Int J Health Policy Manag 5:165–172
7. Nian K, Zhang H, Tayal A, Coleman T, Li Y (2016) Auto insurance fraud detection using unsupervised spectral ranking for anomaly. J. Fin Data Sci. 2:58–75
8. Bell M, Is analytics the underwriting we know?, https://insurance-journal.ca/article/is-analytics-changing-theunderwriting-we-know. Accessed 16 December 2016
9. Fang K, Jiang Y, Song M (2016) Customer profitability forecasting using big data analytics: a case study of the insurance industry. Comput Ind Eng 101:554–564
10. Cummins J, Smith B, Vance R, Vanderhel J (2013) Risk classificaition in life insurance, 1st edn. Springer, New York
11. Bhalla A (2012) Enhancement in predictive model for insurance underwriting. Int J Comput Sci Eng Technol 3:160–165
12. Dwivedi S, Mishra A, Gupta A (2020) Risk prediction assessment in life insurance company through dimensionality reduction method. Int J Sci Technol Res 9:1528–1532
13. Maier MC, Hayley Sanchez F, Balogun S, Merritt S (2019) Transforming underwriting in the life insurance industry. Assoc Adv Artif Intell 31:9373–9380
14. Noorhannah B, Jayabalan M (2019) Risk prediction in life insurance industry using supervised learning algorithms. Complex Intell Syst 4:145–154
15. Batty M, Tripathi A, Kroll A, Sheng Peter Wu C, Moore D, Stehno C, Lau L, Guszcza J, Katcher M (2010) Predictive modeling for life insurance. Deloitte Consulting LLP
16. Hutagaol BJ, Mauritsius T (2020) Risk level prediction of life insurance applicant using machine learning. Int J 9:2213–2220
17. Mustika WF, Murfi H, Widyaningsih Y (2019) Analysis accuracy of XGBoost model for multiclass classification—a case study of applicant level risk prediction for life insurance. In: 5th International Conference on science in information technology (ICSITech), pp 71–77. IEEE, Yogyakarta, Indonesia
18. Jain R, Alzubi JA, Jain N, Joshi P (2019) Assessing risk in life insurance using ensemble learning. J Intell Fuzzy Syst 37:2969–2980
19. Boodhun N, Jayabalan M (2018) Risk prediction in life insurance industry using supervised learning algorithms. Complex Intell Syst. 4:145–154

20. Biddle R, Liu S, Tilocca P, Xu G (2018) Automated underwriting in life insurance: Predictions and optimisation. In: Wang J, Cong G, Chen J, Qi J (eds) Databases Theory and Applications: 29th Australasian Database Conference, ADC 2018, vol 10837. Lecture Notes in Computer Science. Springer, Cham, pp 135–146
21. Prabhu T, Darshana J, Dharani kumar M, Hansaa Nazreen M (2019) Health risk prediction by machine learning over data analytics. Int Res J Eng Technol 6:606–611
22. Franch-Pardo IM, Napoletano B, Rosete-Verges F, Billa L (2020) Spatial analysis and GIS in the study of COVID-19. A review. Sci Total Environ 739:140033
23. Saran S, Singh P, Kumar V, Chauhan P (2020) Review of geospatial technology for infectious disease surveillance: use case on COVID-19. J Indian Soc Remote Sens 48:1121–1138
24. Zhong H, Xiao J (2017) Enhancing health risk prediction with deep learning on big data and revised fusion node paradigm. Scientific Programming

BioBERT-Based Model for COVID-Related Named Entity Recognition

Govind Soni, Shikha Verma, Aditi Sharan, and Owais Ahmad

Abstract In natural language processing, information extraction from textual data is an important task. Named entity recognition, is the most popular task of information extraction, especially in the context of the medical domain. The entity extraction task aims to identify the entities and categorize them (Cho and Lee in BMC Bioinform 20:1–11, 2019) into predefined categories. With the emergence of COVID-19, COVID-related digital resources increased drastically and new types of entities are being introduced, which are semantically similar; also, new entities are being introduced which were earlier unknown. Now due to this, the task of entity extraction has become more challenging for which well-defined models developed earlier are not suitable for extracting such entities. Earlier research suggested that the state-of-the-art models were generic and focused less on domain-specific knowledge. Thus, it becomes important that the research progresses in a direction that considers biomedical domain knowledge for named entity recognition. The paper thus aims to identify the entities of biomedical nature specifically on the COVID benchmark (Cho and Lee in BMC Bioinform 20:1–11, 2019) dataset which was released by the University of Illinois. The experiments were performed using the biomedical domain-specific model BioBERT. Further, we have compared different versions of pre-trained weights on the BioBERT model and the experimental results show that the (Lee et al. in Bioinformatics, 2019) BioBERT-Base v1.1 (+PubMed 1M) weighted version outperforms the other models.

Keywords Deep learning · Named entity recognition · Information extraction · CORD-19 · BioBERT

G. Soni (✉) · S. Verma · A. Sharan
School of Computer and System Sciences, Jawaharlal Nehru University, New Delhi, India
e-mail: sonigovind07@gmail.com

S. Verma
e-mail: shikha51scs@jnu.ac.in

O. Ahmad
Thoucentric, Bangalore, India

© The Author(s), under exclusive license to Springer Nature Singapore Pte Ltd. 2023

A. Mishra et al. (eds.), *Advances in IoT and Security with Computational Intelligence*, Lecture Notes in Networks and Systems 755,
https://doi.org/10.1007/978-981-99-5085-0_32

1 Introduction

The rapid increase in data has led to significant growth in structured, semi-structured, and unstructured data. Unstructured data, which includes text, images, audio, and video, may initially appear as debris, but it holds valuable information once extracted and analyzed. Information extraction (IE) has become increasingly prevalent in various domains, including biomedical, agriculture, and meteorology. Information extraction aims to extract meaningful information from textual documents, enabling us to understand semantics, identify patterns, and recognize entities within the text. Named Entity Recognition is a crucial task in Natural Language Processing and plays a significant role in information extraction. In the biomedical domain, named entity recognition has been employed for various applications, such as drug identification, protein identification, and DNA identification. With the recent outbreak of the COVID-19 pandemic, a vast amount of medical literature has become available. Medical practitioners and academic researchers strive to extract valuable insights from this growing corpus of literature.

Therefore, our paper focuses on the objective of applying NER to the biomedical benchmark dataset CORD-19, released by the University of Illinois [3]. While state-of-the-art NER models are generally generic and lack domain-specific knowledge, it is crucial to advance research in the direction of the biomedical domain to address tasks relevant to the medical field [4]. In this study, we propose the utilization of the BioBERT model for entity extraction tasks on the COVID-related NER benchmark dataset. The BioBERT model is a domain-specific model pre-trained on a biomedical corpus. We consider various pre-trained weights corresponding to different versions of BioBERT, fine-tuning them separately to compare their performance. This paper aims to bridge the gap in domain-specific knowledge for applications in the biomedical domain, specifically targeting the CORD-19 dataset, which consists of over 30,000 scholarly articles related to the COVID-19 pandemic. The subsequent sections of this paper are structured as follows.

Section 2 provides background work related to the topic.

Section 3 presents the proposed model, focusing on the utilization of BioBERT.

Section 4 elaborates on the results obtained and provides discussions.

Finally, Sect. 5 concludes the study, summarizing the key findings and their implications. By conducting this research, we aim to leverage the domain-specific knowledge of the biomedical field and apply it effectively to extract valuable information from the CORD-19 dataset, contributing to the advancement of NER techniques in the medical domain.

2 Background

In biomedical corpus, NER aims to assign the same category to similar entities. However, the complicated links between the words in the biomedical corpus make it difficult to identify semantically similar entities. Moreover, the relation between the entities can either be in the forward or backward direction. The pre-trained BioBERT model is trained to learn these complicated relationships pertaining in both directions.

2.1 BioBERT

BioBERT model architecture is domain-specific and is trained on PubMed and PMC corpora. This BioBERT model is being trained using the weights of BERT architecture. BioBERT is emerging as one of the successful models for biomedical text mining tasks including entity recognition (ER) tasks. BioBERT follows the BERT architecture, so here we discuss the layout of BERT with reference to its usage for domain-specific models [5]. According to the literature a contextualized word representation model called BERT was pre-trained using bidirectional transformers and is based on a masked language model [6]. Previous language models could only combine two unidirectional language models since that is how language modeling works and future words cannot be seen (i.e., left-to-right and right-to-left). BERT may be used to learn bidirectional representations since it makes use of a masked language model that forecasts randomly masked words in a series. Additionally, it achieves cutting-edge performance on the majority of NLP tasks with the least amount of task-specific architecture change [7]. Recent efforts have concentrated [8] on learning context-dependent word representations, in contrast to earlier models that focused on learning context-independent word representations, such as Word2Vec [9] and GloVe [10]. For instance, GloVe applies machine translation to integrate semantic features into document representation, whereas ELMo [11] uses a bidirectional language model.

2.2 Pre-training BioBERT

The BioBERT model is a domain-specific variant of the BERT model that is specifically trained on a biomedical corpus, including articles from PubMed and PMC [12].

The main goal of this specialization is to improve the performance of the model on biomedical text mining tasks [13]. The idea behind this is that domain-specific models are better suited for identifying entities relevant to medical practitioners and researchers. The preprocessing stage in BioBERT involves tokenizing the text using the WordPiece tokenizer. The tokenization process is important for the model to understand the input text; the WordPiece tokenizer is a sub-word tokenization

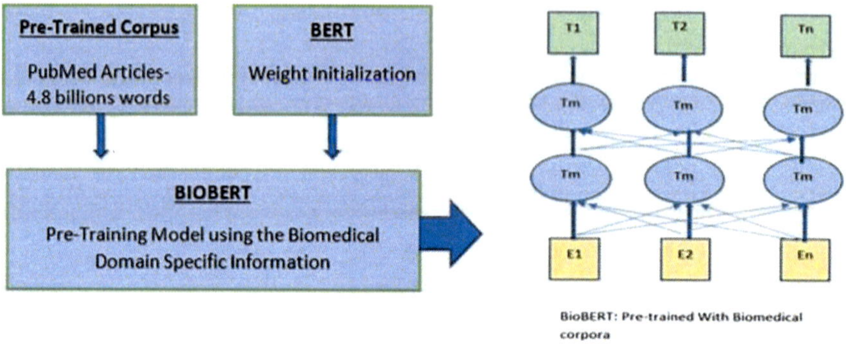

Fig. 1 BioBERT pre-trained model

method that can handle out-of-vocabulary words by breaking them down into the most frequently occurring sub-words. This is important for the model to understand the input text, especially when it encounters new or rare words that are not present in the pre-trained vocabulary.

For creating the BioBERT model, the weights of an existing BERT-based model are initialized with a pre-trained corpus of biomedical text data. This allows the BioBERT model to leverage the pre-trained knowledge of the BERT model while also specializing in the biomedical domain. The process of creating the BioBERT model can be visualized in Fig. 1, where the pre-trained weights of the BERT model are combined with the pre-trained biomedical corpus to create the final BioBERT model.

3 Proposed Model

3.1 Dataset Description

The experiment was conducted using a benchmark dataset consisting of [13] 29,500 documents from the COVID-19 Open Research Dataset Challenge (CORD-19) corpus, which includes meta-data, a full-text corpus, and named entity recognition (NER) results that were collected on March 13, 2020 [14]. The dataset is organized according to a specific file schema, which defines the structure and format of the data. The dataset used in the experiment is structured using a specific format, with several fields such as 'id', 'source', 'doi', 'pmcid', 'pubmedid', 'publishtime', 'authors', 'journal', 'title', 'abstract', 'body', and 'entities'. Each field contains specific information related to the documents, such as the unique identifier, the source of the document, the digital object identifier, the PubMed Central identifier, the PubMed identifier, the time of publication, the authors of the document, the journal in which the document was published, the title of the document, the abstract, and the full text

Table 1 Description of article

Name	Information
Id	Kumar et al. [13] unique identification for each document
Source	It contains the source from which documents are fetched
PMC Id	PMC paper records id
Body	Body contains full-text corpus
Entities	Entities contain CORD-19-NER results from annotation

of the document. The 'entities' field contains information about the named entities that were identified in the text, including the text, start and end positions, and the type of named entity.

For each article present in the dataset, it includes the following items as mentioned in Table 1 [13].

3.2 Framework of the Proposed Model

In this study, the proposed work involves conducting experiments using the COVID-19 Named Entity Recognition (NER) dataset. This dataset is specifically designed for the task of identifying and classifying named entities in the context of COVID-19-related text.

The COVID-19 NER dataset is structured in a way that facilitates the evaluation of the model's performance. It consists of segmented sentences, where each sentence is divided into individual words or tokens. For each word in the sentence, an associated tag is provided to indicate the type of named entity it represents.

During the experiments, the model is trained using the COVID-19 NER dataset, where the words and their associated tags are provided as input. The model learns from this labeled data to understand the patterns and characteristics of different named entities related to COVID-19. By training on this dataset, the model gains the ability to recognize and classify disease names, locations, and other relevant entities within COVID-19 text.

The experiments with the COVID-19 NER dataset aim to assess the performance and effectiveness of the proposed model in accurately identifying and classifying named entities specific to COVID-19. The evaluation of the model's performance using this dataset provides insights into its capability to handle COVID-19-related text and extract meaningful information, which can be valuable for tasks such as information retrieval, data analysis, and decision-making in the context of the ongoing pandemic. The steps involved in Figs. 2 and 3 are listed below.

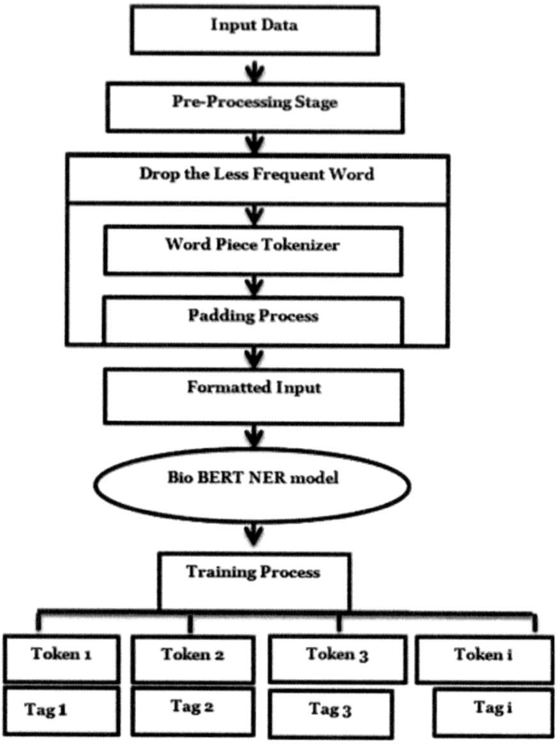

Fig. 2 Part 1 of flowchart of the proposed work

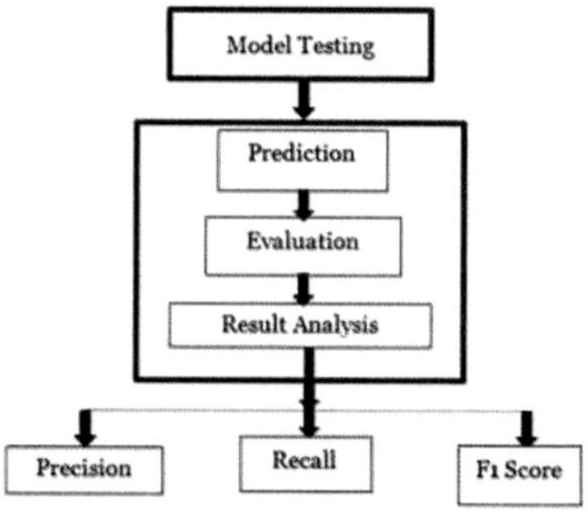

Fig. 3 Part 2 of flowchart of the proposed work

Tokenization and Formatting:

1. Use the WordPiece tokenizer to tokenize the input data.
2. Perform padding on the tokens to ensure a consistent sequence length.
3. Format the tokens as input for further training.

Model Training:

1. Initialize the BioBERT NER model with 12 encoders.
2. Fine-tune the BioBERT NER model during the training phase.
3. Set the sequence length limit to 100 tokens.
4. Choose a batch size (e.g., 10, 16, 32, and 64) and learning rate (e.g., 5e-5, 3e-5, and 1e-5) based on previous studies.
5. Train the model using the formatted input data.

Model Evaluation:

1. Evaluate various performance parameters on the trained model using the evaluation dataset.
2. Calculate metrics such as precision, recall, and F1-score to assess the model's performance.

Testing:

1. Provide unseen biomedical input sentences as test data.
2. Tokenize the test data using the WordPiece tokenizer.
3. Feed the tokenized data into the trained BioBERT NER model.
4. Predict the entity tags for each token in the test data.
5. Output the predicted entity tags for analysis and evaluation.

Performance Analysis:

1. Analyze the performance of the BioBERT NER model on the test data.
2. Compare the predicted entity tags with the ground truth tags.
3. Calculate performance metrics such as precision, recall, and F1-score to evaluate the model's accuracy.

After the model has been trained, various performance parameters are evaluated on the data to evaluate the model's performance. The BioBERT model is then tested for a given sample of test data, or unseen biomedical input sentences as shown in Fig. 4. The model then performs the prediction of the word tokens and predicts the tags for each token of the unseen data provided for testing. For example, when the unseen data is provided, the model would predict the tags for each token in the text, such as identifying named entities like B-Gene, I-Chemical, and Protein.

3.3 Fine-Tuning

In this study, we focused on evaluating the performance of the BioBERT model in the context of fine-tuning the CORD-19 dataset, which is a collection of biomedical articles and research papers. Fine-tuning is an essential step in leveraging pre-trained models like BioBERT for specific tasks or domains. By fine-tuning the model on the biomedical data, we aimed to adapt it to identify medical keywords or proper nouns as entities and extract meaningful information from the text.

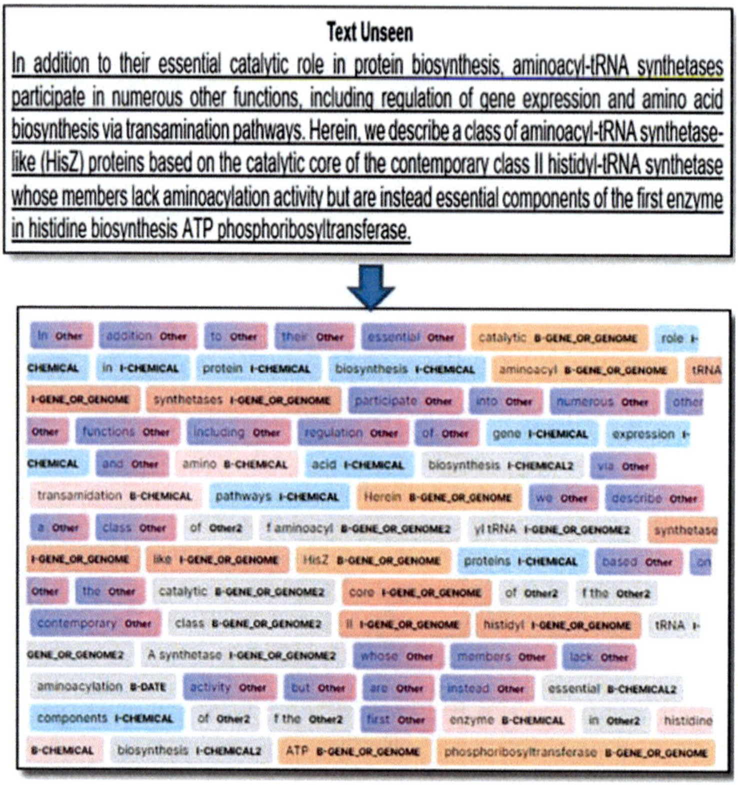

Fig. 4 Unseen entity to tagged entity

The process of fine-tuning the BioBERT model on the biomedical data involves training it on the CORD-19 dataset, which contains a wide range of biomedical texts. The model learns from the dataset by analyzing the relationships between words and their corresponding entity boundaries. This process allows the model to understand the specific context and semantics of the biomedical domain.

To illustrate the functioning of the BioBERT model, Fig. 5 provides an example that showcases its ability to identify entities and assign appropriate tags. In this example, the model recognizes the word 'Amino' as a specific entity and assigns it the tag 'B-Chemical,' indicating the beginning of a chemical entity. Similarly, the word 'acids' is identified as part of the same chemical entity and tagged as 'I-Chemical' denoting its position within the entity. This demonstrates how the BioBERT model effectively identifies medical keywords and assigns relevant tags to them.

The capability of the BioBERT model to accurately identify and tag entities in the biomedical text is crucial for various applications, such as biomedical information retrieval, biomedical named entity recognition, and biomedical text mining. By fine-tuning BioBERT on the CORD-19 dataset, we aim to enhance its performance and enable it to extract relevant information from biomedical texts more effectively.

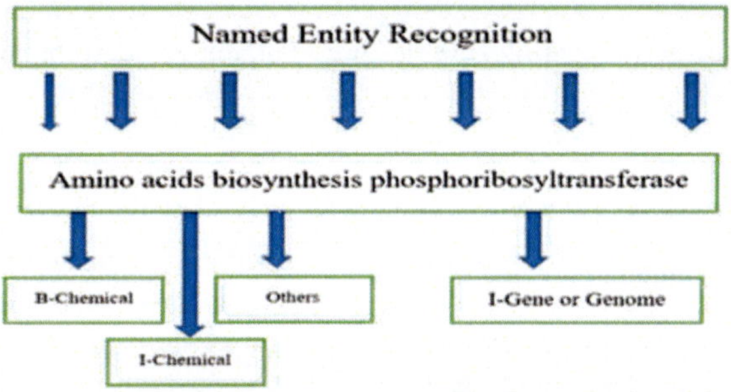

Fig. 5 Fine-tuning of BioBERT on CORD-19 dataset

Overall, Atliha [15] the evaluation of the BioBERT model's performance in fine-tuning the CORD-19 dataset is expected to provide insights into its suitability for biomedical entity recognition tasks and contribute to advancing biomedical research and information extraction in the field.

The study was conducted on the CORD-19 dataset, and the results of the fine-tuning task were analyzed to determine the model's accuracy in identifying entities and extracting semantics from the biomedical data. The study aimed to demonstrate the usefulness of the BioBERT model in fine-tuning for the purpose of information extraction from biomedical literature, thereby contributing to the field of natural language processing and its application in the biomedical domain.

4 Results and Discussions

In our study, we have performed fine-tuning using the BioBERT model on the CORD-19 dataset. Lyman et al. [12] The BioBERT model is a domain-specific model [16] that is pre-trained on a biomedical corpus, which consists of words from PubMed and PMC articles. The BioBERT model is based on the BERT base model architecture, but it is pre-trained on a different combination of corpora to improve its performance on biomedical text mining tasks.

4.1 Evaluation and Metric

Evaluation metrics are an essential part of any research paper that involves predictive models or classification tasks. In NER, Zhang et al. [17] precision, recall, and F1-score are the most commonly used evaluation metrics. These metrics help to measure the accuracy and performance of a model, and Zhong et al. [18] provide a comprehensive picture of how well the model is doing. Precision is defined as the number of true positive predictions made by the model divided by the total number of positive predictions made by the model. Precision measures the proportion of correctly identified entities out of all entities that the model has predicted as entities [19]:

$$\text{Precision} = \text{TP}/(\text{TP} + \text{FP})$$

Recall, on the other hand, is defined as [20] the number of true positive predictions made by the model divided by the total number of actual entities in the dataset. Recall measures the proportion of entities that the model has correctly identified out of all entities present in the dataset:

$$\text{Recall} = \text{TP}/(\text{TP} + \text{FN})$$

The F1-score is a harmonic mean of precision and recall and is a better indicator of the overall performance of the model. The F1-score is defined as the harmonic mean of precision and recall, and it [21] balances both precision and recall to give an overall picture of the model's performance. The F1-score ranges from 0 to 1, with 1 being the best possible score:

$$\text{F1 Score} = (2 * \text{Precision} * \text{Recall})/(\text{Precision} + \text{Recall})$$

It is important to note that there is often a trade-off between precision and recall, and choosing the right metric depends on the specific requirements of the task and the dataset. In NER, it is often necessary to achieve a good balance [22] between precision and recall, as high precision but low recall, or vice versa, may not be satisfactory for the task.

In our research paper, we will use precision, recall, and F1-score as our evaluation metrics, and we will report these metrics for each model we experiment with. This will allow us to compare the performance of different models, and select the best one for our task.

4.2 Result

We have used several versions of [23] pre-trained weights for the BioBERT model, such as BioBERT v1.0 (+PubMed + PMC), which contains a Wikipedia, books, PubMed abstract, and PubMed central full text, and it is trained for 270k steps. BioBERT-Base v1.0 (+PubMed 200K) contains a Wiki, book, and PubMed data, and it is trained for 200 thousand steps. BioBERT-Base v1.0 contains a Wiki, book, and PubMed central full-text data, and it is trained for 270k steps. We also used other versions of BioBERT like BioBERT-Large v1.1 (+PubMed 1M) and BioBERT-Large v1.12 (+PubMed 1M) which are trained for 1M steps and contain a PubMed corpus.

We fine-tuned the BioBERT model for the Named Entity Recognition (NER) task using these versions of pre-trained weights. For evaluation purposes, we computed the precision, recall, F1-Score, and support values for each entity for all versions of

BioBERT models. Among all the models, we obtained the best result for BioBERT v1.1 (+PubMed 1M). Table 2 shows the results for the BioBERT v1.1 (+PubMed 1M) model.

Table 3 shows the precision values for all five pre-trained weight models. Among these models, the BioBERT BASE CASED v1.1 on CORD-19-NER demonstrated superior performance. In our analysis, we evaluated the performance of individual entities as well as their overall performance. This examination revealed that certain entities are relatively easier to comprehend compared to others that exhibit greater complexity.

Furthermore, we observed that the support of an entity plays a significant role in determining its learnability. Entities with higher support tend to be more easily learned by the model, whereas entities with lower support may present challenges in terms of accurate recognition. This insight highlights the importance of considering the frequency and prevalence of entities within the dataset when assessing their learnability and overall performance. By examining the precision values for each entity, we gained insights into the varying levels of difficulty associated with different entity types. Some entities demonstrated consistently high precision scores, indicating that they were accurately identified by the models. On the other hand, certain entities exhibited lower precision values, suggesting the presence of more nuanced patterns or contextual dependencies that make them more challenging to recognize accurately.

This comprehensive evaluation of the precision values for individual entities, along with the collective performance of the models, provides valuable insights into the strengths and limitations of the pre-trained weight models. These findings can guide future research and development efforts to improve the recognition and understanding of entities within the biomedical corpus.

Table 2 Performance measures on BioBERT v1.1 (+PubMed 1M) model

Entities	Precision	Recall	F1-score	Support
B-Cardinal	0.82	0.83	0.83	2681
B-Chemical	0.80	0.79	0.80	9435
B-Coronavirus	0.99	01.00	0.99	7837
B-Date	0.78	0.77	0.78	2478
B-Disease or syndrome	0.81	0.81	0.81	3426
B-Gene or genome	0.77	0.81	0.79	8084
I-Chemical	0.65	0.69	0.67	3238
I-Gene or genome	0.72	0.73	0.72	4342
Other	0.91	0.88	0.90	15,766
Pad	1.00	1.00	1.00	78,413
Accuracy			0.93	135,700
Macro avg	0.83	0.83	0.83	135,700
Weighted avg	0.93	0.93	0.93	135,700

Table 3 Precision scores for all pre-trained weighted model

Entities	BioBERT v1.1 (PubMed 1M) (%)	BioBERT large v1.1 (+PubMed 1M) (%)	BioBERT base v1.0 (+PubMed 200K) (%)	BioBERT base v1.0 (+PubMed 200K + PMC 270K) (%)	BioBERT base v1.0 (+PMC 270K) (%)
B-Cardinal	82	76	69	67	67
B-Chemical	80	70	62	60	62
B-Coronavirus	99	98	96	97	96
B-Date	78	79	75	74	72
B-Disease or syndrome	81	68	70	66	68
B-Gene or genome	77	64	48	48	47
I-Chemical	65	58	39	39	36
I-Gene or genome	72	69	37	37	34

In Fig. 6, an overall comparison of the validation accuracy of all models is presented. The graph shows the performance of several different versions of the model, each of which was run for a specific [24] number of epochs. The x-axis of the graph represents the number of epochs, while the y-axis represents the validation accuracy. The different lines on the graph correspond to different versions of the model, with each line representing the validation accuracy of a particular model over the course of its run.

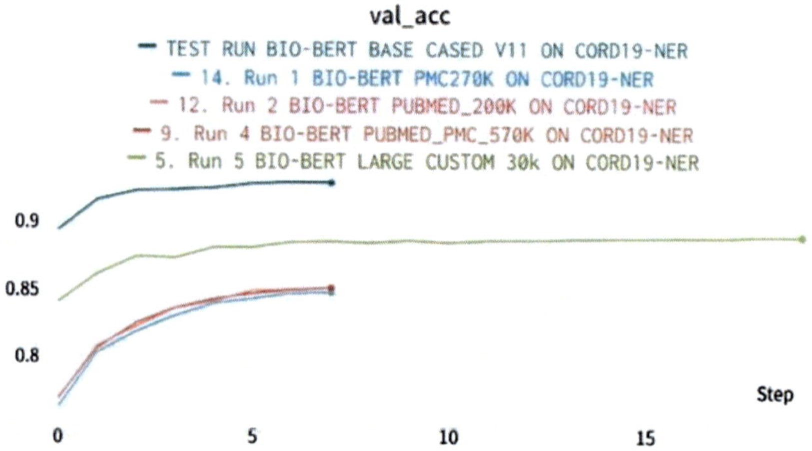

Fig. 6 Overall comparison of validation accuracy of all experiments

The various versions of the model are distinguished by different colors and/or markers. However, the rate at which the validation accuracy increases varies between the different versions of the model. Some models converge to a high validation accuracy more quickly than others. Additionally, it is found that the models that performed the best on the validation set had the highest accuracy. Among all the models, the BioBERT BASE CASED v1.1 model performed the best, achieving the highest validation accuracy. The validation accuracy provides a good overall indication of the performance of the model and can be used as a starting point for further analysis.

5 Conclusion

The paper presented a study on fine-tuning the BioBERT model for the Named Entity Recognition (NER) task using the CORD-19 dataset. Five different pre-trained weights of the BioBERT model were experimented with for the NER task, including BioBERT-Base v1.0 with PubMed 200K, BioBERT-Base v1.0 with PMC 270K, BioBERT-Base v1.0 with purposes 200K and PMC 270K, BioBERT-Large v1.1 with PubMed 1M, and BioBERT-Base v1.2 with PubMed 1M.

The performance of all the pre-trained weights was compared and it was found that the BioBERT-Large v1.1 with PubMed 1M and BioBERT-Base v1.2 with PubMed 1M models performed the best for the NER task. These models achieved an accuracy of 93%, which is significantly better than the other models discussed in this work. One possible reason for the improved performance of these models is the larger pre-training dataset. The BioBERT-Large v1.1 and BioBERT-Base v1.2 models were pre-trained for 1M steps, which may have allowed the model to learn a greater amount of information from the dataset and generalize better to the NER task.

In conclusion, the study demonstrates that fine-tuning a pre-trained BioBERT model on a domain-specific dataset can lead to significant improvements in performance for the NER task. The BioBERT-Large v1.1 with PubMed 1M and BioBERT-Base v1.2 with PubMed 1M models are well-suited for the NER task on the CORD-19 dataset and can be used as a reliable model for extracting entities from biomedical text.

References

1. Cho H, Lee H (2019) Biomedical named entity recognition using deep neural networks with contextual information. BMC Bioinform 20:1–11
2. Lee J, Yoon W, Kim S, Kim D, Kim S, So CH, Kang J (2019) BioBERT: a pre-trained biomedical language representation model for biomedical text mining. Bioinformatics. https://doi.org/10.1093/bioinformatics/btz682
3. Wang X, Song X, Li B, Guan Y, Han J (2020) Comprehensive named entity recognition on cord-19 with distant or weak supervision. arXiv preprint arXiv:2003.12218

4. Zhao D, Li J, Feng Y, Ji H (2015) Natural language processing and Chinese computing. Springer
5. Das D, Katyal Y, Verma J, Dubey S, Singh A, Agarwal K, Bhaduri S, Ranjan R (2020) Information retrieval and extraction on covid-19 clinical articles using graph community detection and bio-BERT embeddings. In: Proceedings of the 1st workshop on NLP for COVID-19 at ACL 2020
6. Catelli R, Gargiulo F, Casola V, De Pietro G, Fujita H, Esposito M (2020) Crosslingual named entity recognition for clinical de-identification applied to a covid-19 Italian data set. Appl Soft Comput 97:106779
7. Arguello-Casteleiro M, Maroto N, Wroe C, Torrado CS, Henson C, Des-Diz J, Fernandez-Prieto M, Furmston T, Fernandez DM, Kulshrestha M, et al (2021) Named entity recognition and relation extraction for covid-19: explainable active learning with word2vec embeddings and transformer-based BERT models. In: Artificial intelligence XXXVIII: 41st SGAI international conference on artificial intelligence, AI 2021, Cambridge, UK, December 14–16, 2021, Proceedings 41. Springer, pp 158–163
8. Lee J, Yoon W, Kim S, Kim D, Kim S, So CH, Kang J (2020) BioBERT: a pre-trained biomedical language representation model for biomedical text mining. Bioinformatics 36(4):1234–1240
9. Mikolov T, Chen K, Corrado G, Dean J (2013) Efficient estimation of word representations in vector space. arXiv preprint arXiv:1301.3781
10. Pennington J, Socher R, Manning CD (2014) Glove: global vectors for word representation. In: Proceedings of the 2014 conference on empirical methods in natural language processing (EMNLP), pp 1532–1543
11. Peters ME, Neumann M, Iyyer M, Gardner M, Clark C, Lee K, Zettlemoyer L (2018) Deep contextualized word representations. corr abs/1802.05365. arXiv preprint arXiv:1802.05365 (1802)
12. Lyman CA, Anderson C, Morris M, Nandal UK, Martindale MJ, Clement M, Broderick G (2019) When the how outweighs the what: the pivotal importance of context. In: 2019 IEEE international conference on bioinformatics and biomedicine (BIBM). IEEE, pp 2149–2156
13. Kumar S, Sahu A, Sharan A (2022) Deep learning based architecture for entity extraction from covid related documents. In: Proceedings of 4th international conference on information systems and management science (ISMS) 2021. Springer, pp 419–427
14. Wang LL, Lo K, Chandrasekhar Y, Reas R, Yang J, Eide D, Funk K, Kinney R, Liu Z, Merrill W, et al (2020) Cord-19: the covid-19 open research dataset
15. Atliha V (2023) Improving image captioning methods using machine learning approaches. PhD thesis, Vilniaus Gedimino technikos universitetas
16. Brockmeier AJ, Ju M, Przybyla P, Ananiadou S (2019) Improving reference prioritisation with PICO recognition. BMC Med Inform Decis Mak 19:1–14
17. Zhang Y, Lin H, Yang Z, Wang J, Sun Y, Xu B, Zhao Z (2019) Neural network-based approaches for biomedical relation classification: a review. J Biomed Inform 99:103294
18. Zhong N, Bradshaw JM, Liu J, Taylor JG (2011) Brain informatics. IEEE Intell Syst 26(5):16–21
19. Srinivasan P, Qiu XY (2007) Go for gene documents. BMC Bioinform (BioMed Central) 8:1–15
20. Ganeshkumar M, Ravi V, Sowmya V, Gopalakrishnan E, Soman K, Chakraborty C (2022) Identification of intracranial haemorrhage (ICH) using ResNet with data augmentation using CycleGAN and ICH segmentation using SegAN. Multimed Tools Appl 81(25):36257–36273
21. Jha PK, Valekunja UK, Reddy AB (2023) SlumberNet: deep learning classification of sleep stages using residual neural networks. bioRxiv, 2023–05
22. Hong G, Kim Y, Choi Y, Song M (2021) BioPREP: deep learning-based predicate classification with SemMedDB. J Biomed Inform 122:103888
23. Zhang R, Hristovski D, Schutte D, Kastrin A, Fiszman M, Kilicoglu H (2021) Drug repurposing for covid-19 via knowledge graph completion. J Biomed Inform 115:103696
24. Liang Y, Kelemen A (2005) Temporal gene expression classification with regularised neural network. Int J Bioinform Res Appl 1(4):399–413

Detecting Object Defects for Quality Assurance in Manufacturing

Mohit Varshney, Mamta Yadav, Mamta Bisht, Kartikeya Choudhary, and Sandhya Avasthi

Abstract The manufacturing industries have been searching for fresh ideas for improving product quality while cutting down on expenses and production time. Techniques for finding defects take a lot of time in production and manual quality control checks. A key element in assuring the quality of the final product is finding defects throughout the production process. It is essential to identify issues or flaws as soon as possible and take the necessary action to reduce operational and quality-related costs. Human operators are unable to accurately conduct defect detection due to their distraction and inattentiveness. The software system will cut down on human labor and the amount of time needed to find broken goods. On both a large and local scale, our system is useful. To build the application, a deep learning model along with the YOLO algorithm for better accuracy is utilized. YOLO algorithms perform really well for objects of similar color, without printing code, and visually difficult to distinguish. To evaluate the system results, Precision, Recall, and inference time metrics are utilized, the obtained value of these metrics clearly indicates the improved performance as compared to other existing methods.

Keywords Industry 4.0 · Object identification · Defect detection · Deep learning · YOLO algorithm · CNN · OpenCV

1 Introduction

The main characteristic of Industry 4.0 is intelligent production, which involves completing activities as efficiently as possible and quickly adapting to changing conditions. After the emergence of Industry 4.0, many organizations began to focus on automating regular tasks. Everything was automated, including deployment,

---

M. Varshney · M. Yadav · M. Bisht · K. Choudhary · S. Avasthi (✉)
Department of CSE, ABES Engineering College, Ghaziabad, UP, India
e-mail: sandhya_avasthi@yahoo.com

M. Varshney
e-mail: mohit.19b101121@abes.ac.in

© The Author(s), under exclusive license to Springer Nature Singapore Pte Ltd. 2023
A. Mishra et al. (eds.), *Advances in IoT and Security with Computational Intelligence*, Lecture Notes in Networks and Systems 755,
https://doi.org/10.1007/978-981-99-5085-0_33

product development, and test automation services. The incorporation of automated technologies enhanced the factory floor and assisted workers in avoiding tedious or hard tasks. Artificial intelligence (AI) and machine learning have made it possible to replace manual operations in fault testing. By speeding and automating the process of extracting insights from visual data, training AI to recognize objects might be helpful. Quality and pace are the two cogs of the manufacturing industry of this time and age and they are constantly on the lookout for time-efficient and accurate methods of making their products [1, 2]. This makes defect detection a crucial step in the post-production stage in any industry and helps reduce operational costs as well as plausible damage to the trust in the client-vendor dynamic. Therefore, we aim to create an application that can perform this defect detection precisely and in a way that blends with the manufacturing process, quality assurance should not add but rather shave time from the production process.

The intelligence, specificity, speed, and effectiveness of defect inspection algorithms [1] have been considerably improved in terms of their scope, specificity, speed, and efficiency, especially as a result of the introduction of cutting-edge deep learning technology. As a result, thorough quality monitoring of the production process is crucial for boosting production efficiency and reducing financial losses [2]. Object defect detection is an important subfield of computer vision. In the past, two-stage object detectors were extremely common and effective. Relative to the great majority of two-stage object detectors, single-stage object detection, and its underlying algorithms have profited from recent developments. The idea is to reduce the percentage of error that is undeniably present in manual defect detection [3, 4]. Also, the software will certainly work at a faster pace than a human assembly line. Our application can be used by large- and small-scale industries.

After detecting the object, it will be added to its dataset and then the defect in that object will be identified and displayed. Our system has both features, it will first identify the object, and then it will detect defects that will be present in that [5–7]. The training step utilized multitask learning and offline approaches for mining hard samples. The proposed framework for defect identification was evaluated using both a self-constructed dataset and an already labeled data with 'defect' and 'ok', and the results of the experiments showed its efficacy. Our technology can precisely identify flaws in images, hence resolving the technical application challenge of pin defect recognition in transmission lines. The main advantage of applying object and defect detection processes in large-scale manufacturing industries is saving time, and overall, the manufacturing process faster. Further, the defect detection model can reduce labor work hours and the need for human intervention at various stages. The overall productivity of the manufacturing unit will increase due to the defect detection process in quality checks.

2 Literature Review

Recent improvements in image processing technology, especially in the field of machine vision, have led to the development of automated ways to find defects in objects or object surfaces [5]. Because of the work of *several* scholars, it is now easier to find a defect on the surface. Ground Penetrating Radars (GPR) were used in Ref. [7] to measure the damage caused by water to pavement bridges. It used mixed deep convolutional neural networks (CNNs) to pull out features and made three datasets with different levels of detail. Structured light, deep learning, and two laser sensors were used to find and measure surface cracks in concrete structures [8]. The size measurement was more accurate when a custom-built fixture module, a distance sensor, and a laser alignment correction algorithm were all used together.

2.1 Defect Detection in Objects

Defect detection is used in a variety of industries, such as manufacturing, construction, and healthcare. A literature survey might explore how defect detection is being applied in these industries, and identify challenges and opportunities specific to each industry [8–11]. Traditional visual inspection methods rely on human inspectors to visually check materials or products for flaws. These techniques can be efficient but they can also be subjective, prone to mistakes, labor-intensive, and time-consuming [9]. Machine learning and AI methods use machine learning algorithms to automatically identify defects in products or materials. These methods can be more objective and accurate than human inspection but can require large amounts of labeled training data and computational resources to train [10].

2.2 Applications of Automatic Defect Detection

There are many potential applications for defect detection, beyond the traditional applications in manufacturing and quality control. These could include areas such as food safety, structural health monitoring, or medical imaging. Image-based quality control and defect detection technologies are used and discussed for food and beverage packaging and containers. Examples include testing the quality of food storage containers [12] and inspecting bottle closures [13]. Also covered are systems for spotting flaws in porcelain manufacturing [14]. This indicates that the full process can be finished in real-time. This system uses a robotic computer vision architecture. Rapid picture processing is combined with an auto-learning system. Additionally, a real-time computer vision system for spotting mistakes in the manufacturing of ceramic tiles is planned. In Industry 4.0, visual equipment can access a lot of data that can be used right away to find problems in products and fix them quickly and

Fig. 1 Generic object categories (image url https://images.app.goo.gl/h5w2wvxBgoypHfYH6) for object identification

effectively. In [14], the authors proposed using photos captured by a drone to identify issues with power line insulators. They trained a cascading architecture of CNN-based detectors that functioned effectively and could be utilized for additional air quality tests. Some general categories of objects in any object identification system are shown in Fig. 1.

3 Methodology for Defect Identification

Before pre-learning, OpenCV [15] is used to pre-process the input image. YOLO is used to determine the photograph's intended subject. The ROI, or region of interest for the item, is derived from the extracted region's data. Then, (3) an error is identified in accordance with the error determination standard that conforms to the process standard that each manufacturer has already developed. It is possible to reduce the financial and technical burden of a small organization by allowing the model to be used to identify any problems that can only be observed through external photographs taken without a scale or other equipment. Figure 2 describes the process flow of working the defect detection system, the method goes through three steps, image pre-processing, object recognition, and defect identification. YOLO (You Look Only Once) divides the input image into grids of varying sizes by adding scale-dependent feature layers. This facilitates the detection of targets of various sizes. However, there are numerous little targets for faults on any surface or object, and because YOLO is

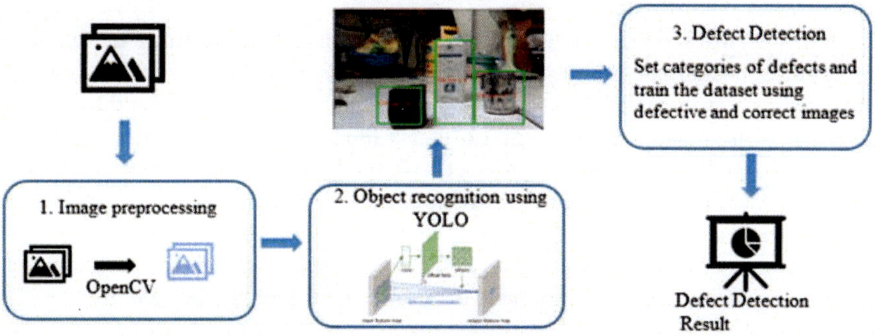

Fig. 2 A model to detect defects based on the YOLO algorithm

not adept at identifying small targets, it is easy to miss them. Therefore, YOLO must be modified in certain ways to improve its ability to locate minor fault sites.

3.1 OpenCV Tools

OpenCV, commonly referred to as open-source computer vision, is a software library for machine learning. It was first created by Intel, and Willow Garage later provided support. It expands the application of artificial perception. This cross-platform library comes with a source BSD license and is available for purchase. Nearly 3000 algorithms are now included in OpenCV, and the methods are being effectively tuned. OpenCV supports applications for real-time vision. Modern computer vision algorithms and machine learning are the categories under which all algorithms are categorized. Java, MATLAB, Python, C, C++, and other programming languages can quickly implement any algorithm on an operating system like Windows, Mac OS, Linux, or Android. Full-featured CUDA and OpenCL are used to advance technologies. More than 500 different algorithms contain functions that compare or support those methods. OpenCV uses a templated interface that is built in C++ and neatly integrates with STL.

3.2 Python Library NUMPY

Large multi-dimensional arrays and matrices are provided by the Python software package known as NumPy [6]. Various open-source software interfaces and contributors are present in NumPy. It also has the following things: An array object with powerful N-dimensions. Data size and type can be properly defined in NumPy. NumPy allows providing quick integration with a variety of databases. Although there are significant limitations, it is also published under the BSD license. This research paper aims to identify objects by using various techniques from the surface of a complex background image. Using image processing techniques, it can remove items from trees more quickly and easily than by hand, such as apples and bananas. Language identification steps. Python offers a library called the Python Imaging Library (PIL) that may be used to interact with images. There are numerous image processing functions available in the Python Imaging Library. We used PIL modules to carry out a few simple tasks. The OpenCV library has been implemented in Python 2.7 with the aid of NumPy, and object identification may be investigated in the same way that a virtual Artificial Neural network can be built with the Sci-Kit Tools.

3.3 Implementation of the YOLO Algorithm

Regression rather than classification is the new terminology used by the authors of YOLO [16] to reframe the object detection problem. An image's bounding boxes and class probabilities are predicted by a convolutional neural network for each object shown there. You Only Look Once was chosen as the name for this technique because it uses bounding boxes to identify items and their locations by only ever glancing at the image once (YOLO). Large, multi-dimensional arrays and matrices are provided by the Python software package known as NumPy [6]. Various open-source software interfaces and contributors are present in NumPy. It also has the following things: An array object with powerful N-dimensions. The field of defect identification has seen success using the YOLO method.

YOLO works by dividing an image into a grid and using a single neural network to predict the bounding boxes and class probabilities for objects within each grid cell. This allows it to process the entire image in one go, rather than using multiple passes or sliding windows, which makes it much faster than other object detection algorithms [15, 16].

We can see in Fig. 3 how an input image is divided into $S \times S$ Grid, and each grid it predicts B bounding boxes. The equation for encoding is $S \times S \times (B * 5 + C)$, where C represents the number of classes of objects. The algorithm's ability to quickly and accurately identify defects in images has significantly improved the efficiency and reliability of the defect detection process. Object detection is essential for security and authentication purposes if we wish to physically identify a person or read his behavior. A variety of biological characteristics are utilized, including fingerprints, hand geometry, the pattern of the retina and iris, etc. [17–20].

Benefits of using YOLO Algorithm—To identify similar colors and size of objects which are having no specific printing code and visually difficult to distinguish YOLO algorithm performs well on these datasets. The currently used object detection methods are R-CNN, YOLO, and SSD. In R-CNN, detecting speed is not fast enough for any real-time application. SDD works moderately in between the other two algorithms. YOLO has high detecting speed and great real-time performance. In busy places like pharmacies, it requires a high detecting speed, so YOLO will work more efficiently [21–24].

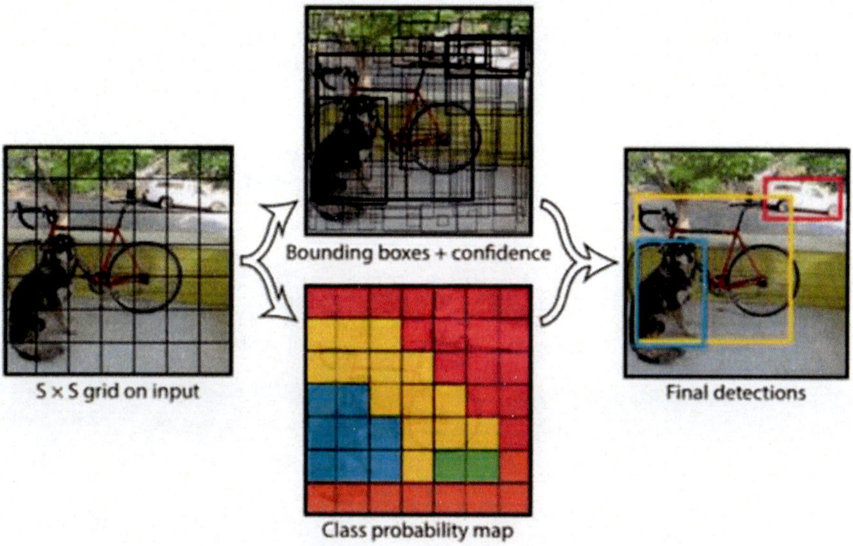

Fig. 3 Example to illustrate the YOLO algorithm

4 Methodology for Defect Identification

The primary goal of the web-based application used for the object detection system in photos is to identify several items from diverse image kinds. Shape and edge features from the image are extracted to accomplish this purpose. For accurate object detection and recognition, a sizable image database is used. Numerous strategies are employed to find defects in detected items. The basic functions of object detection include segmentation, localization, and object classification.

4.1 Dataset for Training

The ImageNet 1000-class competition data are used to train our convolutional layers [21, 25, 26]. The first 20 convolutional layers are used for pretraining, followed by an average-pooling layer and a fully linked layer. Similar to the GoogLeNet models in Caffe's Model Zoo [22], this network takes around a week to train and delivers a top-five accuracy of 98% on the ImageNet 2012 validation set. The second dataset used for defect identification is the casting product image dataset for quality inspection, which is publicly available on Kaggle [23, 27, 28].

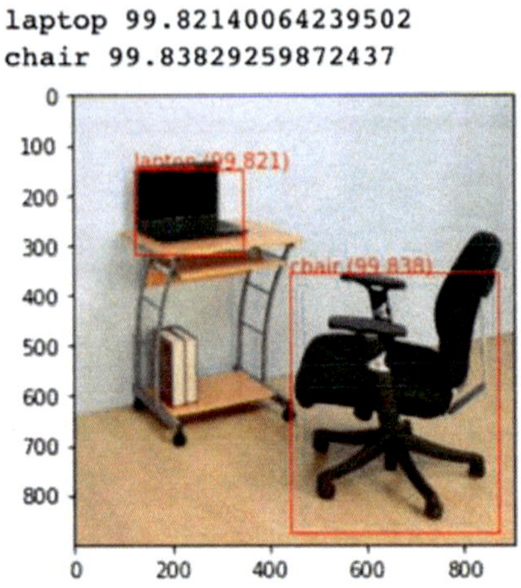

Fig. 4 Identifying chairs and laptops through an identification system

Table 1 Accuracy summary of YOLO for the identification process

Metric	YOLOv4	YOLOv5
Precision	0.949	0.886
Recall	0.958	0.946
mAP@0.5	0.978	0.958
Inference time (ms)	12	11

4.2 *Training and Validation of Identification of Objects*

The only thing we need to do to train adds an option. If this is the first time a new configuration is trained, the annotation will also be parsed, making training simple. Figure 4 shows the output generated through our system when we provided the image of the laptop and chair to the classification model. Table 1 summarizes the precision, recall, and inference time for a sample of 1000 images.

4.3 *Detection and Identification of Defects*

For Defect detection and identification experiments, the second dataset from Kaggle is utilized. The original dataset on Kaggle contained 7348 images, and each image was 300 × 300 pixels in size, and grayscale. There are two categories: Defective front (def-front) and Ok front (ok-front). Two samples each from the defective front surface and the correct front surface are shown in Fig. 5. The number of training def-front images was 3758, and the number of training of-front images was 2875. In the test set, the number of def-front images was 453, and the number of ok-front

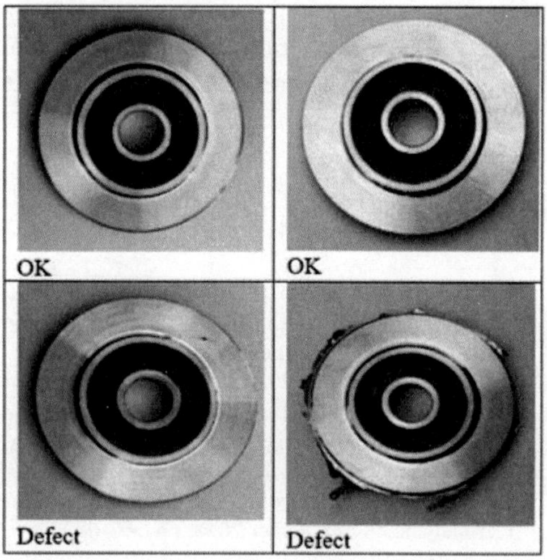

Fig. 5 Sample of two good quality objects and two defective objects

images was 262. The defects in the images are categorized as 'pinholes', 'scratches', and 'deformations'.

5 Conclusion

Traditionally, defect detection and quality assurance have been important in the manufacturing industry because they have the potential to reduce scrap rates, production time, and energy use, which increases productivity. Automated image-based systems try to reduce rejection rates and meet quality criteria in highly automated operations. Human operators are unable to accurately conduct defect detection due to their distraction and inattentiveness. Our software will cut down on human labor and the amount of time needed to find broken goods. The field of defect identification has seen success using the YOLO method. The effectiveness and dependability of the detection process have greatly increased thanks to its capacity to swiftly and precisely identify flaws in images. It was made clear in the previous sections that developing reliable and resilient algorithms with the help of a camera sensor had a specific property to handle common issues and deliver a reliable output. Each of the mentioned methods has advantages and disadvantages, and they may work admirably in certain cases while utterly failing in others. The system described in the research can recognize an object and then look for any potential defects in it with precision and recall scores of 88% and 94%, respectively. The main advantage of applying object and defect detection processes in large-scale manufacturing industries is saving time, and overall, the manufacturing process faster. Further, the defect detection model can reduce labor work hours and the need for human intervention at various stages. The overall productivity of the manufacturing unit will increase due to the defect detection process in quality checks.

References

1. Tsoli A, Argyros AA (2018) Joint 3D tracking of a deformable object in interaction with a hand. In: Proceedings of the European conference on computer vision (ECCV), Munich, Germany, 8–14 September 2018, pp 484–500
2. Ge L, Ren Z, Yuan J (2018) Point-to-point regression PointNet for 3D hand pose estimation. In: Proceedings of the European conference on computer vision (ECCV), Munich, Germany, 8–14 September 2018, pp 475–491
3. Cai Y, Ge L, Cai J, Yuan J (2018) Weakly-supervised 3D hand pose estimation from monocular RGB images. In: Proceedings of the European conference on computer vision (ECCV), Munich, Germany, 8–14 September 2018, pp 666–682
4. Wu XY (2019) A hand gesture recognition algorithm based on DC-CNN. Multimed Tools Appl 1–13
5. Wu W-H, Lee J-C, Wang Y-M (2020) A study of defect detection techniques for metallographic images. Sensors
6. Zhang J, Yang X, Li W, Zhang S, Jia Y (2020) Automatic detection of moisture damages in asphalt pavements from GPR data with deep CNN and IRS method. Autom Constr 113:103119
7. Song EP, Eem SH, Jeon H (2020) Concrete crack detection and quantification using deep learning and structured light. Constr Build Mater 252:119096
8. Peng C, Xiao T, Li Z, Jiang Y, Zhang X, Jia K, Yu G, Sun J (2018) Megdet: a large mini-batch object detector. CVPR
9. Girshick RB, Felzenszwalb PF, Mcallester DA (2011) Object detection with grammar models. In: NIPS, p 1
10. Bodla N, Singh B, Chellappa R, Davis LS (2017) Soft-NMS improving object detection with one line of code. In: ICCV, p 5, 6
11. Law H, Deng J (2018) Cornernet: detecting objects as paired keypoints. In: ECCV, pp 1, 2, 3, 4, 6, 7
12. Pavlakos G, Zhou X, Chan A, Derpanis KG, Daniilidis K (2017) 6-dof object pose from semantic keypoints. In: ICRA, p 3
13. Maninis KK, Caelles S, Pont-Tuset J, Van Gool L (2018) Deep extreme cut: from extreme points to object segmentation. In: CVPR, pp 2, 4, 5, 7, 8
14. Tao X, Zhang D, Wang Z, Liu X, Zhang H, Xu D (2018) Detection of power line insulator defects using aerial images analyzed with convolutional neural networks. IEEE Trans Syst, Man, Cybern: Syst 50(4):1486–1498
15. Russakovsky O, Deng J, Su H, Krause J, Satheesh S, Ma S, et al (2015) Imagenet large scale visual recognition challenge. Int J Comput Vis 115(3):211–252
16. Saengseangfa T, Tantrairatn S (2021, November) Design and development of a vision system for identifying a workpiece quality and defects using object detection. In: 2021 25th international computer science and engineering conference (ICSEC). IEEE, pp 240–245
17. Avasthi S, Chauhan R, Acharjya DP (2022) Information extraction and sentiment analysis to gain insight into the COVID-19 crisis. In: International conference on innovative computing and communications. Springer, Singapore, pp 343–353
18. Avasthi S, Chauhan R, Acharjya DP (2022) Topic modeling techniques for text mining over a large-scale scientific and biomedical text corpus. Int J Ambient Comput Intell (IJACI) 13(1):1–18
19. Avasthi S, Chauhan R, Acharjya DP (2022) Extracting information and inferences from a large text corpus. Int J Inf Technol 1–11

20. Avasthi S, Sanwal T (2016) Biometric authentication techniques: a study on keystroke dynamics. Int J Sci Eng Appl Sci (IJSEAS) 2(1):215–221
21. Gupta A, Avasthi S (2016) Image based low-cost method to the OMR process for surveys and research. Int J Sci Eng Appl Sci (IJSEAS) 2(7)
22. Mishkin D (2015) Models' accuracy on Imagenet 2012 val. https://github.com/BVLC/caffe/wiki/Models-accuracy-on-ImageNet-2012-val. Accessed 02 Oct 2015
23. https://www.kaggle.com/datasets/ravirajsinh45/real-life-industrial-dataset-of-casting-product. Accessed 16 Dec 2022
24. Li Y, Huang H, Chen Q, Fan Q, Quan H (2021) Research on a product quality monitoring method based on multi scale PP-YOLO. IEEE Access 9:80373–80387
25. Ren Z, Fang F, Yan N, Wu Y (2022) State of the art in defect detection based on machine vision. Int J Precis Eng Manuf-Green Technol 9(2):661–691
26. Kumar A (2008) Computer-vision-based fabric defect detection: a survey. IEEE Trans Industr Electron 55(1):348–363
27. Lowe MJ, Alleyne DN, Cawley P (1998) Defect detection in pipes using guided waves. Ultrasonics 36(1–5):147–154
28. Li F, Xi Q (2021) DefectNet: toward fast and effective defect detection. IEEE Trans Instrum Meas 70:1–9

Model Explainability for Masked Face Recognition

Sonam

Abstract With the incessant increase of COVID-19 and related variants, various regulating agencies have emphasized the importance of face masks, especially in public areas. The face recognition systems already deployed by various organizations have to be recalibrated to be able to detect subjects wearing a face mask. The modern-day face recognizers composed of various face detection and classification algorithms appear as a black box to the end user. The nature of such systems becomes suspicious when highly sensitive scenarios are concerned. This consequently raises trust issues towards the model deployed in the background. The behavior of the image classification model can be interpreted for this purpose using LIME (Local Interpretable Model Agnostic Explanations). It can give a clear idea of the features or super-pixels that are in charge of making a particular prediction. This work aims to investigate the local features of a target image that helps the classifier to make a prediction using LIME. A Vanilla CNN model has been selected, which was trained with 7553 face images. The model exhibits a classification accuracy of 98.19%, and it is revealed from the heatmaps that the model works by learning the structure of a face with and without masks for making accurate predictions.

Keywords Explainable artificial intelligence (XAI) · COVID-19 · LIME · Image classification · CNN model

1 Introduction

The advent of technology has enabled the processing of large amount of data with high efficiency, thus allowing us to carry out intense processing tasks with relative ease [1]. This includes, but not limited to the data scientists, relying on deep learning techniques to extract, process, and analyze useful information from the raw

Sonam (✉)
Department of Electronics & Communications Engineering, Indraprastha Institute of Information Technology Delhi, New Delhi, India
e-mail: sonam.octer@gmail.com

© The Author(s), under exclusive license to Springer Nature Singapore Pte Ltd. 2023
A. Mishra et al. (eds.), *Advances in IoT and Security with Computational Intelligence*, Lecture Notes in Networks and Systems 755,
https://doi.org/10.1007/978-981-99-5085-0_34

data in a time-efficient manner. Numerous data-intensive tasks, including instance or semantic segmentation, and image classification, have been carried out in a real time with relative ease using deep learning techniques. Such systems have been deployed worldwide for seamless integration with the existing techniques. Despite, having state-of-art systems and technologies, the safety issues were declared a major concern during the COVID-19 pandemic. While the regulatory bodies insisted on wearing face masks to avoid the transmission, there were certain groups that took advantage of the grim situation for their benefits. Thus, in order to minimize exposure and avoid unnecessary risks, there was a requirement for the recalibration of the existing state-of-the-art face recognition systems to accurately detect and classify subjects wearing a face mask [2]. Most of the modern systems work on a traditional approach such as the feature selection, extraction, and its subsequent classification [3]. The classification may make sense when fewer features are extracted and a method like the decision tree is applied. It is nearly impossible to demonstrate the contributions of all features in deep learning models using the decision tree classifier. Three main approaches—numerical, visual, and rule-based—are used to explain the models [4]. Using a technique like information gain, the numerical methods determine the contributions of all inputs from one to zero or the opposite [5]. Numerical methods can identify significant features by manually experimenting with inputs to see how they affect classification outcomes. Decision Tree and Random Forest are two examples of rule-based methods that use information gain to determine the significance of inputs for classification. However, due to parameter size, they cannot be used in large models like deep neural networks to build an intelligible structure.

For deep neural models, visual methods such as a class activation map is used. The significance of each pixel in an image can be displayed over the image by using an activation map that has been created (also known as a heat map). On the other hand, Deep Learning has come under fire for being used to perform computer vision-related tasks. Neural networks have been criticized for being "black boxes," which essentially means that we teach them with expectations of the results by providing them with a ton of data. However, we are unsure of how it accomplishes our goals. As a result, understanding how a neural network accomplishes what it is intended to accomplish is essential to comprehending how an algorithm recognizes objects and their characteristics from an image. By doing so, researchers will be able to troubleshoot in addition to convincingly explaining how it works. A schematic diagram to represent the problem statement has been depicted in Fig. 1.

This manuscript is structured as follows: A brief description of related work on face mask detection is compiled in Sect. 2. Section 3 presents a brief description of all the algorithms employed in this work. Section 4 presents the methodology of face mask detection using the LIME XAI. Experiments and results are presented in Sect. 5, and Discussion is given in Sect. 6, respectively. In Sect. 7, the conclusion and future works are explained.

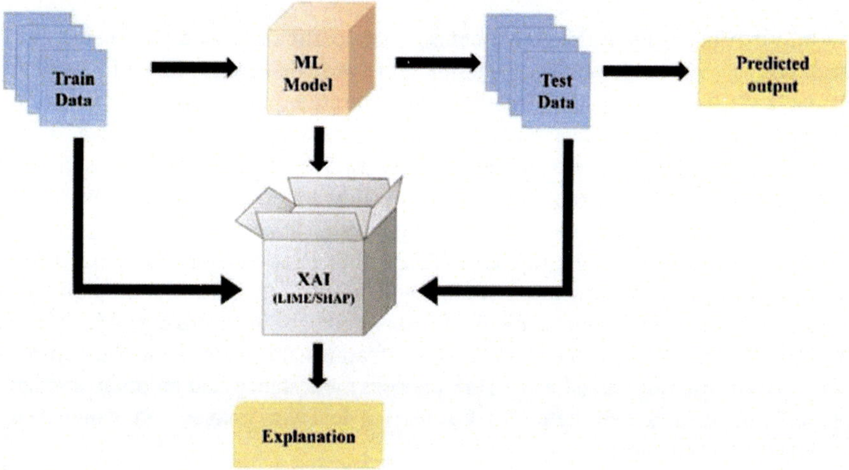

Fig. 1 Schematic diagram to depict explainable artificial intelligence (XAI)

2 Related Work

Numerous studies using deep learning are conducted to identify COVID-19, but these studies focus on clinical outcomes [6–8]. A system that detects whether people are wearing masks can be created by utilizing deep learning. Hybrid deep learning [9] and machine learning model for detecting face mask images have been proposed. Support vector machines (SVM) are used to classify the extracted features after deep learning is used to extract the features. In various datasets, the SVM classifier achieved testing accuracies of 100%. By fusing image classification and super-resolution techniques, a recognition system is created, which is able to identify if the face mask has been used [10]. Four processes make up their algorithm: data preprocessing, face detection and extraction, super-resolution, and face mask identification. Using the deep learning technique, the proposed methodology achieved 98.70% accuracy. Ejaz et al. [11] demonstrate a masked-face recognition system based on Principal Component Analysis (PCA). According to their report, the system performs better when the subject is not wearing a mask; their respective average accuracy rates were 95.68% and 70.53%. Convolutional-based solutions to the face detection problem offer high efficiency, courtesy of deep learning techniques [12–14] including the modified YOLOv3, application of softmax function, and dimension reduction tools to improve the detection efficiency. Kong et al. [15] have demonstrated a mask detection model based on edge detection that works in real-time scenarios. The proposed methodology relies on image restoration, followed by face and mask detection, providing an accuracy of 94.90%.

For explaining the predictions of any model, Ribeiro et al. [16] introduce a new algorithm called Local Interpretable Model-Agnostic Explanations (LIME). The prediction of an ML model, "x" dependent upon the prediction function $f(x)$, which is in turn dependent upon the constituents of any particular model, is treated as a "black box" by LIME. It works on attempting to provide explanations of how $f(x)$ behaves in x's environment. This gives locally accurate explanation of the prediction instance. Simple models with a reputation for being easy to understand, such as decision trees, falling rule lists, etc., are used to generate the explanations. Considering that LIME only explains specific examples and not the model as a whole, the model must be fully explained using a number of examples as demonstrated by Rajpal et al. [17]. SP-LIME (Selective pick LIME) and RP-LIME (Random pick LIME) are two techniques used to choose the instances that will be shown to the user, both of which aim to explain the entire model using the fewest number of instances possible. While SP-LIME chooses the instances so that all the features are covered and the number of instances is also minimized, RP-LIME chooses the instances at random to present a predetermined number of them. LIME is used to deal with images [18], which deal with huge useful features.

3 Preliminaries

This section briefly discusses the algorithms used in the proposed face mask recognition along with model explainability using LIME.

3.1 Local Interpretable Model-Agnostic Explanations (LIME)

In order to gain additional insights into the prediction outcomes of a model, LIME is used post-hoc to evaluate, study, and analyze the features that help the model to make a particular decision as shown in Fig. 2. LIME works by displaying or highlighting specific events in an input data (such as strings of words or locally disjoint patches in an input image), which helps the model to make a particular decision [16]. In terms of machine learning knowledge, it is crucial that explanations be simple enough for a layperson to understand as well. Another crucial factor is local fidelity, which states that an explanation must at least match the behavior of the model close to the instance being predicted in order to be considered meaningful [16]. LIME treats the model under investigation as a "black box," meaning it is unaware of the model's structure. This makes it easier to explain any model as well as upcoming classifiers.

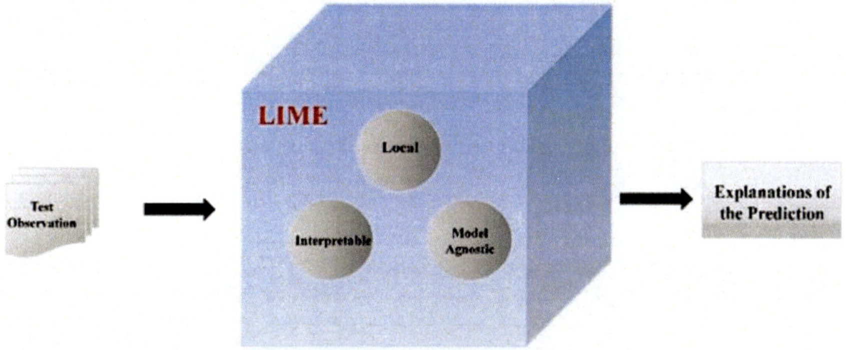

Fig. 2 Schematic diagram depicting LIME architecture

3.2 *LIME for Images*

LIME for images operates differently from LIME for text and tabular data. Since many pixels contribute to each class, it would seem counterintuitive to alter individual pixels [16]. The predictions would most likely not be significantly altered by randomly changing individual pixels. In order to create different versions of the images, the image is divided into "super-pixels" and those super-pixels are either turned on or off. Super-pixels are groups of related-colored interconnected pixels that can be disabled by changing each one to a custom color. A super-pixel's likelihood of being turned off in each permutation can also be specified by the user.

3.3 *Vanilla CNN*

A CNN is a Deep Learning algorithm that uses learning filters to automatically extract complex features from inputs, allowing it to take in a multi-dimensional input and distinguish it from another. Beginning with a Vanilla CNN network that uses CNNs as its layers, the network has six layers, including three forward layers, one max pooling layer, and two CNN layers as shown in Fig. 3.

4 Methodology

This section of the manuscript provides detailed insights into the proposed Explainability of Masked-face recognition. The work has been done using Jupyter Notebook where the face mask dataset is taken from Kaggle [19] shown in Fig. 4. Figure 5 shows the flowchart of the proposed system for model explainability.

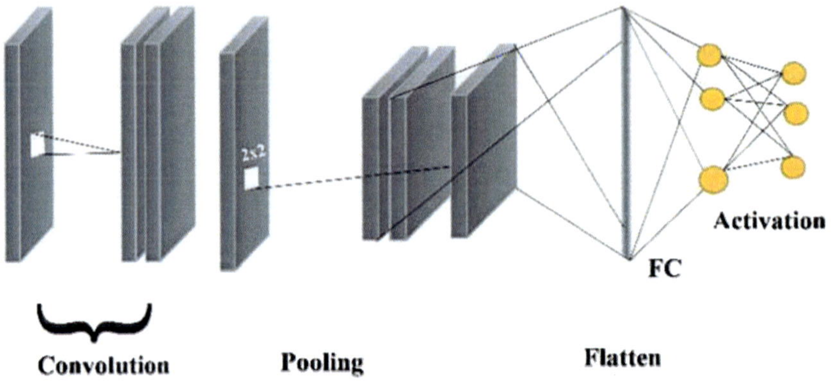

Fig. 3 Schematic diagram depicting a vanilla CNN network

Fig. 4 Sample face images from the dataset taken from Kaggle [19]

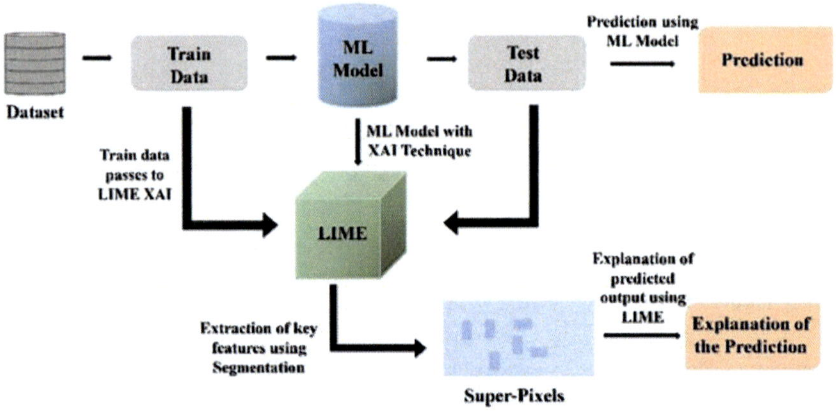

Fig. 5 Block diagram of the proposed work

4.1 Process of Explainability

This section explains the stepwise process of Explainability for face mask recognition. The process includes (i) the generation of training and testing data. The dataset is randomly split into training and testing dataset to avoid ambiguity. This is followed up by (ii) building and training the CNN model. The developed model is then (iii) used for the prediction using the testing dataset. It predicts the class for unseen data by choosing an image randomly from the testing data as shown in Fig. 6a. The next step includes (iv) analyzing the prediction of CNN model using LIME. In LIME, super-pixels are used in feature extraction by applying the Segmentation algorithm (Fig. 6b). Next, (v) perturbed samples are created by randomly turning ON and OFF some super-pixels (Fig. 7a) and calculating the distance between the original and perturbed images (Fig. 7a). This forms the basis for (vi) a local linear model, which is used to select key features. Next, (vii) top features are selected (Fig. 7b), and a weighted local linear model is fitted to explain the prediction made by the "black-box" in the immediate vicinity of the original image. Finally, (viii) LIME works step by step to explain the prediction, which shows the top 10 key features in the image and shows whether a person wears a mask or not.

5 Experiments and Results

The results of the predicted mask or unmask face using the face mask dataset shown in Fig. 4 are compiled in this section. Super-pixels are achieved by using segmentation as shown in Fig. 6b. After the training, 7553 new images (6799 with masks, 754 without) were taken from Kaggle to test the model using actual images. Classification results are shown in Table 1. The top key features that contribute to the classification of face mask recognition and as predicted by LIME are shown in Fig. 7b. Heatmap helps to show the visualization of the key features and tell how much they are contributing to the prediction of face mask recognition with the scale as shown in Fig. 7c. Different color indicates the importance and contribution of that part. Orange color indicates the less contribution whereas, dark blue color indicates the high contribution or key feature for the prediction.

6 Discussion

Deep neural networks, such as CNN with LIME XAI, can be used to implement classification with high performance, but their complexity prevents them from providing an explanation for why a model is placed into a particular class. That is why these models are referred to as "black box" models. Vanilla CNN model was chosen and trained to demonstrate what a CNN learns from images with and without face masks.

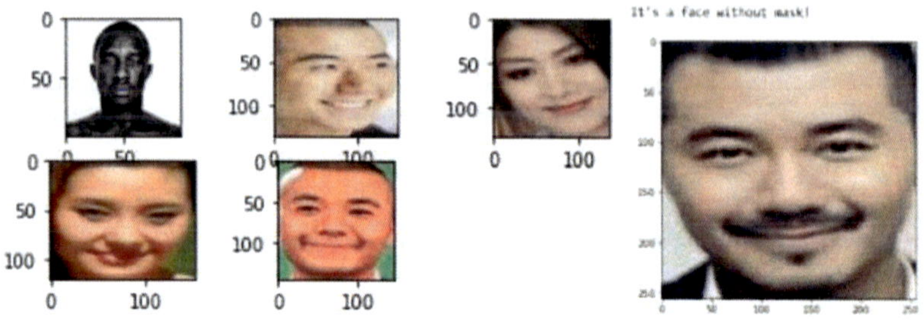

(a)

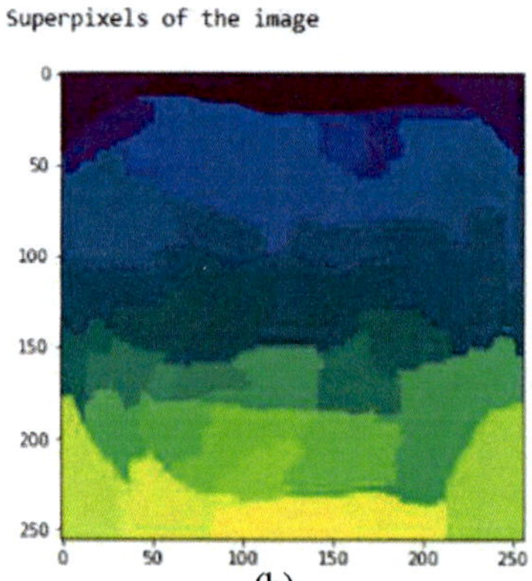

(b)

Fig. 6 **a** Prediction using CNN model and **b** heatmap of key features of a sample

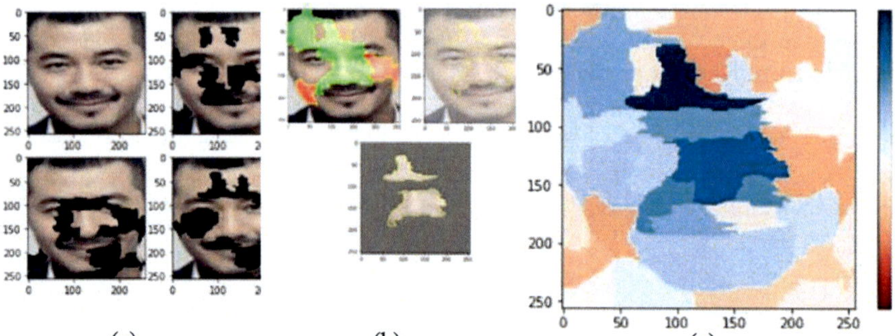

(a) (b) (c)

Fig. 7 **a** Perturbed image; **b** key features of a prediction using LIME; **c** heatmap of prediction

Table 1 Results compiled after classification

Dataset composition	Number of images
Images with mask	6799
Images without mask	754
Correct classification of images with mask	6787
Correct classification of images without mask	751
Accuracy	98.19%

LIME XAI is used to explain the predicted results which come from the CNN model. The heatmap technique was used to highlight regions the model learned from the images. The results demonstrate that the LIME XAI is able to explain the prediction outcome of a model. LIME with CNN model successfully classifies images into the mask and no mask classes. Additional insight is provided by the LIME as to why a particular image has been classified into mask or unmasked class by highlighting the local features of an input image that contributed towards the positive hit.

7 Conclusion and Future Work

With the unprecedented development in the computational facilities, AI and neural networks have observed unparalleled progress in the last decade. However, these techniques have failed to capture the interests of most business and social organizations dealing with sensitive issues. This is a direct consequence of the lack of confidence in the prediction outcomes of the state-of-the-art AI and neural network techniques courtesy of their "black box" nature. This highlights the need for making neural networks more interpretable. To bridge this gap, this work focuses on LIME model explainability to build a model whose prediction outcomes are comprehensible to a wider audience. This will help in establishing confidence and easy deployment of state-of-the-art neural network models for solving most of the issues in a precise and time-efficient manner.

Acknowledgements The author would like to acknowledge the Computer Science and Engineering Department, Indraprastha Institute of Information Technology Delhi for providing necessary tools and financial assistance for the completion of this work. Author is also thankful to his senior Mr. Khushwant Sehra, Department of Electronic Science, University of Delhi South Campus for his guidance and support during this work.

References

1. Ornek A, Celik M, Ceylan M (2021) Explainable artificial intelligence: how face masks are detected via deep neural networks. Int J Innov Sci Res Technol 6(9):1104–1112
2. Brienen NC, Timen A, Wallinga J, Van Steenbergen JE, Teunis PF (2010) The effect of mask use on the spread of influenza during a pandemic. Risk Anal: Int J 30(8):1210–1218
3. Uddin MF, Lee J, Rizvi S, Hamada S (2018) Proposing enhanced feature engineering and a selection model for machine learning processes. Appl Sci 8(4):646
4. Gilpin LH, Bau D, Yuan BZ, Bajwa A, Specter M, Kagal L (2018) Explaining explanations: an overview of interpretability of machine learning. In: 2018 IEEE 5th international conference on data science and advanced analytics (DSAA). IEEE, pp 80–89
5. Raileanu LE, Stoffel K (2004) Theoretical comparison between the Gini index and information gain criteria. Ann Math Artif Intell 41(1):77–93
6. Huang L, Han R, Ai T, Yu P, Kang H, Tao Q, Xia L (2020) Serial quantitative chest CT assessment of covid-19: a deep learning approach. Radiol: Cardiothorac Imaging 2(2):e200075
7. Oh Y, Park S, Ye JC (2020) Deep learning covid-19 features on CXR using limited training datasets. IEEE Trans Med Imaging 39(8):2688–2700
8. Ismael AM, Şengür A (2021) Deep learning approaches for covid-19 detection based on chest X-ray images. Expert Syst Appl 164:114054 (2021)
9. Loey M, Manogaran G, Taha MHN, Khalifa NEM (2021) A hybrid deep transfer learning model with machine learning methods for face mask detection in the era of the covid-19 pandemic. Measurement 167:108288. https://www.sciencedirect.com/science/article/pii/S0263224120308289
10. Qin B, Li D (2020) Identifying facemask-wearing condition using image super resolution with classification network to prevent covid-19. Sensors 20(18):5236
11. Ejaz MS, Islam MR, Sifatullah M, Sarker A (2019) Implementation of principal component analysis on masked and non-masked face recognition. In: 2019 1st international conference on advances in science, engineering and robotics technology (ICASERT). IEEE, pp 1–5
12. Hjelmås E, Low BK (2001) Face detection: a survey. Comput Vis Image Underst 83(3):236–274
13. Kumar A, Kaur A, Kumar M (2019) Face detection techniques: a review. Artif Intell Rev 52(2):927–948
14. Li C, Wang R, Li J, Fei L (2020) Face detection based on YOLOv3. In: Recent trends in intelligent computing, communication and devices, pp 277–284
15. Kong X, Wang K, Wang S, Wang X, Jiang X, Guo Y, Shen G, Chen X, Ni Q (2021) Real-time mask identification for covid-19: an edge computing based deep learning framework. IEEE Internet Things J
16. Ribeiro MT, Singh S, Guestrin C (2016) "Why should I trust you? Explaining the predictions of any classifier. In: Proceedings of the 22nd ACM SIGKDD international conference on knowledge discovery and data mining, pp 1135–1144
17. Rajpal A, Sehra K, Bagri R, Sikka P (2022) XAI-FR: explainable AI-based face recognition using deep neural networks. In: Wireless personal communications. Springer, pp 1–18
18. del Castillo Torres G, Riog-Maimó MF, Mascaró-Oliver M, Amengual-Alcover E, Mas-Sansó R (2022) Understanding how CNNs recognize facial expressions: a case study with LIME and CEM. Sensors 23(1):1–13
19. Kaggle Mask Dataset. https://www.kaggle.com/datasets/omkargurav/face-mask-dataset. Accessed 10 Jan 2022

A Research Agenda Towards Culturally Aware Information Systems

Abhisek Sharma, Sarika Jain, Naveen Kumar Jain, and Bharat Bhargava

Abstract Our world is not homogeneous in nature, it has multiple languages, ethnic groups, culture, beliefs, and varying perceptions of things. There is development towards catering experiences based on these varying dimensions by utilizing the available cultural models (such as Trompenaars, Hofstede, and Deal and Kennedy). Though these models can help in the development of globalized information systems that can be adapted for a particular locale (location/language/culture-specific), we don't have these dimensions in a dereferenceable form. This paper is an attempt to put some light on the possible solution for the problems discussed. Here, we discuss research questions that should be addressed while incorporating cultural representation. We also discuss various cultural models, and available usages and probable usecases.

Keywords Knowledge graph · Linguistics · Culture · Cultural models

---

This work is supported by the IHUB-ANUBHUTI-IIITD FOUNDATION set up under the NM-ICPS scheme of the Department of Science and Technology, India.

A. Sharma (✉) · S. Jain
National Institute of Technology, Kurukshetra, India
e-mail: abhisek_61900048@nitkkr.ac.in

S. Jain
e-mail: jasarika@nitkkr.ac.in

N. K. Jain
Zakir Hussain College, University of Delhi, Delhi, India

B. Bhargava
Purdue University, West Lafayette, IN, USA
e-mail: bbshail@purdue.edu

© The Author(s), under exclusive license to Springer Nature Singapore Pte Ltd. 2023

A. Mishra et al. (eds.), *Advances in IoT and Security with Computational Intelligence*, Lecture Notes in Networks and Systems 755,
https://doi.org/10.1007/978-981-99-5085-0_35

1 Introduction

We live in a society with multiple languages. Each language has different syntactical representations and cultural influences [1, 2]. It's still difficult to comprehend diverse languages in their proper context [3]. For language understanding, a majority of computer science researchers have employed techniques that solely take lexical differences into account [4, 5]. Language is assumed to reflect peoples' worldviews in multilingual work. Contrary claims have been made by cognitive and sociolinguistics scholars. They have discovered that linguistic distinctions go beyond vocabulary and grammar and represent differences in their cultural and moral paradigms [6, 7]. The use of explicit opinion vocabulary with regard to a particular topic (e.g., "policies that benefit the poor"), idiomatic and metaphorical language (e.g., "the company is spinning its wheels"), and other types of figurative language, such as irony or sarcasm, are the most prominent ways through which cross-cultural differences manifest themselves.

Cross-cultural representations and linguistic tools can help computers become more contextually aware. To represent and depict different cultural properties, cultural models exist. These cultural models have defined a number of dimensions and properties. There are the following set of cultural models, namely the Trompenaars Cultural Dimensions, Hofstede Cultural Dimensions, GLOBE Study Cultural Dimensions, Deal and Kennedy Culture Model, Human Synergistics Culture Model, Barrett Values Spiral Dynamics Culture Model, the Competing Values Framework, and the OCAI culture survey. Based upon these cultural models, we have identified 25 dimensions and 8 properties to define culture of a region as a unit to specifically address the leadership and the business environment. Although there is some overlap between these dimensions, properties, and areas of motivation, we have not merged them to cover a wide range of domains.

Here, we list down certain research questions catering to the cultural perspective of globalized information systems:

- Do we have a standardized cultural knowledge representation that allows systems to achieve culturally appropriate results?
- Are we able to interpret the effects of cultural variations in decisions made by computer algorithms?
- Are machine learning algorithms getting well-formed cultural information?
- Is cultural knowledge available to policy makers?
- Can we dynamically cater to User Interfaces based on cultural orientation of the user?

The remainder of the paper is organized as follows: Sect. 2 defines different cultural models; previously worked upon usecases using cultural knowledge are explained in Sect. 3; in Sect. 4, we take a business expansion usecase for a better understanding of language and culture when used together; finally, we discuss probable work direction for the future.

2 Cultural Models

Cultural models are the results of many studies over decades. Many versions of cultural models have been developed by many authors: some contain some overlapping information, as well and some have extended previous works. We have looked into some of the known models and enlisted them here:

1. Trompenaars Cultural Dimensions [8]
2. Hofstede Cultural Dimensions [9]
3. GLOBE Study Cultural Dimensions [10]
4. Deal and Kennedy Culture Model [11]
5. Human Synergistics Culture Model [12]
6. Barrett Values Spiral Dynamics Culture Model [13]
7. The Competing Values Framework and the OCAI culture survey [14].

Models have been proposed focusing on different domains. The most common one is related to Organizational operations and behaviour of employees and leaders inside organization(s). In Table 1, we have summarized these models by listing the dimensions the said model has proposed and at what domain the model was targeted.

Other than these models and dimensions, we can also have the cultural properties defined by Erin Meyer in [15]. Though we found similarities in the definition of these properties with the dimensions defined by the above models, we feel that they can provide more insights to the culture at hand.

3 Previous Usages (of Culture) and Probable Usecases

Culture has been identified as beneficial by some of the previous usecases in a number of domains. Some of these are policy making, decision-making, user interface, teaching, and business. Further expansion in these domains is possible as the current implementations are rather limited and have scope for development. Other than these domains, multilingual cultural knowledge can be used in usecases like Question-Answering, NLP-related applications (such as contextual tokenization), deep learning-related applications (such as classifications), knowledge management, and query processing.

3.1 Policy Making

Policies are one of the crucial elements of society, companies, and governments. They drive change and development of company and country. Below are some of the works that have used cultural properties in tasks related to policy development and use. The authors [16] have provided an outline of the implications of differences in

Table 1 Different cultural models

No.	Model	Dimensions	Focus
1	Trompenaars cultural dimensions	1. Universalism versus particularism 2. Individualism versus communitarianism 3. Neutral versus emotional 4. Specific versus diffuse 5. Achievement versus ascription 6. Sequential versus synchronous time 7. Internal direction versus external direction	Trompenaars was to inquire about employee's behaviour in both work and leisure time
2	Hofstede cultural dimensions	1. Power distance index (high versus low) 2. Individualism versus collectivism 3. Masculinity versus femininity 4. Uncertainty avoidance index (high versus low) 5. Long-versus short-term orientation 6. Indulgence versus restraint	Hofstede's main aim was to evaluate work values across countries
3	GLOBE study cultural dimensions	1. Uncertainty avoidance 2. Power distance 3. Future orientation (degree to which society values the long term) 4. Assertiveness orientation (masculinity) 5. Gender egalitarianism (femininity) 6. Institutional 7. Societal collectivism 8. Performance orientation (degree to which societies emphasize performance and achievement) 9. Humane Orientation (extent to which societies places importance on fairness, altruism, and caring)	Organizational behaviour
4	Deal and Kennedy culture model	1. History 2. Values and beliefs 3. Rituals and ceremonies 4. Stories 5. Heroic figures 6. The cultural network—The informal network within an organization is often where the most important information is learned. Informal players include – Storytellers – Gossipers – Whisperers – Spies – Priests and priestesses	Understanding rites and rituals in corporate culture
5	Human synergistics culture model	1. Ideal culture (values) 2. Causal factors (levels of change) 3. Operating culture (OCI norms) 4. Outcomes (effectiveness)	In the organizational context on how people work and interact with each other
6	Barrett values spiral dynamics culture model	1. Ensuring viability 2. Supporting relationships 3. Achieving excellence 4. Courageously evolving 5. Authentic expression 6. Cultivating communities 7. Living purpose	Areas of human motivations
7	The Competing Values Framework and the OCAI culture survey	1. Internal-external dimension 2. Stability-flexibility dimension	Assessing organizational culture

geographical and cultural clusters in relation with future development of AI applications. It is done through the analysis of national AI policy documents of 25 countries through topic modelling, clustering, and reverse topic-search. They have provided an overview of topics discussed in the documents. The authors have also analysed the frequency of 11 ethical principles across the corpus.

3.2 Decision-Making

Any process that has to make a choice from a number of choices is decision-making. While making a decision, the decision maker has to be aware of all the variables and in most situations culture plays a role. As one of the examples of decision-making, the authors of [17] have taken Bangladesh's case. Bangladesh is highly influenced by the social institutions of Bangladesh like high-power distance, collectivism, masculinity, and high uncertainty avoidance. The high-power distance and division among the groups persuade to maintain gaps between the mass people and the decision makers and not to express their opinion. The authors have also discussed the rationale behind decision-making using Hofstede cultural dimensions by taking Bangladesh's example of how still hierarchical/centralized administration dominates over decentralized structure. It also discusses how cultural norms determine the level of participation of the stakeholders.

3.3 User Interface

User interface (UI) is one of the crucial parts of user's experience when interacting with a software or computer system. Different user's have different preferences for the way they want to interact with the system, and it has to do with how they perceive things which is directly related with their culture and their experiences. Below are some works that have used cultural dimensions to make UI catered to the background of the user.

As stated by [18], users prefer content that is accurate in terms of context and represents cultural values and properties. UI designs that take into account the preferences, requirements, and variances that exist within various communities, this could improve the system's usability. Through Hofstede cultural dimensions used on Arab users, they have generated design guidelines which the authors then evaluate using 78 participants and stated high satisfaction among the users.

Accommodating four out of five cultural dimensions of Hofstede into the user interface design will result in an increase in usability for all users [19]. The aim was to develop user interface with cultural effectiveness. In this version of the tests, the results show the approach is somewhat viable. Although the work presented by Ford G, KotzÃ© P (2005) [19], is a hypothesis, in which the authors have stated that it needs to be further researched.

3.4 Teaching

The research is founded on the idea that different people learn in different ways, and cultural diversity may also influence how they learn and are taught.

Intelligent Learning Environments (ILE) and AI can be used to achieve individually tailored learning experiences [20]. Users' cultural background and preferences may not align with most mainstream educational systems. Through the use of ILE and AI, mainstream systems can be culturally tailored.

Using Hofstede cultural dimensions, the authors of [21] are investigating whether ERP training can accommodate cultural orientation of the user. After the analysis, they found that the current ERP learning system didn't accommodate Thai culture and is based on Western culture without accounting for other cultures. It was stated that this disassociation has affected the effectiveness of the training outcome.

In [22], communicative experiences of professors from South Africa and students from Sudan during a two-year master's course in Computers in Education was monitored. The purpose of the research was to determine the extent to which Hofstede's static quantitative research could be used as a basis for an essentially qualitative dynamic interpretation. The authors found that power distance and uncertainty avoidance tend to amplify each other, though it results in less inclination towards individualism and more towards collectivism. They mentioned that 3 elements play a key role when working with different versions: (1) construction of shared meaning, (2) reduction of communicative uncertainty, and (3) appropriate use of technology.

Hofstede cultural dimensions are used to find the suited dimensions for the localized software. Software that supports multiple languages is evaluated to allow more diversity than just changing to local language and date format [23].

3.5 Business

Businesses of today are operating worldwide and have to deal with audience/customers from a variety of nations. With this much variety, businesses constantly have to deal with issue of catering products and experiences that are suitable for all the varied users. Culture is the most significant factor that influences the business model when expanding business to new markets [24]. This paper aims to provide supporting analysis for decision-making using Hofstede cultural dimensions. The study was based on secondary data of Canada, South Korea, Germany, and Morocco. The paper concluded that it is necessary to get inclined with cultural values before entering the market.

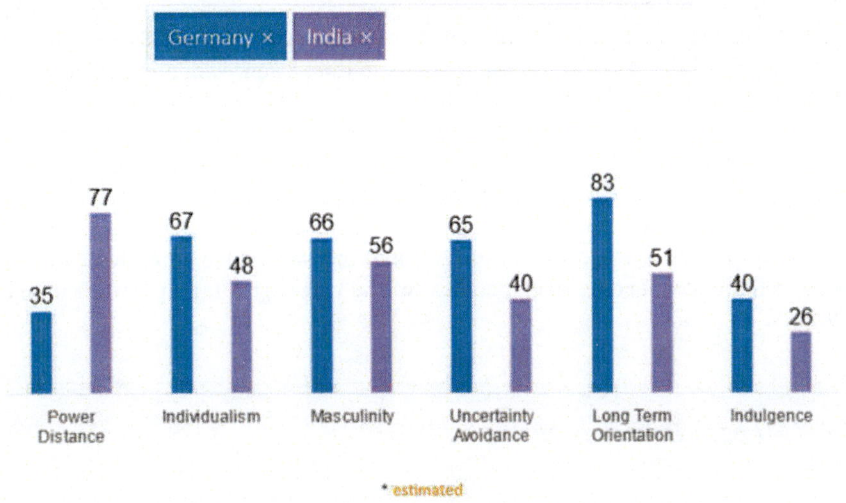

Fig. 1 Cultural difference between India and Germany (www.hofstede-insights.com)

4 Language and Culture

4.1 A Business Expansion Usecase

Let's take a look into how cultural properties can be used in the context of business expansion. Business expansion is when a business is expanding to new markets, often to new countries. At this point, it is really crucial for people responsible for the expansion to be aware of the language as well as cultural differences their current country and the target country have. Let's suppose a German company is planning to expand in India, then the company must account for differences specially in dimensions like power distance, individualism, uncertainty avoidance, and long-term orientation. Other dimensions are also present but their difference is comparatively minimal. We are aware of the language differences as both of them have different base languages. German in case of Germany. In case of India, there are many languages, though Hindi is the most commonly used language. Figure 1 shows the cultural difference as defined by Hofstede cultural dimensions.[1]

As we can see in Fig. 1, India has a score of 77 in power difference that means a more steep top-down structure in society and organizations compared to Germany. Germany's score in individualism is 67, meaning they focus more on parent-children relation rather than aunts and uncles. Though it is different in the case of India, along with direct relations it also focuses on other relations (like aunt, uncle). Meaning of Uncertainty avoidance is defined as "The extent to which the members of a culture feel threatened by ambiguous or unknown situations". In this, Germany is on the

[1] www.hofstede-insights.com/country-comparison/germany,india.

higher end with a score of 65, so their preference for uncertainty avoidance is higher than that of India. Last is long-term orientation dimension, which means how the members of the society refer past experiences to deal with current situations. In this, India has a considerably lower score than Germany, which means Germany's people are more connected with their past for handling current situations than India.

Similar distinction between cultures of other countries can also be found from the studies conducted by the researchers who have devised these cultural models, and each one of these models provides insights into one way or the other that can help in the better understanding of the context of the text depending upon the cultural diversity.

4.2 A Legal Usecase

The law has language at its heart; therefore, it is not unexpected that some aspects of the legal profession have used natural language processing software for a long time. However, during the past few years, there has been a growth in interest in using contemporary methods to solve a larger variety of issues [25].

There are generally five legal systems considered in the world today: civil law, common law, customary law, religious law, and mixed legal systems. The summarization from university of South Carolina's Law Library[2] shows a legal system adopted based on countries. These differences that we see are not only because of the language difference, they are because each and every nation has a different history which shapes the culture of the nation. Which in turn creates differences between nations' view towards law and order, and how they try to solve societal disturbances with it.

5 Discussion and Future Plans

Though there are many usage examples of cultural dimensions in the context of computer technologies and other domains, the literature is still rather limited and the usage and association of cultural dimensions have to be further assessed to understand the significance of it in various domains. In future, a knowledge graph of some sort can be helpful in incorporating cultural information in a computer understandable form that can enable information systems to take culturally aware decisions.

[2] https://guides.law.sc.edu/c.php?g=315476&p=2108388.

6 Conclusion

This paper briefly highlights the current state and lack of language understanding in the correct context in the present information system's tools and techniques. This paper also tries to answer questions that can lead to the development of culturally and contextually aware language understanding in information systems. Contextual awareness of text has more that just the lexical properties of the language; it also has cultural influence over it. The cultural models and the dimensions/properties defined by them have been derived over various studies and have the capability to represent different cultures. This coexistence of culture and language has been tested in some domains like policy making, decision-making, etc. and was found to be effective. Further development of language tools with culture has potential and can improve the contextual understanding of text.

References

1. Jain S, Kysliak A (2022) Language-agnostic knowledge representation for a truly multilingual semantic web. Int J Inf Syst Model Des (IJISMD) 13(1):1–21
2. Jain S (2021) Multilingual and multimodal access. In: Understanding semantics-based decision support. Chapman and Hall/CRC, pp 117–130
3. Sharma A, Jain S (2021) Multilingual semantic representation of smart connected world data. In: Smart connected world: technologies and applications shaping the future, pp 125–138
4. Sharma A, Patel A, Jain S (2022) LSMatch and LSMatch-multilingual results for OAEI
5. Sharma A, Jain S, Trojahn C (2022) Multifarm11-extending the multifarm benchmark for Hindi language
6. Kövecses Z (2004) Introduction: cultural variation in metaphor. Eur J Engl Stud 8(3):263–274
7. Lakoff G, Wehling E (2012) The little blue book: the essential guide to thinking and talking democratic. Simon and Schuster
8. Smith PB, Dugan S, Trompenaars F (1996) National culture and the values of organizational employees: a dimensional analysis across 43 nations. J Cross Cult Psychol 27(2):231–264
9. Hofstede G (1980) Culture's consequences: national differences in thinking and organizing. Sage, Beverly Hills, California
10. House RJ, Hanges PJ, Javidan M, Dorfman PW, Gupta V (eds) (2004) Culture, leadership, and organizations: the GLOBE study of 62 societies. Sage Publications
11. Deal TE, Kennedy AA (1983) Corporate cultures: the rites and rituals of corporate life. Addison-Wesley (1982). ISBN: 0-201-10277-3. Business Horizons 26(2):82–85
12. https://www.humansynergistics.com/
13. https://www.valuescentre.com/barrett-model/
14. https://www.ocai-online.com/about-the-Organizational-Culture-Assessment-Instrument-OCAI
15. Meyer E (2014) The culture map: breaking through the invisible boundaries of global business. Public Affairs
16. van Berkel N, Papachristos E, Giachanou A, Hosio S, Skov MB (2020) A systematic assessment of national artificial intelligence policies: perspectives from the Nordics and beyond. In: Proceedings of the 11th Nordic conference on human-computer interaction: shaping experiences. Shaping Society, pp 1–12
17. Dutta B, Islam KM (2016) Role of culture in decision making approach in Bangladesh: an analysis from the four cultural dimensions of Hofstede. Bangladesh e-J Sociol 13(2)

18. Alsswey A, Al-Samarraie H (2021) The role of Hofstede's cultural dimensions in the design of user interface: the case of Arabic. AI EDAM 35(1):116–127
19. Ford G, Kotzé P (2005) Designing usable interfaces with cultural dimensions. In: IFIP conference on human-computer interaction, pp 713–726. Springer, Berlin, Heidelberg
20. Mohammed PS, Nell'Watson E (2019) Towards inclusive education in the age of artificial intelligence: perspectives, challenges, and opportunities. In: Artificial intelligence and inclusive education. Springer, Singapore, pp 17–37
21. Chayakonvikom M, Fuangvut P, Cannell S (2016) Exploring education culture by employing Hofstede's cultural dimensions to evaluate the effectiveness of the current ERP training approach in Thailand. J Educ Train Stud 4(10):79–89
22. Cronjé JC (2011) Using Hofstede's cultural dimensions to interpret cross-cultural blended teaching and learning. Comput Educ 56(3):596–603
23. Ping TP, Chan CP, Sharbini H, Julaihi AA (2011) Integration of cultural dimensions into software localisation testing of assistive technology for deaf children. In: 2011 Malaysian conference in software engineering. IEEE, pp 136–140
24. Al-Alawi A, Alkhodari HJ (2016) Cross-cultural differences in managing businesses: applying Hofstede cultural analysis in Germany, Canada, South Korea and Morocco. Elixir Int Bus Manag 95(2016):40855–40861
25. Dale R (2019) Law and word order: NLP in legal tech. Nat Lang Eng 25(1):211–217

Convolutional Neural Network-Based Quality of Fruit Detection System Using Texture and Shape Analysis

Om Mishra, Deepak Parashar, Harikrishanan, Abhinav Gaikwad, Anubhav Gagare, Sneha Darade, Siddhant Bhalla, and Gaurav Pandey

Abstract This research describes an excellent method for detecting fruits' quality using convolutional neural networks. Fruit grading is done by inspections, experiences, and observation. To rate the quality of fruits, the proposed system employs machine learning techniques. Shape and color-based analysis methods are used to grade two-dimensional fruit depictions. Different fruit photos may have identical color, size, and shape qualities. As a result, utilizing color or form property analysis methods to identify and differentiate fruit photos is ineffective. As a result, we combined a size, shape, and color-based method with a CNN to improve accuracy and precision of fruit quality recognition. The advisable system begins the process by selecting the fruit images. The image is then sent to the rectification stage, where fruit sample properties are retrieved. Subsequently, fruit images are trained and tested using a CNN. The convolutional neural network is used in this proposed paper to abstract out colors, size, and shape of the fruits and results achieved with a combination of these features are quite promising.

Keywords CNN · Image processing · Image classification · Raspberry Pi

1 Introduction

Agriculture is one of India's most important economic sectors, and it is crucial to the country's economic success. Human experts continue to evaluate fruits in India as traditional. Mankind wastes a substantial amount of money and time checking the quality of the fruits in the fields [1].

In this study, we look into a reliable and pocket-friendly method for determining fruit quality based on size, shape, and color. Because fruits are highly breakable,

O. Mishra (✉) · D. Parashar · Harikrishanan · A. Gaikwad · A. Gagare · S. Darade · S. Bhalla
Symbiosis Institute of Technology, Symbiosis International (Deemed) University, Pune, India
e-mail: om.mishra83@gmail.com

G. Pandey
Netaji Subhas University of Technology, Delhi, India

© The Author(s), under exclusive license to Springer Nature Singapore Pte Ltd. 2023
A. Mishra et al. (eds.), *Advances in IoT and Security with Computational Intelligence*, Lecture Notes in Networks and Systems 755,
https://doi.org/10.1007/978-981-99-5085-0_36

they should be tested in non-damaging ways. When it comes to fruit sizing, the most important physical attribute is its hue, which offers visual property.

As an outcome, the classification of fruit quality is critical for expanding market share and creating higher performance criteria. If classification and grading are done manually, the procedure will be slow and occasionally prone to errors. Humans classify the freshness of fruit based on its color, size, and other characteristics [2].

If these freshness measures are mapped into an automatic system using the Python, work will be completed more quickly and errorless. This increases the speed and decreases the cost of the fruit sorting process [3]. The shape and color of the fruit are critical for perceptible checking. Structure independent of structure for fruit grading based on quality must be capable of identifying both criteria accurately. The shape of the fruits may be clearly determined from their digital photograph [4]. Color identification, on the other hand, combines many mental and physical concepts, making it challenging to effectively represent and process color in that picture. There are numerous degrees of color systems available for classifying fruits depending on color. Software development is critical in this color-based classification system [5].

2 Literature Survey

In the previous work for exact fruit identification, features are found out from segmented images. To capture noticeable and near IR images, Ref. [1] built and experimented with a multi-spectral camera. This paper aimed toward the replacement of a manual inspection system using image processing.

The method [2] used object recognition of standard of fruit by seeing bad and excellent supported quality fruit by using MATLAB in image processing by generating accuracy. This paper aimed toward the maturity of disease and maturity using ANN in the fruit domain.

In other work, researchers used Quality Control of Fruit using Hyperspectral imaging of calyx and stem discrimination on apple, i.e., it helps to avoid the error detection produced by the presence of stems and calyxes, which are most trending problem in control of food quality. The Apple defects were detected and evaluated by Hyperspectral image processing algorithms by the images of the VIS region [3]. This paper aimed toward Quality Control of Fruit using Hyperspectral imaging of calyx and stem discrimination on apple.

To enhance the robustness of the detection method, Ref. [4] used LBP-SVM that helped filter out the false-positive detections generated by background chaff interferences. This paper aimed toward the tomato grading and also sorting by using machine vision approach based on a microcontroller using classifier. The proposed system for the classification of tomatoes consists of three stages, utilizing digital images of the samples captured in experimental setups and deployed using microcontrollers. The system, along with a comparative analysis of some similar methods, demonstrates the efficacy of the proposed approach in grading and sorting tomatoes.

From the literature survey, we found that Deep learning algorithms are more effective in the fruit industry, particularly for applications in fruit quality monitoring.

In the proposed method to evaluate the color, size, and shape of the fruit, the entire structure is created using the neural network approach [6]. Color is the most essential aspect in fruit classification, yet many fruits have similar colors, thus their size helps us solve this type of difficulty. The size and color-based classification include physically observing the surface of the fruit to obtain the essential data and classifying it to the appropriate grade.

The paper is described in different sections where Sect. 1 and 2 explained Introduction and Literature survey, respectively. The proposed methodology has been discussed in Sect. 3. The experimental results and conclusions have been explained in Sects. 4 and 5, respectively.

3 Methodology

The Conventional Neural Network: A CNN is a network designed for DL algorithms that are primarily used for pixel data processing applications and image recognition. There are various forms of neural networks (NN) in deep learning, but Convolutional Neural Network (CNN) is the most preferable network architecture for detecting and identifying objects. The proposed method is illustrated in the block diagram shown in Fig. 1.

The convolutional NN is utilized to save spatial relationship between pixels of varying sizes. Mathematical operations are stacked on top of each other to form network layer. CNN algorithm works by avoiding overfitting and it enhances the generalization of networks, i.e., the number of trainable network parameters is reduced in it. CNN is on the extracted features the output of model will be highly reliant and organized because of classification and feature extraction.

In the methodology, we involve the following steps, mentioned in the flowchart in Fig. 2.

1. Firstly, detect the presence of fruit using an IR proximity sensor.
2. After object detection we will be capturing the image of fruit whose quality we want to detect.
3. Compare the image of the fruit with the dataset.
4. The fruit picture will be resized to the standard size.
5. Then convert the RGB image into Grayscale.
6. Then by using a gas sensor we will capture the smell of the fruit.
7. Then if all the parameters indicate "freshness" then it will show that "Normal".
8. If the fruit is rotten, it will trigger a buzzer for around 1–2 s, and a red LED will light up along with the buzzer. A comparison of the proposed method with existing methodologies is provided in Table 1.

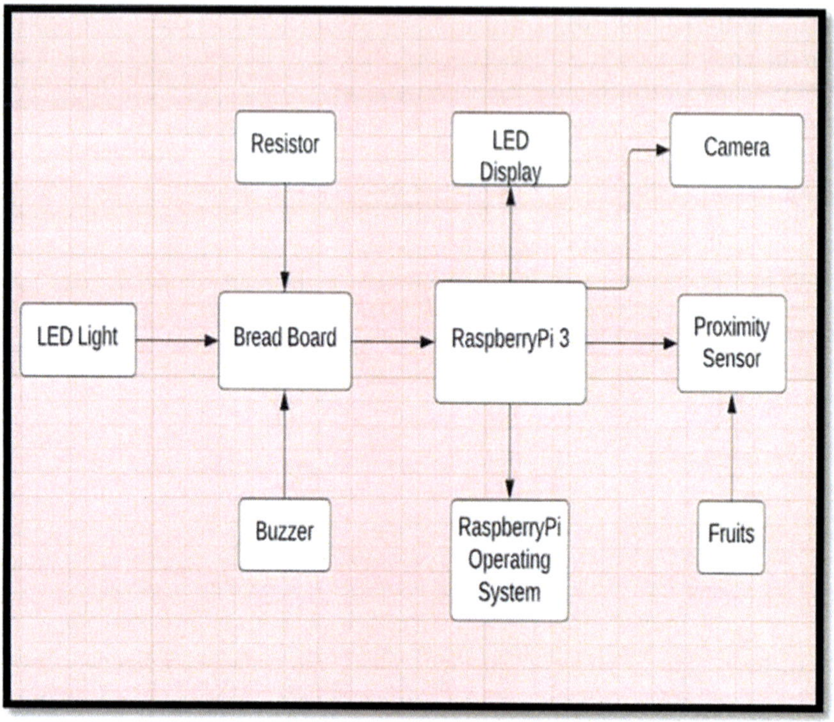

Fig. 1 The block diagram of the proposed method

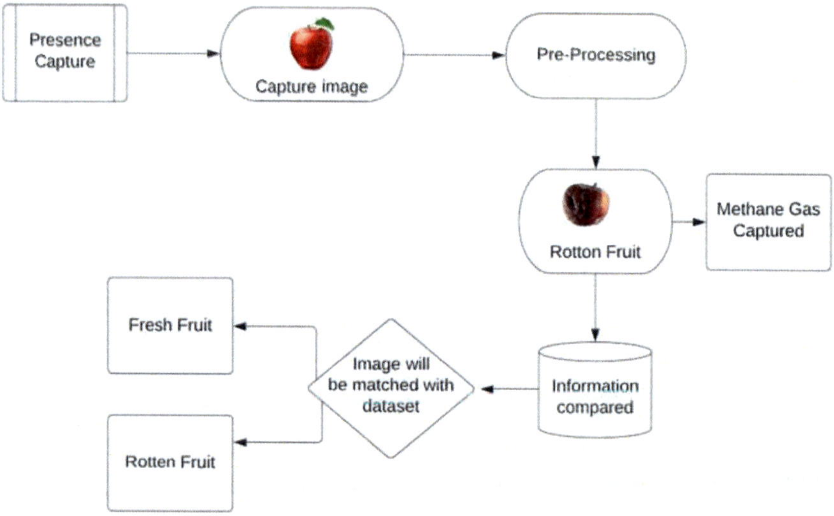

Fig. 2 Flowchart of the proposed methodology

Table 1 Comparison of the proposed method with existing methodologies

Existing methods	Proposed method
It's been observed from the literature survey that the methodologies used for a similar project were based on "SVM" [7–9]	For our project on the Fruit quality detection system, we have implemented the latest ML (Machine Learning) technologies in the place of old ML tech
The use of deep learning wasn't found	Here we used advanced deep learning methods and made prominent use of CNN (convolutional neural network)
Upon survey, it has been observed that the projects made before were made using the ANN (artificial neural network) method	For better implementation and working of a similar project, we used the CNN method with a customized and advanced code that is compatible and very quick at the resulting efficiency even at Raspberry Pi 3
The image processing part is the most important aspect of any quality detection system, upon which we realized that it isn't that efficient, which gave a low accuracy [10, 11]	For the image processing part, we made use of digital image processing, which gave us the picture quality the clarity which all together helped in the processing of the image. This was later compared to our model where the model consisted of sample images, the digital image processing helped us with getting visualization, recognition, sharpening, restoration, and pattern recognition
The use of gas sensor was nowhere reported in the previous models, which is a very important component in the detection system of any fruit, vegetable, or any consumable quality detection system	Considering all the literature survey's here before we made prominent use of the gas sensor which is used for detecting methane. Methane gas is mostly present and released from rotten fruits, which tells us whether the fruit is rotten or not

4 Experimental Results

In this project, our primary goal is to present new prototypes for detecting the freshness in fruits. Today, people sometimes do not consider how fresh the fruits they are consuming actually are. Thus, this prototype is made with the several sensors such as a Proximity sensor, and a Gas sensor, Raspberry Pi 3. Operating System used is Raspberry Pi OS with scripting language Python 3.7. In this, we have used deep learning library TensorFlow 1.15.2. Terminal Emulator used is PuTTy. The text editor used is Jupyter Notebook and Anaconda, while the remote PC software used is VNC Viewer. The state chart diagram is shown in Fig. 3.

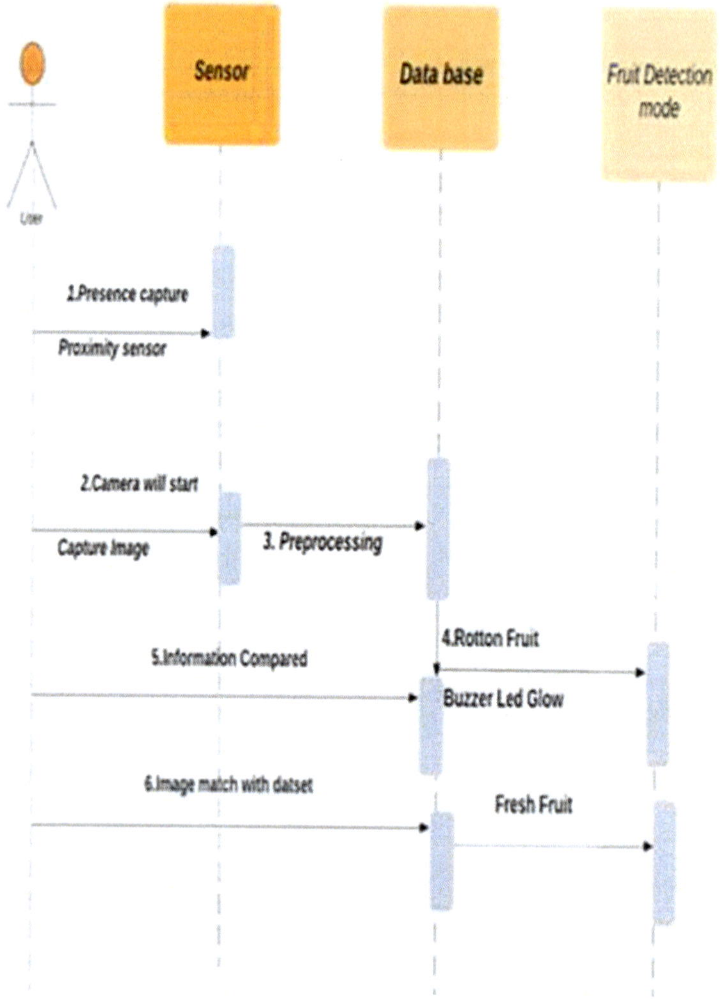

Fig. 3 Statechart diagram

4.1 Raspberry Pi 3

We used Raspberry Pi, a series of small single-board computers developed in the UK in partnership by the Raspberry Pi Foundation and Broadcom. The Raspberry Pi is used in different fields such as weather monitoring, patient monitoring, crop monitoring, etc. The main reason for the popularity of the Raspberry Pi is that it reduces the cost of the system significantly. It also has simple architecture. Jupyter Notebook and Anaconda are often used by computer and electronics hobbyists due

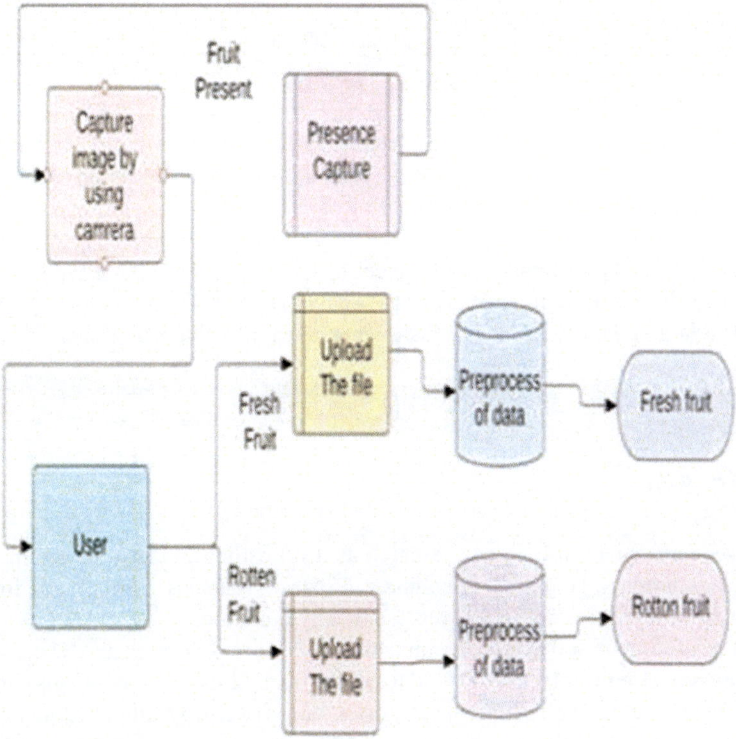

Fig. 4 Data flow diagram (a): level 1

to their support for HDMI and USB standards. The dataflow diagram is shown in Fig. 4.

4.2 *Proximity Sensor*

A transmitter and receiver are included in E18-D80NK Adjustable IR Sensor Proximity Switch 3–80 cm Range photoelectric switch sensor. According to demand, the detection distance can be changed. It has 3–80 cm detecting ranges. The Adjustable Infrared Sensor Switch is a versatile device that can be used in interactive media, industrial assembly lines, a robot to avoid obstacles, and many other situations. It is compact, affordable, simple to use, and simple to assemble. Depending on the obstructions, different switching signal outputs are produced. When there are no impediments, it stays high, and when there are difficulties, it stays down. The probe has a bright light to detect the range of 3–80 cm.

4.3 Gas Sensor

Measuring the amount of methane gas in the air, whether at home or at work, is mostly done with a gas sensor of the Metal Oxide Semiconductor type, or MQ4. This sensor has a stainless steel mesh-encased aluminum-oxide foundation and tin dioxide coating on a ceramic-based detecting element. Every time gas makes contact with the sensing device, its resistance changes. The change is then measured in order to determine the gas concentration. Its detecting range of 300–10,000 PPM is suitable for gas leak detection. Methane burns in a highly exothermal manner and, if ignited, releases a large quantity of heat, which, when used in a controlled manner, can be helpful.

4.4 Dataset

It plays a very crucial role in any research as it is costlier and hard to obtain. That is why our primary goal was to engineer a dataset that has similar fruit images containing different feasible fruits represented by n numbers.

At a resolution of 360 pixels, the images were captured through an HD camera's, from a practical point of views. The dataset has many images to compare with the fruits. The dataset has two different categories, one is of fresh fruits and another is of rotten fruits. In both categories, the images of fruit have been captured from different angles so that when some fruit is placed at any angle it can understand the quality of the fruit. The dataset also contains information about the smell of rotten fruit. According to a scientific fact when fruit gets rotten, they produce a special kind of gas called Ethylene.

This section details the fruit sample trials and quality detection, as well as a previously unknown fruit. We collected two distinct fruit samples ranging in size, color, and form and compared them to the system's output to get a better understanding of the project. On the basis of the experimental results, the merits of the proposed method are as follows. (1) Anyone who consumes edibles is aware of quality of the fruit before they are going to buy them. (2) Waste-related pollution can be decreased. (3) It can quickly determine freshness of any edibles. The proposed method can quickly determine the freshness of any edibles and assist food businesses in obtaining high-quality fruits for the production of jams and ketchup. Fresh and rotten apple fruit, Implementation setup, and Output are shown in Figs. 5, 6, and 7, respectively.

Fig. 5 Fresh and rotten apple fruit

Fig. 6 Implementation setup

5 Conclusion and Future Scope

This research paper basically focuses on use of the CNN in the agriculture and food industry. Appearance is a highly important characteristic of agricultural products.

For restoring experimental examination, CNN is very useful because it gives us actual classification. It is an advisable method to recognize fruit quality with a number of real-world problems in the dataset to increase functionality. Our results showed that the system is perfect. We can also modify the design in future by adding a conveyor belt in the project, by adding it we can increase the accuracy and fruit

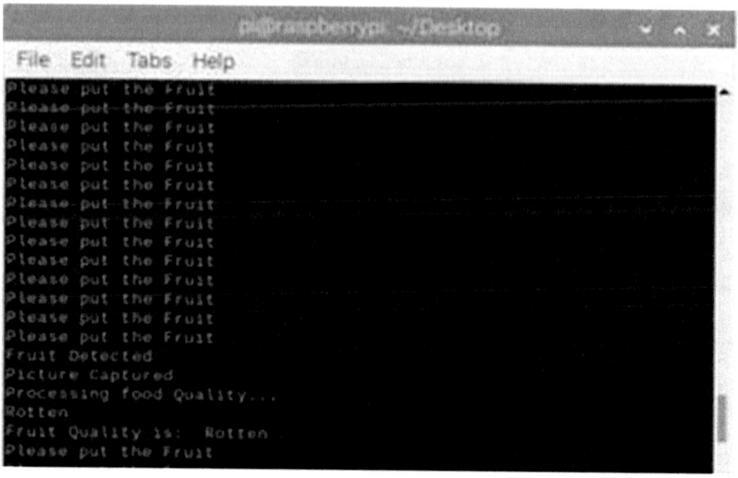

Fig. 7 Output

quality inspection we can do. By doing this we can differentiate between the hybrid colors from original colors. We can also count the number of fruits passed. It could be also our future work. By using data association, we will distinguish new image frame and previous image frames, to identify the fruit we will use tracking, feature matching, association tech., if the fruit is already seen from the frame or not, we can do this task.

References

1. Samajpati BJ, Degadwala DS (2016) Hybrid approach for apple fruit diseases detection and classification using random forest classifier. In: International conference on communication and signal processing (ICCSP), Melmaruvathur, India, pp 1015–1019
2. Satpute MR, Jagdale SM (2016) Automatic fruit quality inspection system. In: International conference on inventive computation technologies (ICICT), Coimbatore, India, pp 1–4
3. Chandini AA, Maheswari B (2018) Improved quality detection technique for fruits using GLCM and multiclass SVM. In: International conference on advances in computing, communications and informatics (ICACCI), Bangalore, India, pp 150–155
4. Behera S, Sangita S, Rath A, Sethy P (2019) Automatic classification of Mango using statistical feature and SVM. Springer Nature Singapore Pte. Ltd., pp 137–189
5. Kaur S, Girdhar A, Gill J (2018) Computer vision-based Tomato grading and sorting. J Gill RIMT IET, Mandi Gobindgarh, India © Springer Nature Singapore Pte. Ltd. Kolhe ML et al (eds) Advances in data and information sciences. Lecture notes in networks and systems, vol 38
6. Fruit maturity and disease detection using artificial neural network. E-ISSN: 2582-5208, volume:02/issue:09/September (2020)
7. Patel H, Prajapati R, Patel M (2019) Detection of quality in Orange fruit image using SVM classifier. In: 3rd international conference on trends in electronics and informatics (ICOEI), Tirunelveli, India, pp 74–78

8. Dubey S, Jalal A (2012) Detection and classification of Apple fruit diseases using complete local binary patterns. In: Third international conference on computer and communication technology, Allahabad, India, pp 346–351
9. Komal K, Sonia S (2019) GLCM algorithm and SVM classification method for Orange fruit quality assessment. Int J Eng Res Technol (IJERT) 8(9)
10. Pineda I, Alam N, Gwun O (2019) Calyx and stem discrimination for Apple quality control using hyperspectral imaging. Springer Nature Switzerland AG, pp 274–287
11. Dhakshina S, Esakkirajan S, Keerthiveena A (2020) Microcontroller based machine vision approach for tomato grading and sorting using SVM classifier. Microcontroll Microsyst 76:103090

A Hybrid Approach for Spontaneous Emotion Recognition in Valence–Arousal Space

Gyanendra K. Verma

Abstract Affect analysis has been dominated by two seemingly opposing theories: the categorical approach, which is based on six universal discrete emotions. The dimensional approach is based on 2D/3D valence, arousal, and dominance. This paper proposes a hybrid emotion recognition approach using ResNet152 and eXtreme Gradient Boosting on the DEAP dataset. Firstly, the EEG data has been transformed into a scalogram using a discrete wavelet transform. We propose an enhanced emotion classification model using ResNet150 and eXtreme Gradient Boosting (DLXGB) on the DEAP database of EEG signals. We transformed the EEG signals into a Scalogram, a spectral presentation of EEG signals. The features are extracted using the ResNet150 version two model and combined with a robust gradient boosting classifier. We have experimented with twelve emotions: Cheer, Depress, Happy, Sad, Love, Shock, Exciting, Hate, Fun, Melancholy, Mellow, and Terrible. We have obtained promising results with the highest accuracy of 85%. The results show the potential to use the proposed method for emotion classification using scalograms.

Keywords Spontaneous emotion · Scalograms · ResNet150 · XGBoost · Valence · Arousal

1 Introduction

Affective computing deals with the systems and devices that can improve Man–Machine interaction by interpreting and processing emotions. The wide applications of affective computing made it viable for various applications, from behavior analysis to decision-making. Emotions are essential in humans due to their linkage to neurophysiologic features related to a coordinated set of reactions, including verbal, behavioral, physiological, and neurological processes.

G. K. Verma (✉)
Department of Information Technology, National Institute of Technology Raipur, Raipur, India
e-mail: gkverma.it@nitrr.ac.in

© The Author(s), under exclusive license to Springer Nature Singapore Pte Ltd. 2023

A. Mishra et al. (eds.), *Advances in IoT and Security with Computational Intelligence*, Lecture Notes in Networks and Systems 755,
https://doi.org/10.1007/978-981-99-5085-0_37

Emotions can be represented in two ways, namely categorical and dimensional approaches. The categorical approach deals with discrete emotions that can be universally detected through facial emotions; however, dimensional emotions are based on emotion primitives such as valence, arousal, and dominance. The most commonly used 2D/3D emotions are valence–arousal and dominance. Valence denotes pleasantness from highly positive (pleasure) to extremely negative (displeasure). However, arousal represents the intensity of emotion, i.e., eagerness (positive) to calm (negative). Dominance is the degree of control that a stimulus exerts [1].

The electroencephalogram (EEG) measures brain activity using electrodes mounted on the head that measures the potential difference in a conductive medium [2]. According to current research, human emotion may be detected via peripheral physiological signals such as electroencephalograms. An EEG signal is a 1D signal; however, we can transform an EEG signal into a spectral representation called a 'Scalogram.' A scalogram is a time–frequency representation of an EEG Signal [3]. A few samples of scalograms are illustrated in Fig. 1.

This paper proposes a hybrid approach for affect analysis using ResNet and XGBoost model. We have used DEAP [4], a benchmark database for spontaneous emotion recognition. The performance of both models is compared during the training, validation, and testing phases.

The remainder of the paper is structured as follows: The relevant work in a related field is presented in Sect. 2. The information and procedures used in this investigation are thoroughly described in Sect. 3. The results and comments from the experiment are given in Sect. 4. The conclusion of the work and future endeavors are presented in Sect. 5.

2 Related Work

C-RNN, a hybrid deep learning model that combines CNN and RNN, was introduced by Xiang Li et al. [5] for emotion recognition. They utilized continuous wavelet transform and frame construction as pre-processing steps before applying deep learning architectures. The performance for the arousal and valence aspects in this experiment is 74.12% and 72.06%, respectively.

Deep CNN end-to-end learning on the DEAP dataset was used by Lin et al. [6] to present a strategy for categorizing emotional states [6]. The input data is converted into a grayscale image format that includes time and frequency domain data, and the features were retrieved before the AlexNet model was trained on them. In this investigation, the accuracy of the arousal and valence models was 87.30% and 85.50%, respectively.

Li et al. [7] used the DEAP dataset together with a combination of CNN and LSTM RNN (CLRNN) for their emotion recognition test. Before processing, the dataset is transformed into a multi-dimensional image series. With the hybrid neural networks described in the study, the average accuracy of each person's emotion categorization is 75.21%.

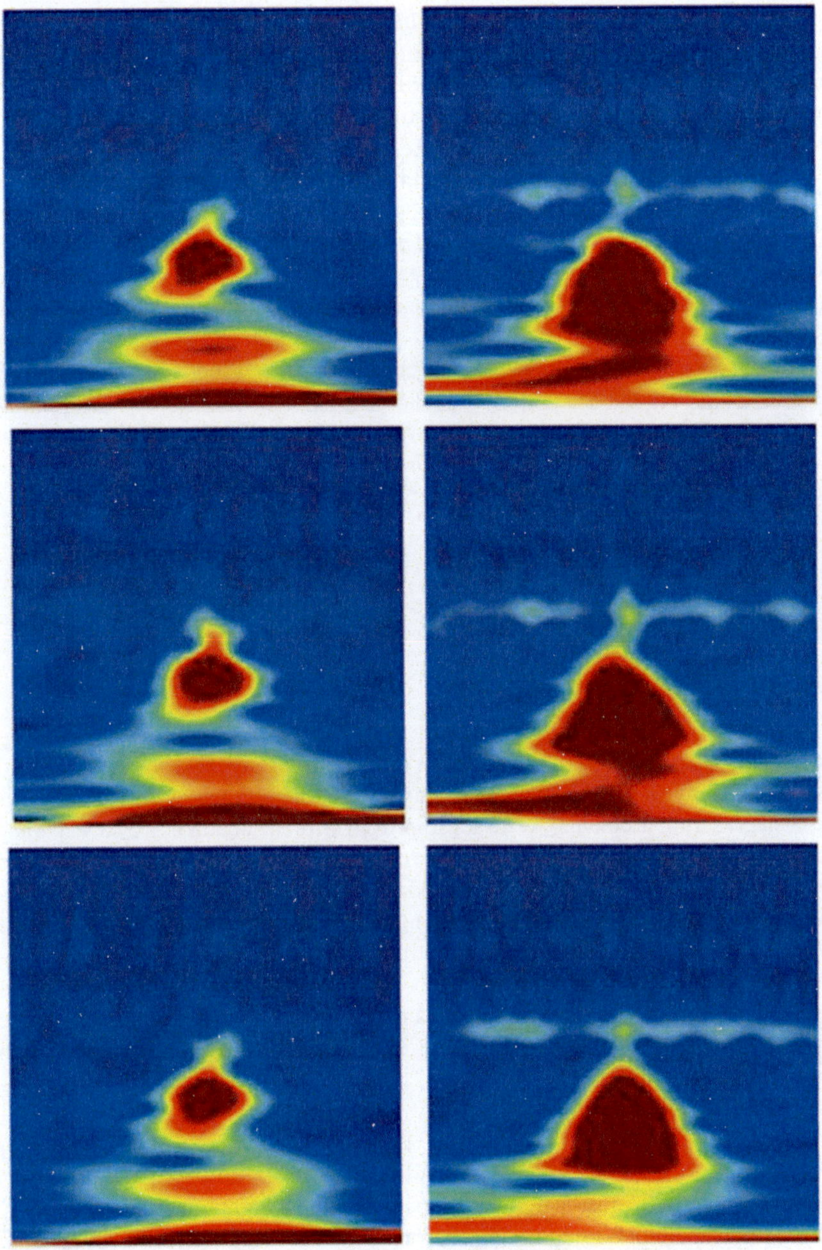

Fig. 1 Samples of scalograms

Using statistical learning approaches like Bayesian Classification, Chung and Yoon (2012) [8] concentrated on classifying DEAP data into Valence and Arousal classes. Two or three groups have been created from the valence and arousal data. High/low valence and high/low arousal distinguish two classes, but low/normal/high valence and arousal distinguish three classes.

The EEG signal was shown by Rozgic et al. (2013) [9] as a series of overlapping segments. A new non-parametric Nearest-Neighbor model combined segment-level data with response-level characteristics. They have employed Kernel PCA dimensionality reduction as a pre-processing step. Numerous techniques were used to classify the data, including NBN Neighbors, Nearest Neighbor Voting, and RBF SVMs.

When analyzing EEG signals, Candra et al. (2015) [10] looked at how window size affected the results. According to the authors' conclusions, a vast window will increase the information load, so the feature needs to be varied with alternative information. Similarly, if the temporal window is too small, it will not be easy to mine the information associated with emotion.

3 Materials and Methods

3.1 ResNet Architecture

The ResNet architecture [11] is effective in promoting hyper-DNN training and improving accuracy. The CNN depth is where the ResNet function evolves, and this causes degradation problems. Accuracy improved with the depth; however, accuracy decreased downsides with the depth improvement. The error would be ignored when a deeper network does not provide the maximum training sample error, suffers from saturation accuracy, and is composed of congruent mapping layers. The fundamental concept of using congruent mapping is to transfer the current result to the following layer. ResNet uses the residual block to fix the vanishing gradient and degradation problems in CNNs. These ResNet blocks carry out the remaining operations while considering the residual block's input and output. The following gives the expression for the residual function:

$$y = F(x, W) + x \tag{1}$$

where x designates the residual block's input, W is the weight of the residual block, and y denotes the block's output. The ResNet network comprises different residual blocks with different convolution kernel sizes. Some of the classical variants of ResNet are Ret-Net18, RcstNct50, and RestNet101. Figure 2 illustrates ResNet150 architecture.

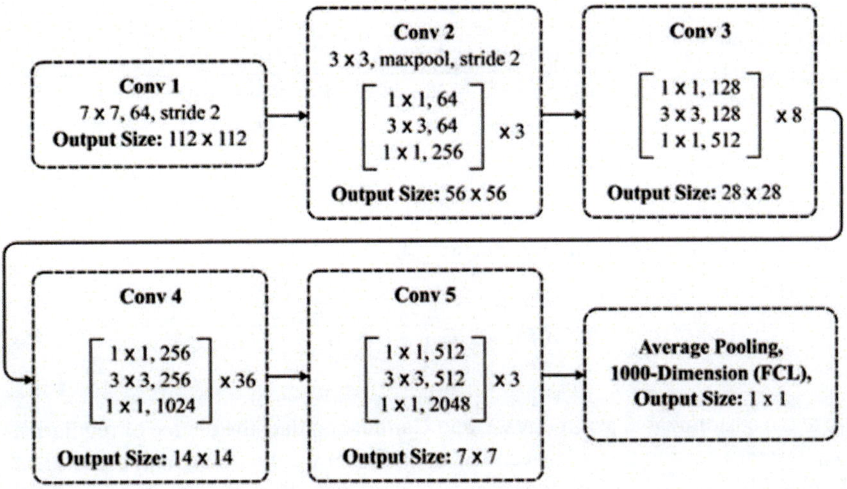

Fig. 2 ResNet150 architecture [12]

3.2 Extreme Gradient Boosting

An emerging tree-based method called Extreme Gradient Boosting (XGBoost) has gained favor for classifying data. It has been demonstrated to be a very successful strategy for classifying data [13]. When performing classification and regression tasks using machine learning, the XGBoost is a highly scalable end-to-end tree-boosting method [14]. The XGBoost classifier has been used in place of the ResNet150's Fully Connected Layer (FCL). The creators of this method, Chen and Guestrin, have provided a detailed explanation of their conceptual framework. We have summarized the computations and definitions in the following section because this method is new.

The first is a classification and regression tree ensemble approach (CARTs) with k nodes. The class label Yi's final prediction output is computed based on the overall prediction scores at a leaf node fk for each tree Kth. Below is presented Eq. 2:

$$\hat{y}_i = \varphi(x_i) = \sum_{k=1}^{K} f_k(x_i), \quad f_k \in F, \tag{2}$$

where F is the collection of all K scores for all CARTs, and Xi is the training set. The results are then improved by applying a regularization step, as indicated in Eq. 3:

$$L(\varphi) = \sum_i l(\hat{y}_i, y_i) + \sum_k \Omega(f_k), \tag{3}$$

where l stands for the differentiable loss function, which is defined by comparing the target Yi and Yi-hat class labels' respective error values, and the second section

Table 1 Dataset structure

Array name	Array shape	Array contents
Data	40 × 40 × 8064	Video/trial × channel × data
Labels	40 × 4	Video/trial × label (V, A, D, L)

applies OMEGA penalization to the model complexity to prevent over-fitting issues. Equation 4 calculates the penalty function (3):

$$\Omega(f) = \gamma T + \frac{1}{2}\lambda \sum_{j=1}^{T} w_j^2, \tag{4}$$

where the customizable parameters Y and Gamma regulate the degree of regularization, T stands for the tree's leaves, and w holds the weights assigned to each leaf. The classification problem is then successfully solved using Gradient Boosting (GB), which is expanded by a second Taylor expansion.

4 Proposed Model

4.1 Dataset

In this work, all the experimentation was performed on the DEAP database [4], a benchmark database of physiological and peripheral signals like EEG, EMG, EOG, etc. The dataset is compiled through 32 subjects by showing video clips of different emotions. The subjects rated the video in terms of arousal, valence, likes and dislikes, dominance, and familiarity on a continuous scale. The DEAP dataset organization is given in Table 1.

4.2 Proposed Methodology

To acquire a better training outcome, we must first analyze the DEAP dataset before feeding it into our suggested model. Our pre-processing data methods will be detailed in depth in the following sections. This part will go through the specifics of the datasets utilized in this study and how the data will be categorized. The recommended approach for the experiment will then be thoroughly explored. The proposed hybrid framework is illustrated in Fig. 3.

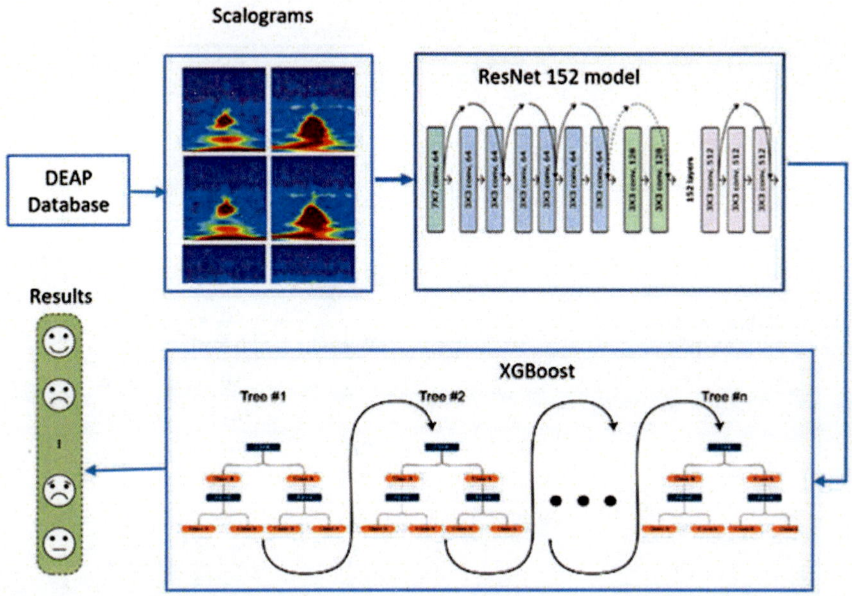

Fig. 3 Proposed methodology

5 Experiments and Results

It is crucial to test our model to comprehend the system's quantitative performance to demonstrate the efficacy of our proposed system for emotion analysis. Standard statistical metrics have been employed, including the F1 score, precision, recall, and accuracy. Every experiment was run on a scalogram made from the DEAP database. Images were used to train and evaluate the suggested hybrid model for multi-class categorization. For classification, we have chosen four categories of emotions.

The emotions are grouped into three sets for the experimentation by considering four different emotions in each set. The database structure with the number of Scalograms used in training and testing is illustrated in Table 2.

The experiments are performed with three sets of databases, and corresponding results are shown in Tables 3, 4, and 5. All the results shown are with weighted average performance. Confusion matrix for different test sets are presented in Figs. 4, 5, and 6.

Table 2 Dataset organization

Set	Emotions	#Scalograms (Train)	#Scalograms (Test)
Set-1	Cheer, Depress, Happy, Sad	128	50
Set-2	Love, Shock, Exciting, Hate	128	40
Set-3	Fun, Melancholy, Mellow, Terrible	128	40

Table 3 Hybrid model performance with Test set-1

	Precision	Recall	F1-score
1	0.84	0.79	0.81
2	0.84	0.81	0.82
3	0.81	0.88	0.84
4	0.85	0.85	0.85
Weighted Avg	0.83	0.83	0.83

Table 4 Hybrid model performance with Test set-2

	Precision	Recall	F1-score
1	0.88	0.85	0.86
2	0.89	0.75	0.81
3	0.82	0.90	0.86
4	0.81	0.88	0.84
Weighted Avg	0.85	0.85	0.85

Table 5 Hybrid model performance with Test set-3

	Precision	Recall	F1-score
1	0.76	0.75	0.76
2	0.80	0.63	0.71
3	0.76	0.87	0.81
4	0.74	0.71	0.73
Weighted Avg	0.77	0.76	0.76

A comparative performance with the proposed approach and existing studies reported in the literature is given in Table 6.

We obtained 85% accuracy with Test set-2, the highest among other test sets. The accuracies with Test set-1 and Test set-3 are 83% and 77%, respectively. A performance graph of accuracy is shown in Fig. 7.

Our suggested strategy outperformed the prior approaches described in the related works section regarding overall performance. Comparative performance with existing studies is given in Table 6.

6 Conclusion and Future Scope

We have presented a hybrid emotion classification system based on ResNet150 and XGBoost that classifies the four classes of emotions. The experiments were performed on a benchmark DEAP database. We used Scalograms for the experimentation, which is a transformation of EEG signal into an image format. The four classes of emotions are considered by creating four groups of different emotions. We

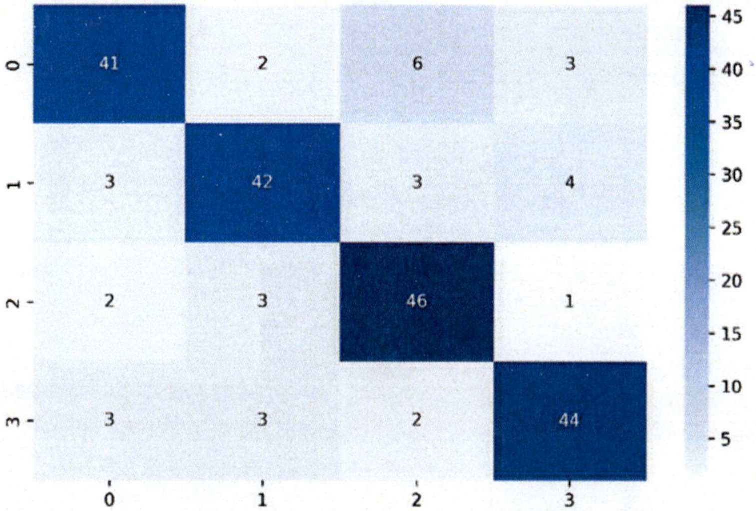

Fig. 4 Confusion matrix for Test set-1 (0: Cheer; 1: Depress; 2: Happy; 3: Sad)

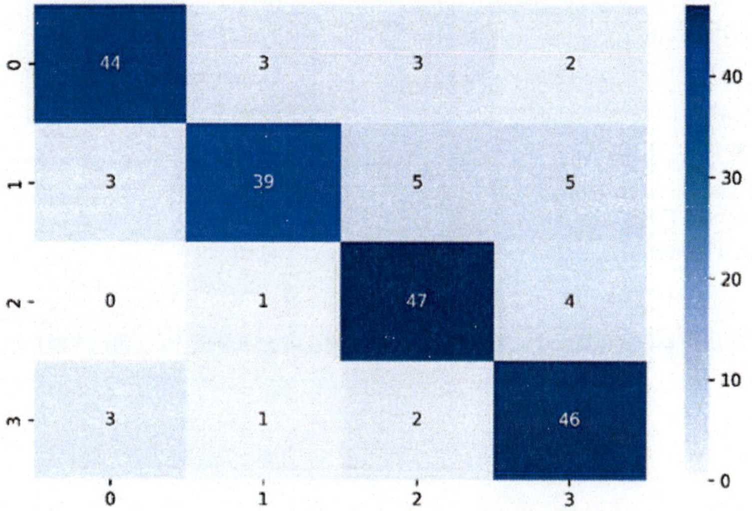

Fig. 5 Confusion matrix for Test set-2 (0: Love; 1: Shock; 2: Exciting; 3: Hate)

have achieved promising results by combining deep learning ResNet architecture and XGBoost classifier. We have achieved a comparable performance with the highest accuracy of 85%, and that too with four classes of emotions.

The study intends to investigate a new model by combining ResNet150 and XGBoost. In the future, we will perform experimentation with a higher number of classes. Therefore, in the future, we want to retrain and test our model utilizing

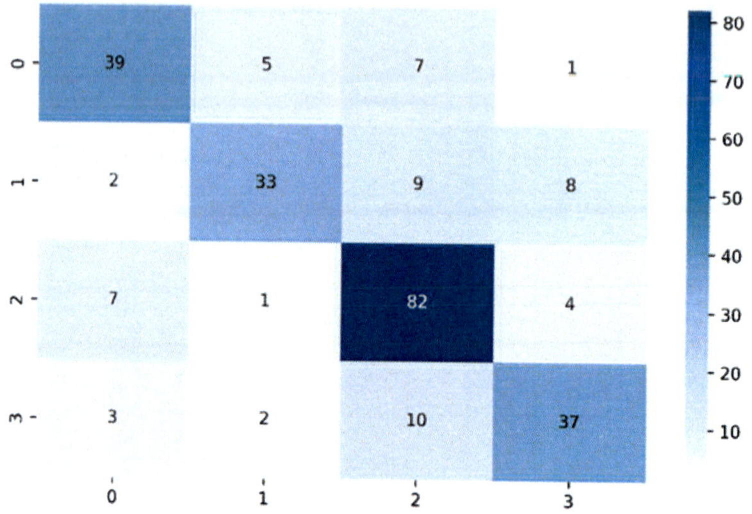

Fig. 6 Confusion matrix for Test set-3 (0: Fun; 1: Melancholy; 2: Mellow; 3: Terrible)

Table 6 Comparison with existing studies

Study	Methodology	Classification accuracy (in %)
Alhagry et al. [15]	LSTM	85.65
Lin et al. [6]	CNN-LSTM	75.21
Li et al. [7]	FFT, CNN	84.70
Acharya et al. [16]	FFT, LSTM	81.91
Proposed model	***ResNet150 + XGBoost***	**85.00**

other EEG datasets utilized in emotion identification research, including DREAMER and AMIGOS.

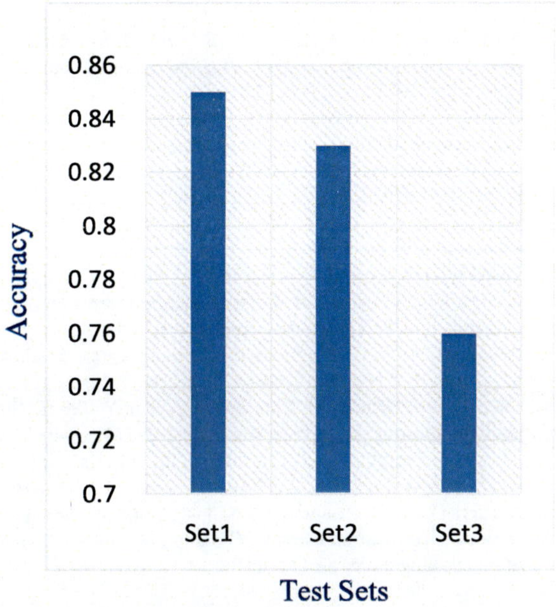

Fig. 7 Overall accuracy of the model

Acknowledgements We would like to thank National Institute of Technology Raipur (NIT Raipur) who have provided funds for this research under SEED GRANT project no. No./NITRR/Dean(R&C)/2022/83 dated March 9, 2022.

References

1. Alarcao SM, Fonseca MJ (2017) Emotions recognition using EEG signals: a survey. IEEE Trans Affect Comput 10(3):374–393
2. Teplan M (2002) Fundamentals of EEG measurement. Meas Sci Rev 2(2):1–11
3. Garg D, Verma GK (2020) Emotion recognition in valence-arousal space from multi-channel EEG data and wavelet based deep learning framework. Procedia Comput Sci 171:857–867
4. Koelstra S, Muhl C, Soleymani M, Lee JS, Yazdani A, Ebrahimi T, Patras I et al (2011) Deap: a database for emotion analysis; using physiological signals. IEEE Trans Affect Comput 3(1):18–31
5. Li X, Song D, Zhang P, Yu G, Hou Y, Hu B (2016) Emotion recognition from multi-channel EEG data through convolutional recurrent neural net-work. In: 2016 IEEE international conference on bioinformatics and biomedicine (BIBM). IEEE, pp 352–359
6. Lin W, Li C, Sun S (2017) Deep convolutional neural network for emotion recognition using EEG and peripheral physiological signal. In: International conference on image and graphics. Springer, Cham, pp 385–394
7. Li Y, Huang J, Zhou H, Zhong N (2017) Human emotion recognition with electroencephalographic multidimensional features by hybrid deep neural networks. Appl Sci 7(10):1060
8. Chung SY, Yoon HJ (2012) Affective classification using Bayesian classifier and supervised learning. In: 2012 12th international conference on control, automation and systems. IEEE, pp 1768–1771

9. Rozgić V, Vitaladevuni SN, Prasad R (2013) Robust EEG emotion classification using segment level decision fusion. In: 2013 IEEE international conference on acoustics, speech and signal processing. IEEE, pp 1286–1290
10. Candra H, Yuwono M, Chai R, Handojoseno A, Elamvazuthi I, Nguyen HT, Su S (2015) Investigation of window size in classification of EEG-emotion signal with wavelet entropy and support vector machine. In: 2015 37th annual international conference of the IEEE engineering in medicine and biology society (EMBC). IEEE, pp 7250–7253
11. He K, Zhang X, Ren S, Sun J (2016) Deep residual learning for image recognition. In: Proceedings of the IEEE conference on computer vision and pattern recognition, pp 770–778
12. Pustokhin DA, Pustokhina IV, Dinh PN, Phan SV, Nguyen GN, Joshi GP (2020) An effective deep residual network based class attention layer with bidirectional LSTM for diagnosis and classification of COVID-19. J Appl Statist 1–18
13. Parashar J, Rai M (2020) Breast cancer images classification by clustering of ROI and mapping of features by CNN with XGBOOST learning. Mater Today Proc
14. Chen T, He T, Benesty M, Khotilovich V, Tang Y, Cho H, Zhou T et al (2015) Xgboost: extreme gradient boosting. R package version 0.4–2, 1(4), 1–4
15. Alhagry S, Fahmy AA, El-Khoribi RA (2017) Emotion recognition based on EEG using LSTM recurrent neural network. Int J Adv Comput Sci Appl 8(10)
16. Acharya D, Jain R, Panigrahi SS, Sahni R, Jain S, Deshmukh SP, Bhardwaj A (2021) Multi-class emotion classification using EEG signals. In: International advanced computing conference. Springer, Singapore, pp 474–491

Author Index

© The Editor(s) (if applicable) and The Author(s), under exclusive license to Springer Nature Singapore Pte Ltd. 2023

A. Mishra et al. (eds.), *Advances in IoT and Security with Computational Intelligence*, Lecture Notes in Networks and Systems 755,
https://doi.org/10.1007/978-981-99-5085-0